Cyrielle Garson

Beyond Documentary Realism

CDE Studies

Edited by
Martin Middeke

Volume 30

Cyrielle Garson

Beyond Documentary Realism

Aesthetic Transgressions in British Verbatim Theatre

DE GRUYTER

ISBN 978-3-11-069669-1
e-ISBN (PDF) 978-3-11-071576-7
e-ISBN (EPUB) 978-3-11-071586-6
ISSN 2194-9069

Library of Congress Control Number: 2020946262

Bibliographic information published by the Deutsche Nationalbibliothek
The Deutsche Nationalbibliothek lists this publication in the Deutsche Nationalbibliografie; detailed bibliographic data are available on the Internet at http://dnb.dnb.de.

Printing and binding: CPI Books GmbH, Leck

www.degruyter.com

À mon père

Acknowledgements

This research into contemporary British verbatim theatre has been conducted under the aegis of the generous award of a full-time "French Ministry of Higher Education Research Studentship" (2011–2014) by the *École Doctorale* 537 "Culture et Patrimoine" at Avignon University in France. In the first three years of this project, I have been very fortunate to be surrounded by a stimulating and supportive group of colleagues within the ICTT research team of Avignon University. Above all, I thank Professor Madelena Gonzalez for her quick and early support of this project, her intellectual rigour and for helping see it through to conclusion. I would also like to express my deepest gratitude to Elisabeth Angel-Perez, Chris Megson, Martin Middeke, Annette Pankratz, Kerstin Schmidt, Eckart Voigts and Christina Wald for providing formal feedback on portions of the manuscript.

A number of people and events in the course of this research project have been invaluable in the development of my thinking in the field of contemporary British theatre, political and documentary performance as well as verbatim theatre more specifically. In particular, my work on contemporary British verbatim theatre has been greatly enriched by my association with the *théâtre(s) politique(s)* group (Université Paris Nanterre, France). I would like to also thank the RADAC team (*Recherches sur les arts dramatiques anglophones contemporains*) and the *Atelier Scènes Anglophones* (SAES society) for inviting me to speak and contribute to their exciting research activities in France, fortifying my knowledge of contemporary theatre in crucial ways. Special thanks are due to CDE, which inspired this project in the context of its PhD Forum in 2012 (Monning, Mülheim/Ruhr), and whose members have been instrumental in making me feel part of an outstanding international collective. It is also thanks to its generous biannual award received in 2018 that I am finally able to turn this somewhat unwieldy doctoral thesis into a monograph for a wider academic readership.

I am also greatly indebted to all the libraries and librarians, archives and archivists that have provided me with practical assistance and service. Deepest thanks to *Bibliothèque Universitaire d'Avignon* and its highly efficient interlibrary loan service, the British Library, the National Theatre Archives and the V&A National Archive of Performance in London. In addition, I have been driven by insights generously offered by practitioners and I extend my gratitude to all those who agreed to let me interview them, who allowed me access to recordings of their work and unpublished manuscripts. I would like to mention by name, however, both Louise Wallinger and Jo Harper for allowing me to publish an archival

https://doi.org/10.1515/9783110715767-002

picture of Non-Fiction Theatre as well as Jo Fisher for letting me reproduce the poster of her 2016 play *And Then You Kissed Me:* thank you! There are also many scholars – too many to mention individually here – in Europe, America and Asia who have helped in more subtle, but meaningful ways.

During the final part of this lengthy process, I am very grateful as well for the understanding of my former colleagues at Eton College between 2014 and 2016.

Finally, I would most of all like to thank my family for their continuous support over the years and in particular my brother Gilles Garson for helping me on the technical front at the eleventh hour. Thanks to all my friends and colleagues who were there from the beginning and stayed the distance, those who are here at the end: Sarah Ablett, Charlotte Agnew, Rosemary Agnew, Vanessa Barbaro, Johanna Bielher, Maria Elena Capitani, Emilie Chehilita, simon David, Johanne Keiding, Charlotte Lamailloux, Lucile Lesage, Framboise Ollivier, Hana Pavelková, Laurence Pons-Wood, Laure Porta, Domi Pritchard, Dominique and Huguette Py, Ellen Redling, Eleanor Rodda, Verónica Rodríguez, Haroon Shirwani, Jan Suk, Clare Summerskill and Agathe Torti-Alcayaga.

Copyright Acknowledgements

Earlier versions of material used in Chapters 3.2, 3.3 and 4.2 appeared in the following articles, reprinted with permission from the publishers:

Garson, Cyrielle. "Remixing Politics: The Case of Headphone-Verbatim Theatre in Britain." *Journal of Contemporary Drama in English,* vol. 2, no. 1, 2014, pp. 50–62.

Garson, Cyrielle. "Does Verbatim Theatre Still Talk the Nation Talk?." *Journal of Contemporary Drama in English,* vol. 6, no. 1, 2018, pp. 206–19.

Garson, Cyrielle. "Playing the Truth Game 'Verbatim' : from Tribunal Plays to Experimental Crime Plays on the Contemporary British Stage." *Coup de Théâtre*, vol. 32, 2018, pp. 157–79.

Abstract and Preface

Verbatim theatre, a type of performance based on actual words spoken by "real people," has been at the heart of a remarkable and unexpected renaissance of the genre in the United Kingdom since the mid-nineties. Simultaneously, contemporary British theatre-makers claimed to have found a renewed avenue to politics through this particular medium. In spite of recent shifts in the British theatrical landscape as well as the much-vaunted postmodern culture of anti-realism, documentary realism is generally conceded to have remained the normative mode of presentation for verbatim plays on stage. Through a comparative examination of seven representative verbatim productions, this book argues, however, that there has been an equally persistent strand of verbatim works that involves a move away from realism as the key element in performance. These productions make use of a wide variety of aesthetic experiments that broaden and transgress what we might understand as verbatim theatre. The strategy adopted by this study is thus to read verbatim theatre against the grain of its claim to "authenticity" and "truthfulness" and to suggest the need for a discourse which better articulates an interdependence between its aesthetic imperatives and the possibilities of social engagement. Finally, this volume accounts for the existence of a range of aesthetic variables in present-day verbatim theatre which is aimed at more media-aware contemporary audiences. These are grouped into three theoretical positions, each of them being analysed and discussed in a separate part and illustrated by case studies. It is hoped that, through this argument, the changing aesthetics of verbatim theatre will be illuminated and that – on however small a scale – this book will contribute to a critical history and theoretical formulation of the complexity of this rich field as both a scholarly discipline and a lived practice.

In broad terms, the central aim of the book is therefore to critically explore and account for the relationship between contemporary British verbatim theatre and realism whilst questioning the problematic mediation of the real in these theatre practices. Taken together, the seven case studies in this volume – the Tricycle Theatre's *Justifying War* (2003), Robin Soans' *A State Affair* (2000), David Hare's *Stuff Happens* (2004), Alecky Blythe's *Come Out Eli* (2003), DV8's *To Be Straight with You* (2008), Alecky Blythe and Adam Cork's *London Road* (2011), and Jonathan Holmes' *Fallujah* (2007) – challenge effectively the widespread assumption that verbatim theatre is formally homogenous, aesthetically unadventurous, and/or in decline.[1] In other words, this study, which showcases a wide variety of aesthetic transgressions from the "original" verbatim orthodoxy, questions the frequent assumption that verbatim theatre represents the degree zero of the aesthetic, by showing that even the most seemingly straightforward realist forms, such as tribunal plays, rely, nevertheless, on specific aesthetic practice and process. It is not only that verbatim theatre is interested in stirring up controversy, but that the means it uses to achieve this are highly counterintuitive, as the following pages will demonstrate through a reconsideration of both verbatim theatre's moral imperative and its political efficacy. More specifically and through a tripartite theoretical structure, I set in place a taxonomy of verbatim dramaturgies – documentary realism (Chapter 2), new realism (Chapter 3), and post-realism (Chapter 4) – that reinforces, reflexively interrogates, and/or deconstructs theatre's arbitration and excavation of the documentary real.

1 In April 2018, the Tricycle Theatre was renamed Kiln Theatre. However, as the case studies in this book do not cover this period, the name Tricycle Theatre is kept throughout.

Contents

1 Introduction

> But beware. Verbatim theatre may not be for you. There is no real dialogue, no narrative structure and no conventional dramatic climax. Instead, you get a cast of characters talking directly to the audience instead of to each other. (King 2)

At a theatre in Britain over the past two decades the spectator might have happened on one of the following:

1) A theatre space that painstakingly resembles a tribunal exposing the constellation of past secrets, lies and the poor workings of British public institutions as it scrupulously recreates a past public inquiry that he or she may have witnessed first-hand in its natural habitat.
2) Or, a theatre space committed to a different and more extensive idea of theatre, deceptively dry, simple and empty except for stools or chairs with actors delivering defiant and unalloyed raw perspectives, discontinuous interstitial glitches apparently caught on the move, glimpses of individuals normally sidelined by mainstream society and official culture within the contemporary moment. In this space, these simplicities and tones of verbatim language interact with each other in a constant push and pull, and they multiply, proliferate ceaselessly and are always pointing towards a plausible "us" that speaks directly to the society and the times in which we are living.
3) Or perhaps a variant of number two that mixes these with more familiar conventions of the stage and playwriting so that Janet King's words of warning – quoted in the epigraph – can almost be safely ignored.
4) Or, the spectator might have ecountered a theatre space occupied by a different kind of actor, performers wearing headphones, or earphones, or even in-ear monitors, where the figure of the actor is at once marginal and central and where meaning is assigned to his or her voice's texture and tone, the tiniest haphazard details of speech, emphasing the disjunction between speaker and speech.[1]
5) Or a more sophisticated and impeccably crafted multimedia theatre space announcing that "every word spoken on stage comes directly from interviews" – a stock feature of verbatim theatre, of which more a little later – and challenging or perhaps recalibrating some of the settled beliefs that so pervasively govern the spectator's thinking as it combats the more famil-

1 Considering the figure of the actor as both marginal and central may seem like a contradiction or an impossible feat. As demonstrated in Chapter 3.2 of this book, the aforementioned strand of theatre technically requires the presence of an actor in order to exist and, yet, the performance apparatus attempts to marginalise his or her presence.

https://doi.org/10.1515/9783110715767-003

iar narratives of the mass media's representations of the real. This performance may even be showcasing a physicality that moves the spectator and/or further complicates these matters.[2]

6) Or a piece that is so new that it does not sound or look like anything the spectator has ever seen before. It could well be a completely new genre that experiments with what he or she thought a musical ought to be.
7) Or, finally, a theatre space that is about the quality of engagement, that is used in a completely different way, always in the process of becoming something else, making the spectator experience – in the middle of the action taking place – some profoundly memorable and multifaceted painful emotions, small shifts of perception and perhaps even self-perception. In other words, a practice of holding up a space for something else to happen, imprinting itself on the mind of the spectactor.

I begin with these examples not because they are representative of all contemporary verbatim theatre in Britain, but because their particular concerns resonate with many of the key ideas explored in this study. In fact, each example directly relates to one of the case studies and, in a chronological order, so that (1) corresponds to Chapter 2.2. However, as Chapters 2.4 and 3.3 are not related to the main case studies, this means that (3) corresponds to Chapter 3.1, (6) to Chapter 4.2 and (7) to Chapter 4.3. Chapters 2.4 and 3.3 are transition chapters that encapsulate an urgent re-examination of the premises of the two chapters that have preceded them.

To come back to the examples, such performance works, which oscillate from a kind of rigidity to a dynamic flexibility, indicate a distinctive turn in British theatre history that is somewhat difficult to descipher in terms of aesthetics. In the first two instances (1) and (2), what is involved is a complex pact of performance that insists on its special relationship with "the real world" whilst attempting to erase or minimise the usual markers of theatricality.[3] In the next two

2 This disclaimer is the exact opposite of the one found in fictional works that claims that: "Any names, characters, events or locations depicted are the product of the imagination. Any resemblance to actual persons (living or dead) is entirely coincidental" (Smith, *Preston Bill* 7).

3 This and the term "contract" will be used throughout this book. Arguably, even though these terms express ideas that are rather self-evident within the Western world of theatre and performance studies, it is relatively rare to find these terms in English publications. In line with the French scholarly tradition, I have been accustomed to discuss performances through the notion of "*contrat de spectacle*" which simply means the theoretical contract between theatre-makers and audiences that always frames reception and may be changed throughout the performance. In other words, these are the cultural conventions that have been – perhaps unconsciously – learnt by the theatre-goer who may expect that there is a separation between the performance

instances (3) and (4), this time, the spectator seems to be caught between two incompatible positions as if the performance were itself indecisive and admitting its own vulnerability. Unpacking briefly just these few examples uncovers thorny problems in the contemporary theoretical landscape. But before diving into these messy issues, let us first establish what I exactly mean by "verbatim theatre" so that what is at stake here becomes clearer and more concrete.

1.1 Definition and Context

What is contemporary British verbatim theatre? It is so many things that this entire volume can only give a partial answer. In fact, researcher Emma Cox prefers to use the plural, "verbatim theatres" (*Noncitizenship* 31), in recognition of the reputed unruliness of the field, the richly pluralistic and variegated nature of verbatim theatre, "the divergent ways in which such theatre can manifest" (*Noncitizenship* 31) and the valences the term can encompass in a contemporary British context. In other words, "verbatim theatre" is typical of a simple-sounding concept that turns out to be outrageously complicated.

Roughly speaking, verbatim theatre does not describe what the play or performance is about but *how* it is made, that is to say, the way in which the words have been selected. In this book, the kind of verbatim works to be encountered tends to primarily base its performance text on interviews, transcripts of public inquiries, as well as various documents in the public domain. The temptation may indeed be to expect that such performances are more straightforward and demand less work of their spectators. As British critic Aleks Sierz said in the context of a 2010 lecture entitled "*Blasted* and after: New Writing in British Theatre Today," on the occasion of a meeting of the Society for Theatre Research at the Art Workers Guild in London, "what you see is what you get – and often what you see is all there is."[4] However, as this study will demonstrate, this narrow reading of verbatim theatre deserves to be challenged and a more critical reading of the contemporary verbatim output may shed light on some fascinating aspects that are entirely new to the critical arena of British theatre. With this end in

space and the auditorium, that there may also be a curtain and a dimming of light level to announce the beginning and the end of the piece. Each performance engages with these and may set up some conventions of its own, as is the case of verbatim theatre which prompts its audience to accept that every word has been used by a real person, differentiating the pact of performance from fiction-based shows.

4 This quotation is taken from the full transcript available on *TheatreVoice.com* here: http://www.theatrevoice.com/audio/new-writing-in-british-theatre-today/.

mind, here, verbatim theatre is therefore considered as an inherently fluid and unstable discursive category rather than a definite dramatic genre with a shared documentary project. This is especially the case in the period under scrutiny that saw verbatim theatre seamlessly move away from and come back to its historical foundations.[5] It is indeed my contention that what also matters – as in all works of art – is *what* the creative team does with these words.

Verbatim theatre – as both artistic method and output – has expanded exponentially since the mid-1990s and seems to coincide with the homogenising umbrella of a "renewed interest on the part of contemporary theatre practitioners in addressing political and social issues directly" (Goss, "Postdramatic" 160) in a post-political era, which has arguably come to dominate the British theatrical imaginary by stealth, its vibrant and momentous impact still resounding as these lines are being written.[6] More significantly, since the early years of the 2000s, "verbatim theatre has become the new black" (Edwardes). Acres of print have been expended on verbatim theatre which may appear startling when, in fact, quantatively, it amounts to a lot less than more conventional types of theatre, but this (distorted) perception phenomenon has arguably happened before in British theatre history.[7] As verbatim theatre has surged in power, so its detractors have grown in number and strength over the past two decades, as will be explored in Chapter 2 of this volume. Regardless of this situation, however, verbatim theatre has relentlessly continued to grow and to inspire contemporary artists. It should be stressed, moreover, that verbatim theatre may just be one of the fastest growing types of British theatre since it has, as Andrew Haydon suggested in a recent publication, "touched on almost every possible way of working in modern British theatre" ("2000s" 48). Clearly, we live in an age in which performances designated as "verbatim theatre", either by artists themselves or by the press, are increasingly diverse. They have included monologues as exemplified by Sonja Linden and Christine Bacon's *Asylum Monologues* (2006), audio-verbatim plays such as *Going for a Walk*, musicals (see Chap-

5 As will be described in Chapter 2 of this book, even though the phrase "verbatim theatre" itself is fairly recent, the tradition has been around in the UK for many decades.

6 Interestingly, Deirdre Heddon also proposes that perhaps it is not so much an increase of verbatim performances in Britain as a "public recognition" (*Autobiography* 128). As far as can be ascertained, this is not an exclusively British phenomenon, but this study can only look at its specificity within Britain.

7 For instance, "no more than 5 per cent of Royal Court productions ha[s] ever been verbatim" (Dugdale 16) and, as Ariane de Waal has argued, there has been a "disproportionate emphasis placed upon verbatim theatre" (4). In the 1990s, the emphasis was put on what Aleks Sierz called "In-yer-face theatre" (2001) to the detriment of equally valid and important types of theatre works.

ter 4.2), comedies such as Tamasha Theatre's *The Trouble with Asian Men* (2005), very short plays such as Alecky Blythe's *Voices from the Mosque* (2011) that was part of Headlong's *Decade* on the responses to 9/11 on its tenth anniversary, collages of recreated interviews massaged with physical theatre (see Chapter 4.1), more abstract works such as Dan Rebellato's *Outright Terror Bold and Brilliant* (2005), or even poetry as in Simon Armitage's *Black Roses: The Killing of Sophie Lancaster* (2012) and, finally, verbatim sequels such as Cressida Brown's *Re-Home* (2016) which revisited her *Home* (2006).[8] Before going any further, however, it may be necessary to unpack some of the terminology used in the title of this book so as to expose the network of theoretical presuppositions it sets forth as well as its broader pattern of thinking.

To begin with a brief definition, by verbatim theatre, this book encompasses those performances characterised in their making by a central, or exclusive, reliance on the exact words of real people that have been gathered during a prolonged research phase and then creatively edited. The outcome of that work generally meets an audience and is performed by professional actors in a theatre setting.[9] Typically, verbatim theatre-makers have very little knowledge (or no knowledge at all) as far as a specific issue is concerned and, as a result, they interview people in the hope of finding something that may contradict their own bias or some master narratives currently operating.[10] It is not surprising, then, for these performances to hold "some degree of after-effect" (Goode, *Forest* 250), as the audience may become "newly informed by some verbatim piece" (Goode, *Forest* 250). Most of the time, it is necessary to have a written agreement (typically called a release, permission or consent form) between the people involved, that is to say, between the creative team and the interviewees, similar to what an ethnographer or oral historian would use. If one addresses the notion of a pact of performance in the context of these works, it is usually tied to a pact of authenticity, and more precisely to what this book calls "documentary realism." The subject matter of verbatim pieces is extremely variable, but critics tend to focus solely on the technique itself rather than other aspects of the pro-

---

8 In Dan Rebellato's piece, actors perform the internet messages, including the emoticons, by saying them. For instance ☺ becomes <happy>.

9 Sometimes these are not professional actors but the "real people" themselves, especially in the context of theatre with refugees (see Alison Jeffer's fascinating 2011 book on the subject in the bibliography).

10 Scholar and translator Tania Moguilevskaia calls this the "zero position" in reference to the work of the Russian verbatim artist, Mikhaïl Ougarov. Even if it is theoretically possible to imagine that some artists conduct the interviews in this manner, the subsequent fashioning of a performance text would seem to indicate that such a position is not achievable in practice.

duction, even though it is far from being a new form. Having said that, it is my contention that most verbatim works are about something that is already there whilst nobody is paying attention and wanting it to be more significant. For example, a verbatim piece such as Victoria Brittain's *The Meaning of Waiting* (2010) wants the audience to engage with the often-unseen fallout of the war on terror in Britain through the words of eight prisoners' wives. Furthermore, even though it is exclusively based on the words of real people, verbatim theatre can also convey a desire for something to be in the world that is not already there. One of Clare Summerskill's latest verbatim plays, *Rights of Passage* (2016), on the plight of gay asylum seekers coming to the UK, does this rather successfully.

It is called verbatim theatre, but it is made up of many other things. People call it verbatim theatre because its verbatim components are more important than in other performances. The task of establishing and clarifying its aesthetics is complicated by the fact that no agreed principle really unites such performances beyond their use of verbatim materials/oral testimonies, whose degree is open to dispute. For example, as Elisabeth Angel-Perez explains, British playwright David Hare – whose play *Stuff Happens* (2004) is the case study in Chapter 3.1 – considers that verbatim theatre "is a form of drama made, for more than 60 % of the play, of the words of real people" ("Nation" 65), whereas for Dan Rebellato – in a more strict sense – it "involves plays in which all of the words are taken "verbatim" from existing sources" ("Writing" 156). Additionally, as conversations about verbatim theatre have proliferated since the 2000s (due to its resurgence and popularity), inevitably, so too has the meaning of the term as well as the practices it is understood to describe. In this new context, whereby diverse structures of performance increasingly incorporate verbatim material, one must be particularly weary of not mistaking its functional markers for the verbatim model itself. For this reason, it is necessary to go over some definitions that have had currency at this cultural moment.

First of all, by far the most important definition in the field is the one that Derek Paget provided in 1987, which coincided with the first appearance in print of the term "verbatim theatre". In effect, this was the way playwright Rony Robinson, "a pioneer of the method" (Paget, "Oral History" 317), used the term in his own practice:

> [i]t is a form firmly predicated upon the taping and subsequent transcription of interviews with 'ordinary' people, done in the context of research into a particular region, subject area, issue or event, or combination of these things. This primary source is then transformed into a text which is acted, usually by the performers who collected the material in the first place. (Paget, "Oral History" 317)

Arguably, this definition may appear somewhat dated – and not only because theatre-makers have now access to technologies significantly more advanced than the portable cassette recorder – in light of the current spectrum of practice in Britain that claims the appellation today, for Paget's often-cited definition describes a fairly recent phenomenon at the time, which nevertheless has a longer history. Verbatim theatre of the more radical sort was indeed designed to effect social change. In this sense, it was close to Erwin Piscator's conceptualisation of documentary theatre in 1966: "[i]t is a dramatic art which distils from reality – historical or contemporary – a work of art that meets the requirements of a drama and attains in content a degree of actuality and political force rarely possessed by previous dramatic literature" ("Investigation" 349). This upsurge of verbatim performances has therefore needed new definitions to comprehend its changed parameters. Another important definition has been the one given by Will Hammond and Dan Steward in their *Verbatim, Verbatim: Contemporary Documentary Theatre* in 2008:

> The term *verbatim* refers to the origins of the text spoken in the play [...]. In this sense, verbatim is not a form, it is a technique; it is a means rather than an end. For this reason, *verbatim theatre* can be used to describe plays that are sometimes so dissimilar that the term may appear to be of little value. (9)

This seems to indicate that it has become more difficult to define verbatim theatre as what it means ultimately depends on the theatre-maker. It is also my contention that, at the present time, the range of practices that might be considered as verbatim theatre in Britain is far more diverse, oblique and unstable than ever before. On this note, British theatre critic Michael Billington does not hesitate to write that this type of theatre "has proved itself infinitely flexible" ("V").

In addition to this, it will be noted several times in this book that there is a tendency in theatre criticism, that is sometimes extended to scholarship, to use the very same term(s) to designate different things, and conversely different terms are often used to refer to the very same object. For instance, the term "documentary drama," is used by Eric Bentley to mostly characterise his own documentary practice.[11] Its main rule consists in putting "into people's mouths

11 In a more recent publication, Rib Davis explains that the term "documentary drama" can be used in a great variety of contexts, sometimes distancing itself from current understandings of verbatim theatre and sometimes colliding with it as in what follows: "the term 'documentary drama' can be interpreted to mean a play or film in which not only did all the events occur in reality, but also almost everything spoken is documentary as well – the *exact* words coming either from tape-recorded and transcribed interviews or from written sources such as letters, re-

only words which they had used and which indeed they had placed on the public record. No investigative reporting. No confidential sources. Just what people had said in public, and for the public" ("Political Theatre" 51). In this sense, this practice can be understood as a variant of verbatim theatre. More strikingly, Nicholas J. Long defines "documentary theatre" exactly as I define "verbatim theatre" in this study: "documentary theatre – a category of staged performance in which the actual words of real people are edited into a script and performed on stage by actors" (305). For Tom Cantrell, however, verbatim theatre seems rather like a subfield of documentary theatre that one may call "verbatim documentary theatre".[12] Alison Jeffers also considers that these terms are not exactly the same thing, documentary theatre being used when the performance is drawing on documents and, verbatim theatre more specifically, when "the exact, albeit edited, words of the subjects are inserted into the play" ("Esrafil" 92). Similarly, Jenny Hughes writes that, in the strictest sense of the term, the performance text of a verbatim piece – as opposed to a piece of documentary theatre – is based exclusively on previously uttered words, be it "drawn from documented records, personal testimony [or] interview material" (92). Yet, for Andy Lavender, the divide is only a geographical one: "[t]he incursion of people speaking in public has been marked in theatre by what became known in the US as 'documentary theatre' and in the UK as 'verbatim theatre'" (36), and, for Rib Davis, one of degree as strict verbatim is opposed to "the form of documentary theatre which allows some minor alterations to the words from the tape" ("London Road" 126). More interestingly perhaps, David Williams perceives the distinction in terms of its implications:

> Other people call this 'verbatim theatre', but in my view the word 'verbatim' implies a too-close association with 'truth', something that the theatre, no matter how firm its embrace of real-ness and real people, can never actually deliver. The term 'documentary' acknowledges that these materials are real— most are literally documents of one form or another—but also makes it clear that there is a frame around the presentation of these materials, that, despite their strong truth claims, they have been changed by their transformation into theatre. (115)

ports, council minutes, newspapers, official records or school logs. With minimal alterations and very few additions, the words from the source or a variety of sources are knitted together rather like a patchwork, to form a drama" (*Dialogue* 202).

12 See his excellent definition of verbatim theatre available at Drama Online: http://www.dramaonlinelibrary.com/genres/verbatim-theatre-iid-2551.

Whether or not one agrees with such divisions is again open to question.[13] More often than not, the attentive researcher in effect finds in the literature both terms being used interchangeably, even within the same paragraph such as in the following example:

> This is not necessarily intended as a call for more verbatim theatre, but there is something in the form that masterfully takes advantage of theatre's capacity to humanize the audience. I find the documentary play such a fine way to awaken people to the world and its myriad complexities. Though it shares some qualities with journalism, verbatim theatre is a form that can transcend snappy sound bites. (Kushnir 92)

These definitions and remarks taken together, the term "verbatim theatre" thus applies whenever certain minimum conditions are met and must therefore possess some basic coordinates, beyond conducting interviews and capturing the elusive nature of the human voice as part of the creative process. There are kernels of truth in all these definitions, in the sense that they all capture various important aspects of verbatim theatre, yet none of them have much to do with aesthetics, my primary concern in this study.

1.2 The Stakes

A poster for the verbatim show *And Then You Kissed Me* (2016) by Jo Fisher, inspired by another verbatim piece, Robin Soans' *A State Affair*, the case study in Chapter 2.3, poses the question at the heart of this book: how to articulate verbatim theatre's "latest instalment of realism" as seen on the British stage since the mid-1990s? The terms "latest instalment" presuppose both some sort of lineage in line with stage realism – a concern with the here and now of how ordinary people live with actual problems and facts – and, at the same time, the possibility that it may not be frameable within its conventional schemata. In three main sections, each consisting of three main chapters, this book thus interrogates verbatim theatre's engagement with and departure from the ongoing aesthetic project that bears the name realism. More specifically, this is a book about the contemporary British verbatim theatre phenomenon that narrates the development of a "documentary realist" theatrical style in the work of a gen-

13 Chris Megson, for instance, considers that verbatim theatre includes both "spoken testimony (derived, for example, from personal interviews) and 'found' texts (often collated from a range of documentary sources)" ("Hare" 512).

Figure 1: Jo Fisher's *And Then You Kissed Me* (2016)

eration of British practitioners. As the title of this book advertises, this study also narrates the gradual move away from documentary realist strategies in recent British verbatim theatre without imposing a presumed chronology based on this theoretical paradigm.[14] I contend here that a study of verbatim theatre's relationship with realism is most pertinent to its contemporary British output, as it has shown a striking consistency in the period under discussion. Importantly, this book attempts to provoke questions about the qualitative distinctions between making/viewing a piece of verbatim theatre and making/viewing a

14 A complex term to which I dedicate Chapter 2.1.

piece of fictional drama by retrieving something from that experience which seems worthy of critical analysis.

One might wonder at this stage in what way contemporary British verbatim theatre is seemingly tied to realism. It would be no exaggeration to suggest that one's own presuppositions when encountering the words "verbatim theatre" or "pure verbatim theatre" is that they carry a natural affinity as well as a convincing sense of the orderly and the stable. One expects to find represented in the theatre exactly what one finds in the "real world", that is if one looks hard enough for the "really real." Of course, one also expects some sort of "natural" transformation, due to the change of context and audience in the fleeting passage of performance, a transformation not dissimilar to the one famously formulated in the field of chemistry by Antoine Lavoisier,[15] around the time of the French Revolution: "Nothing is lost, nothing is created, everything is transformed".[16] Paradoxically, such a transformation through the process of verbatim theatre denotes its own lacunae, since verbatim theatre only deals with remains and traces from a past that can never be fully revived or accessed. This is an especially important consideration that relates to aesthetics and its structural role in performance, as it carries varying weight that may shift our experience of these verbatim elements.

As such, verbatim theatre is equally a commitment made to the audience that despite the aforementioned relation to lack, it will appear as plenitude, perhaps even as indistinguishable from what it attempts to recreate. Tracing such a complex pact of performance, Deirdre Heddon rightly reminds us in *Autobiography and Performance* that there is a "realist mode that 'verbatim' encourages and implies" (163), a kind of synergy or aesthetic smoothing over of a glimpse from the past that I am calling here "documentary realism," and which shall remain my primary point of contention. As it would turn out, documentary realism is not to be confused with what is generally termed "theatrical realism" or "stage realism" in the context of non-verbatim plays. It is a realism of a different order that rejects some tenets of theatrical realism, as it directly imports materials from the "real" extra-theatrical world and seems to go all the way to the end of one of theatre's self-contradicting paths. Documentary realism – a concept I shall fully develop in Chapter 2 – thus maintains the illusion of watching an ordinary, unrehearsed, live slice of life and constitutes "what contemporary audiences regard as the governing logic of documentary theatre" (Youker 2015, 219). Alison For-

15 Of course a verbatim theatre production extends far beyond the accumulation of verbatim materials.

16 My translation from the original French "Rien ne se perd, rien ne se crée, tout se transforme".

syth and Chris Megson have rightly identified this strand of performance as "[t]he once trenchant requirement that the documentary form should necessarily be equivalent to an unimpeachable and objective witness to public events" (3). This is perhaps one of the ways in which verbatim theatre differs from oral history performance, if understood as a challenge to "the textual drive toward narrative resolution and the conventional authority of more objective or objectifying modes of knowledge and representation" (Pollock 4).

Above all, against the general flow of British theatre history, these verbatim works stretch and trouble the neat limits of established practices, violate (and thereby subvert) the aesthetic consensus in important respects, for they create an aesthetic context that can accommodate without strain styles deemed antithetical. A couple of examples might put some flesh on these otherwise abstract principles. In Look Left Look Right's *Counted* (2010), a site-specific verbatim play on political apathy and democracy in the UK, the performance engages with realism through the meticulous recreation of the interviews and, yet, the mobility of the audience in this promenade configuration complicates this reading. More obviously, Alecky Blythe and Adam Cork's *London Road* (2011) – a verbatim musical that is also the case study in Chapter 4.2 – features a kind of language that both creates and compromises realism, as it sets real words and speech patterns to music. Finally, my main purpose here is to develop a different way of thinking about contemporary British verbatim theatre, by focusing on some of the concepts that govern its making in this particular period. The three competing aesthetic categories this book proposes (with some overlap and interaction between them) arguably shed light on a concept of contemporary verbatim theatre that, as Lib Taylor reminds us, "rather unhelpfully groups dissimilar devising processes and performance modes together" ("Voice" 379).

1.3 Contemporary

Historical researchers first introduced the tape recorder in the 1930s, but British theatre-makers only began to grapple with the power of the recorded "true story" and the saying of it in the 1960s. In the meanwhile, in the final years of Piscator's life, Hochhuth, Kipphardt, Weiss and others in Europe, Asia and South America were experimenting with a "Revolutionary political theatre," a documentary theatre "on the fringes of the establishment theatre, on the fringes even of the avant-garde theatre, in a type of under-world" (Hainaux 375). Now, to come back to Britain, it is tempting to date the rise of visibility of contemporary verbatim theatre in the early 2000s, as Steve Waters has done in the following: "verbatim theatre [...] had a resurgence shortly after September 11 when politics seemed back in town" ("Political" 139), or even to 2003 in response to the Iraq war. However, it is my contention that this renaissance started in the 1990s. As such, the long-standing term "contemporary" should not be understood here as engaging solely with 21st-century verbatim theatre but, more specifically, as a reflection of the "contemporary period" stretching back to the mid-1990s and forward to the present day.[17] Crucially, this specific period in theatre history is marked by profound changes in the nature of theatre itself and "in the way we view it" (Pavis, Routledge x).

The now-familiar version of the British theatre output in the 1990s is that it is essentially marked by the "In-yer-face" phenomenon, an avant-garde style and sensitivity that resulted in shocking plays; except that it was not really the case because, at the same time, other styles were emerging. It was, for instance, a time that saw the development of new voices thanks to the work of black and Asian companies such as Tamasha, Tara and Talawa, and, for the purposes of this study, it was also the beginning of a renaissance of verbatim theatre. Conditions in Britain were rather favourable to the development of the latter[18] as "In-yer-face" theatre was fading away,[19] state funding for theatre rose under New Labour and the Arts Council of Great Britain changed their policy to include more experimental and devised works.[20] In other words, the need for a new kind of theatre in these theatrically patchy times, one that would directly address the key issues of the day was felt ever more intensely. The Royal Court, the Tricycle Theatre and, perhaps more surprisingly, the National Theatre under Nick Hytner's directorship can be said to have spearheaded a very particular intervention in their audiences by staging, within their core programme, a certain proportion of verbatim plays. Certainly it is the case that we are witnessing the most exciting period for British verbatim theatre, a theatre that has also proven to be particularly receptive to international influences, perhaps against a backdrop of celebrated aesthetic insularity. It is also my contention that this period was equally marked by the ongoing exploration of "a beyond of documentary realism," as will be explained in more detail in Chapters 3 and 4 of this volume.

---

17 Chronologically speaking, the earliest contemporary British verbatim performance mentioned in this book premiered in 1994 and, the most recent one, in 2018.

18 Nevertheless, the occasional verbatim play did reach the British stage in the early 1990s, but it remained relatively ignored by critics who did not recognise it as potentially "the next big thing."

19 In fact, verbatim theatre re-appeared during the high water mark of In-yer-face theatre (see Megson 2006 and my own 2015 article published in *Coup de Théâtre*).

20 The results of the Boyden Report in 2000 injected new money into the sector and generally boosted and favoured the development of new works.

Furthermore, the term "contemporary" also shares a more profound affinity with verbatim theatre since it appears particularly reactive to the world in which it is wrought as it takes audiences behind the headlines, to the point that the resulting playtext is sometimes perceived as unfinished, transformable *ad vitam aeternam*. In short, above all, it seems, through notably its turn to personal and private testimonies, to have its fingers on the pulse. The politically charged period under review that Prem Shankar Jha calls "the descent into systematic chaos" (328) comprises the Stephen Lawrence case and aftermath, the NATO campaign in Kosovo, 9/11, the war on terror, including the British intervention in Afghanistan and the cataclysmic run-up to the invasion of Iraq, as well as the rise of a Manichean rhetoric dividing the world between "good" and "evil", new revolutionary landscapes in the Middle East, the collapse of the distinction between civilian and military space, the London bombings of 7 July 2005, torture at Guantanamo, the Tony Blair and Gordon Brown years of New Labour (1997–2010), the David Cameron/Nick Clegg coalition, the riots, the Lehman Brothers collapse in September 2008, the UK "austerity programme" in 2010[21] and an increasingly naturalised neoliberalism, the referendum on Scottish Independence, a gradual dismantling of the NHS, wide-scale youth unemployment, the housing crisis, ruthless gentrification and rampant precarity, ongoing political apathy (Chou, Gagnon and Pruitt), the ever-increasing commodification of universities, the discourse of crisis in the humanities, a blurring of political agendas, a publicly perceived lack of rigour in journalism (succumbing to last-minute scoops, infotainment, sound-bite reporting and the infamous so-called 'churnalism' and 'clickbait' phenomena), the advent of high-traffic fake news farms, the Levenson Inquiry and the Murdoch scandal, a climate of general insecurity, a growing and spreading arsenal of "Weapons of Mass Destruction" (WMD), the emergence of ISIS, a global refugee crisis, the prospects of an imminent ecological crisis (as a burgeoning population is now exceeding the capacities of the world), the "dematerialization of 'real life' itself" (Žižek, *Desert* 14) and, on some level, a sense of nostalgia for a disappearing Britain.[22]

---

21 This political programme put public services under severe pressure and implemented a growing restriction of benefits.

22 In terms of "political apathy" in Britain, Chou, Gagnon and Pruitt base their claim on recent studies by the Hansard Society. According to them, "few citizens now take part, or recall taking part, in political activities like voting" (607). Worse, more and more citizens feel that their opinions do not ultimately matter and "[i]nstead, the more prevalent sentiment was that everyday citizens have little to no influence on what elected officials do" (608). By commodification, I refer to the meaning given by Liz Tomlin via Colin Leys in a recent book chapter, that is to say, "under the market-driven politics of neoliberalism" ("Academy" 264). More specifically,

But amidst the gloom at this troubled historical juncture, there have been rays of hope. The 2000s saw a reawakening of activism as exemplified by the global Occupy movement, the Northern Ireland peace process, full equality of rights for same-sex couples, greater equality for the "trans" community, more initiatives and actions being taken to close the gender gap, spectacular progress in science, and the advent of open access forever globally changing the landscape of the digital and the information shared online. On the theatre front, an increased diversity started to emerge slowly, especially as regards to BAME – most visible perhaps in the rise to prominence of black British playwrights such as Roy Williams, Kwame Kwei-Armah, Bola Agbaje and debbie tucker green who started to achieve visibility and success in London's mainstream theatres.[23]The period also saw the most significant development for new writing and new work, overtaking revivals in the British repertoire for the first time since records began.[24]It was also a period marked by the launch of the National Theatre of Scotland and National Theatre Wales, a regional revival to combat the industry's London centrism, the Monsterists, a substantial increase in government funding[25] and the creation of National Lottery funds for theatre, opera and ballet.[26]

Tomlin indentifies the top up fees introduced by Blair in 2006 as the seminal turning point for the commodification of higher education in England, a moment in which "[s]tudents were reconfigured as consumers" ("Academy" 269). Before Žižek, Jean Baudrillard in his *Symbolic Exchange and Death* (originally published in French in 1976) already argued that "[t]he principle of simulation governs us now, rather than the outdated reality principle" (2). In his *Simulacra and Simulation* (originally published in French in 1981) he explains that "[s]imulation is no longer that of a territory, a referential being, or a substance. It is the generation by models of a real without origin or reality: a hyperreal" (1). This book does not really extend to the UK general elections in May 2015, June 2017 and December 2019, and does not include Brexit and its aftermath as well as the current Covid-19 pandemic.

23 The term "BAME" designates ethnic minorities in Britain, that is to say "black, Asian and minority ethnic". In the theatre industry, BAME playwrights, actors and directors are now more visible, but there is still a long way to go to achieve a more diverse mainstream.

24 Post war British Theatre can be seen as a history in which new playwrights struggle to have their voices heard. Each decade, a new generation attempts to break through a perceived unjust system that favours dead playwrights over living ones. In the 1990s, new writing started to become something exciting that theatres would like to put on and the 2000s inherited this tendency in the sense that there was a deluge of new writing (also due to better finances in the sector). However, never in recorded history had new writing been reported to be the main activity of theatres across Britain. For more information on this phenomenon, see the British Theatre Consortium's latest publications: British Theatre Repertoire 2013 and 2014.

25 Sadly since 2010, following government cuts to public services more broadly, Arts Council budgets have dropped again. A first report entitled "In Battalions" on the implications of these cuts for the sector was co-produced by Fin Kennedy in 2013 and can be read here:

Last but not least, there is plenty of dismaying evidence to prove how verbatim theatre has moved into the mainstream of the culture in the period under scrutiny. Indeed, the internet user eager to know about the phenomenon of verbatim theatre can now access over 490 000 web pages in less than a second, overtaking the terms "devised theatre" and "In-yer-face theatre."[27] Moreover, verbatim theatre has inspired some novels such as Liam Murray Bell's *The Busker* (2014) where we learn that one of the characters, Emma Scribbens, is a "bright young woman doing research on political theatre, on verbatim theatre specifically" (145) and it has also inspired poetry itself as some poems now call themselves "verbatim" when they find poetry amongst the ordinary words of real people. Verbatim theatre has also had impact on the the world of dance, opera, music, installation art, fine art, musicals and VR to some extent; the world of cinema by setting the template for a new genre of film-making and it now even has its own Twitter account with over 1000 subscribers.[28] Today, it has become increasingly common to go to the theatre to see a verbatim play in which "[a]ll the words that the actors speak are words that [a theatre-maker] ha[s] collected" (Blythe, *Revolution* 5) through interviews, to hear one on the radio or to see one at the cinema or on television.[29] In fact, they now travel to other countries and continents and are even translated into other languages, whilst some others are sold on DVDs.[30] Students can now access well-established training methods in verbatim theatre and it has also become a new way to present research data. In other words, we now accept without question that an invigorated

https://www.scribd.com/document/126273288/In-Battalions. Theatre Uncut was founded in 2010 as a result of this sad situation as regards to public spending. Since then, with Theatre Uncut, leading British playwrights such as Mark Ravenhill, David Greig and Chris Goode have written short plays in response to these cuts that are rights free and for anyone to perform anywhere. These plays have been performed by over 6,000 people across four continents.

26 In 2003, a group of British playwrights, including David Eldridge, Richard Bean, Tanika Gupta and Moira Buffini, gathered and produced a Manifesto (Eldrige 2005) to demand bigger spaces for living writers so that they could produce full-length plays with large casts.

27 These results are based on an experiment I conducted on 9 October 2016 with the Google search engine from a UK location.

28 Besides the film version of Alecky Blythe's *London Road* which will be discussed in Chapter 4.2, Clio Barnard's *The Arbor* (2010) pioneered a new way of approaching cinema by having screen actors lip-sync to the audio verbatim material. See Johnson (2016) for more on this re-appropriation of the verbatim techniques from one medium (theatre) to another (cinema). The Twitter account in question is run by Dr Tom Cantrell, Head of Theatre at the University of York.

29 DV8 Physical Theatre's *John* was filmed and broadcast live to cinemas throughout the UK and the world.

30 The most extreme example of this tendency is Alan Rickman and Katherine Viner's *My Name is Rachel Corrie* which was translated into more than a dozen languages.

technologically-determined verbatim theatre is everywhere and what counts as verbatim performance has changed continually over time to become a moving target, large enough to include (6) and (7) that I mentioned earlier in the second page of this introduction. To name a more concrete example of the way in which verbatim theatre can stretch, one of the latest projects announced by the National Theatre consists in collecting a gigantic verbatim archive on modern Britain following the Brexit vote, which should form the basis of future shows. Taken together, these testify to the extent to which its unmatched accuracy is part of the critically approved British theatre landscape, and this is by no means an exhaustive list.

By failing to place a cultural premium on the idea of the individual, verbatim theatre has provided the germ for impetuous experiments onstage and beyond, weaving together the threads of collective action and art, thereby expanding the British mainstream theatre vocabulary. Its influence has indeed been so pervasive that it has defined the aesthetic profile of entire British theatre companies, tremendously changing in its wake the conventional apparatus of theatre (i. e. distribution of lighting, the stoical artifice of theatre design, the customary "theological" quality of theatrical representation, the curtain call, the suspension of disbelief, dramatic characterisation, etc.), actor-training, rehearsal processes, the role of the playwright, methods of direction and dramaturgy.[31] As such, artists now increasingly define themselves as "verbatim playwrights", "verbatim theatre-makers", "verbatim story-tellers", or "creative editors." In short, the verbatim methodology is now a standard part of the theatre practitioner's armoury. However, in this period, each verbatim practitioner or company is extremely individual in style despite the fact that the words used are those spoken by other people, as British playwright Robin Soans has suggested: "[i]f you gave David Hare, David Edgar, myself and Alecky Blythe this material you would end up with four different plays" (*North* 108). In effect, verbatim theatre is extremely fluid and flexible and therefore particularly suited to the present times, generally characterised by a "multiplicity of contradictory tendencies and incoherent sensibilities" (Nealon x – xi).

What is more, verbatim methods have now entered the sphere of new writing, which indicates that verbatim theatre has prompted new kinds of theatrical experience, notably with regards to a kind of dramatic language at the very edge of semantic legibility and "the form of a series of interlocking monologues, with a couple of short dialogue scenes, that mimic the kind of verbatim theatre that is

---

31 For instance Word for Word Theatre, Hope Theatre, Snippet Theatre, Recorded Delivery, Look Left Look Right, FYSA Theatre, Every Day, Lung and Use a Coaster.

now a staple of contemporary new work" (Sierz "BU21"). This, in turn, automatically makes one wonder whether it is still important to distinguish verbatim theatre from non-verbatim theatre and, by contrast, whether it might be doing something no other theatrical medium can do. Put another way, its act of cultural vandalism along with its indifference to traditional generic distinctions means that it embraces technology in a timely fashion, whilst committing itself to the virtues of simplicity and minimalism, as will be explored in Chapter 2 of this book.

To a striking degree, then, this contemporary renaissance of verbatim theatre – with its inescapable rhetoric of the new – is more permeated by the political and economic context of neoliberalism. As Liz Tomlin writes, and she is quite right to do so, we have moved away from a context that saw its most experimental/radical theatre practices "recuperated by the capitalist economy once its value is acknowledged" to one that sees, from the start, such work being "fertilized and incubated within the economy's ideological predicates and structures" ("Academy", 264). Regurgitating real people's words and/or ripping material straight from the headlines may appear within reach and, to some, economically convenient as it is supposedly quicker to make and easier to sell, thus undermining all artistry and merit,[32] according to the most widespread line of attack that is.[33] Certainly, in mainstream verbatim theatre, it is the "real" component at the mercy of current events that is the decisive factor in raising the capital for a production.

As the subsequent chapters show, the fact that verbatim theatre operates now in the mainstream and is commercially successful has fundamentally changed the ways to encounter and experience it.[34] The issue of the extent to which contemporary verbatim theatre is "new" – new significance, meanings, forms, improbable pairings and functions – along with the criteria at our disposal to assess such a novelty is a complex and important one. Equally important, it seems, is how contemporary verbatim theatre differs from its previous incarnations and I will perhaps give more weight to this argument than required, in the

---

32 Some verbatim theatre-makers – such as Alan Rickman and Katharine Viner with regards to *My Name is Rachel Corrie* – feed the flames of disputes over the artlessness of the form by denying their authorship and claiming that they are mere editors.

33 In a way, it is as if the mediating technology had been conflated with the product. When the first tape recorders were made available to the public at large, the new technology was marketed "as a layman's technology, a device accessible to amateurs" (Olsson 63).

34 Within a British context, the term "mainstream theatre" generally refers to subsidised building-based producing theatres as well as the theatre companies and artists that perform on mid- to large-scale stages within these institutions.

face of an increasing authoritarian de-historicisation of the practice, combined with a disappearance of the prospect of radical change, the seeds of its political and aesthetic neutralisation.

1.4 Past and Current Research

The importance of verbatim theatre and its leading role in the history of contemporary British theatre are now fully recognised.[35] Its resurgence coincided with the explosion of modes of documentary across the wider culture as well as "a growing obsession with the real throughout the world" (Luckhurst, "Introduction" 1). As is now widely acknowledged, there has been a tremendous resurgence of interest and debates in the field of documentary studies, generally in order to get to grips with the current superabundance of "reality" art practices.[36] The profusion of terms to describe these practices crucially suggests that a clear picture of the contemporary verbatim theatre landscape is not easy to come by. As the 2000s advance, it has become evident that verbatim theatre is testimony to an increasingly global "neo-documentary trend, the most recent turn towards *real* people" (Festjens 165). This phenomenon has prompted a series of publications on "documentary theatres" such as the Autumn 2006 edition of *TDR: The Drama Review* edited by Carol Martin and a special issue of *Zeitschrift für Anglistik und Amerikanistik* (56.2) on "Constructing Reality: The New Documentarism" co-edited by Christiane Schlote and Eckart Voigts-Virchow in 2008 and on "reality theatre" and "the theatre of the real" such as Carol Martin's *Dramaturgy of the Real on the World Stage* (2010) and *Theatre of the Real* (2012) as well as Ulrike Garde and Meg Mumford's *Theatre of Real People* (2016), Maryvonne Saison's *Les Théâtres du Réel* (1998), Danielle Merahi's *Théâtres du Réel* (2016) and Jenn Stephenson's *Insecurity: Perils and Products of Theatre of the Real* (2019).

The enthusiasm for the topic means that verbatim theatre is now a significant body of literature that has endured a wide range of critique in order to describe the same artistic scene.[37] The range of critical frames within which verbatim theatre has been couched would seem, to one's dismay, to have left no

---

35 See for instance Siân Adiseshiah and Louise LePage's introduction to *Twenty-First Century Drama: What Happens Now* (2016), p. 3.
36 See David Shield's excellent *Reality Hunger: A Manifesto* (2010).
37 Writing in 2010, Chris Megson already noticed – on the occasion of an academic symposium held at the University of Reading – "[t]he presence of a small army of younger academics and emerging scholars [...], many completing their doctorates in [the field of documentary performance]" (135).

terrain of thought unmapped. They include, but are not limited to, the following[38]: ethics (Luckhurst 2008), jurisprudence (Ward 2011), cognitive neuroscience[39] (McCutcheon 2013), black aesthetics (Kaneko Lucas 2007), mental health (Summerskill 2014), journalism (Bernbaum 2010), pedagogy (Cooper 2015), intersectionality (Lemoine 2013), ethnography (Saldaña 2011), feminism (Speer 2013), social research (Mienczakowki 2006), oral history (Friedman 2006), trauma studies (Little 2015), human rights (Derbyshire and Hodson 2008), activist arts (Paget 2010), dramatherapy (Sajnani 2012), identity politics (Diamond 2012), gerontology (Schweitzer 2007), new wars (Soncini 2015) and translation (Garson 2013).[40]As these lines are being written, many doctoral researchers all over the world are exploring this renaissance of verbatim theatre from a variety of fascinating angles. One such is British playwright, performer and activist Clare Summerskill who is currently researching the role of the contributor in verbatim theatre at Royal Holloway, University of London.[41] These are of course pointers rather than frameworks that aim to encompass contemporary verbatim theatre completely, as will be the work of this book which can only draw attention to one aspect of verbatim theatre and attempt to find patterns.

For the most part, verbatim theatre scholarship tends to focus on the notions of "political theatre" or "political performance." More specifically, due to the fact that this type of performance is known to have "made politicians and thinkers sit up and take notice" (Crompton 2016), verbatim theatre has generated a particular line of scholarly criticism focused on its relationship with politics. However, it is my contention that such a criticism, that is now commonplace to many accounts, must be modulated if one looks beyond the popularised spate of verbatim productions. Indeed, such views do not hinder theatre practitioners who have been

---

38 For practical reasons, one example of each type of criticism will be given.

39 Artist Lexi Strauss is currently working with neuroscientists in an attempt to understand what happens to the actor when they perform in a headphone-verbatim performance (see Chapter 3.2 for an overview of this type of verbatim theatre). An MRI (Magnetic Resonance Imaging) scan along with other investigations might reveal something important about some fascinating aspects of verbatim theatre.

40 More recently, the interdisciplinary Performance of the Real research group at the University of Otago, New Zealand that explores representations and performances of the real, organised an international conference that sought to apply psychoanalytic and Lacanian understandings of the real to the genre.

41 Some documentary theatre companies such as the Australian Urban Theatre use the term "consultant" whilst the German/Swiss collective Rimini Protokoll prefers to call the people interviewed "experts of the everyday" and the South African playwright Pedzisai Maedza calls them "research respondents" (107), and some others in the UK and further afield use the term "verbatim subjects."

exploring other kinds of practices through the verbatim technique. There is equally an important line of inquiry which concerns itself with historicising the phenomenon and tracing its multiple origins in both theory and practice such as Marvin Carlson's *Shattering Hamlet's Mirror: Theatre and Reality* (2016) and also, to some extent, Alison Forsyth and Chris Megson's *Get Real: Documentary Theatre Past and Present* (2009), Lucie Kempf and Tania Moguilevskaia's *Le théâtre néo-documentaire: résurgence ou réinvention?* (2013) and Erica Magris and Béatrice Picon-Vallin's *Les théâtres documentaires* (2019). These seem to indicate that one of the characteristics of verbatim theatre might precisely be its recursive, perhaps even intertextual nature.

In addition, there is now a substantial body of scholarship devoted to the examination of acting in verbatim theatre, which culminated in the publication of Tom Cantrell's *Acting in Documentary Theatre* (2013).[42] Verbatim theatre has also called forth special attention to education, community, therapy and theatre for social change more broadly (for example, the work of Ice and Fire, Age Exchange, Write to Life, Woven Gold and Collective Encounters in the UK). Crucially, verbatim theatre is often discussed within larger frameworks and rarely on its own. For instance, Liz Tomlin in her book on the ideological discourse of "the real" in contemporary performance practice dedicates a chapter to it (2013) and so does Duška Radosavljević in her *Theatre-Making: Interplay Between Text and Performance in the 21st century* (2013). Similarly, Ursula Canton in her 2011 book proposed a functional approach to "biographical theatre" – a term she uses to describe all verbatim and non-verbatim performances based on biographical material – that relies on the concept of "extra-theatrical performances" (48) and Paola Botham, in her 2009 doctoral dissertation, set up "an alternative framework based on the philosophy of Jürgen Habermas" (307) that she applied not only to verbatim theatre but to contemporary British political theatre.[43] The absence of academic monographs focused specifically on contemporary British verbatim theatre, besides Tom Cantrell's *Acting in Documentary The-*

42 This originated with both the AHRC "Acting with Facts" project at the University of Reading led by Derek Paget and focusing on the problematics of acting in fact-based drama across media, and Tom Cantrell and Mary Luckhurst's book of interviews, *Playing for Real: Actors on Playing Real People* (2010). Mary Luckhurst also published some articles on the subject, notably on the "ethical stress" related to performing real people (2010). The "Acting with Facts" project, besides the wealth of interviews they conducted with practitioners, published two special journal issues: *Studies in Documentary Film* 4.3 (2010) and *Studies in Theatre and Performance* 31.2 (2011).

43 In particular, Paola Botham explored Habermas' concepts of "post avant-garde art" and the "public sphere." Verbatim theatre contributes to this political aim of expanding the public sphere, for instance, when it focuses its attention on unheard and marginalised voices.

atre (2013) and Jess McCormack's *Choreography and Verbatim Theatre: Dancing Words* (2018), suggests that a new theoretical framework is needed to gain a better insight into the latest developments in the field, especially as we are now twenty-one years into the 21st century.

As the novelty of the 2000s wears off, the aesthetic ambiguity of contemporary verbatim theatre has not gone unnoticed but its significance and scope have hitherto been mostly restricted to what I call "documentary realism" here. A very persistent determination on the part of academics and critics is therefore to emphasise its problematic promiscuity with non-art practices. In truth, many of the arguments in this book have emerged in the wake of the vigorous debates that verbatim theatre seems to often occasion. While this study is largely sympathetic to that perspective, I would argue that it is also important to discuss in more detail to what extent contemporary verbatim theatre necessitates aesthetic choices that are not typically present in theatre. Put in another way, a discussion of the aesthetics of verbatim theatre needs to be given a prominent place and this particular research intends to take full account of its importance and challenges, especially its interaction with realism. Therefore, this book will endeavour to effect this in a more essential way than has been done before. As a consequence, my argument will take a different tack by addressing the enmeshment of realism and more experimental aesthetic practices in contemporary British verbatim theatre. The conception of verbatim theatre to be developed in these pages – as the case studies have been selected to illustrate – is that of a performance practice that is remarkably more diverse and prone to aesthetic transgressions. Finally, in this book, I shall also cover under-researched fields such as the use of headphones in verbatim performances (Chapter 3.2), the development of site-specific (Chapter 4.3), musical (Chapter 4.2) as well as physical verbatim practices (Chapter 4.1) and the fictionalising tendency most visible in recent years (Chapter 3.1).

1.5 Methodology and Corpus

1.5.1 Methodology

As elaborated below, the narrative of this book seeks to identify what stimuli produce documentary realism and what might frustrate the audience's investment in such a theatre aesthetic. I thus intend to demonstrate how the notion of "documentary realism" offers a fertile ground for an inquiry into contemporary British verbatim theatre. My assessment is informed by interviews I conducted with practitioners, archival research at the National Theatre and at the Victoria & Albert Theatre and Performance Department and also my direct experience as a spectator. The generosity of the practitioners I interviewed or simply contacted via email – Craig Baxter, Alecky Blythe, Jo Harper, Jonathan Holmes, Nicolas Kent, Linda McLean, Richard Norton-Taylor, Mimi Poskitt, Mark Ravenhill, Lexi Strauss, Clare Summerskill and Louise Wallinger enabled me to access documents and information which otherwise would have been unavailable to me as a researcher.

This volume aims to trace the evolution of contemporary British verbatim theatre from the 1990s to the opening decades of the 2000s in a selection of key high profile performances, in order to make the case for a move away from documentary realism. As such, it takes with each chapter a new step outside and beyond documentary realism. Importantly, it also begins to question the "fit" between realism and verbatim theatre, which will be considered as both essential and elusive. Thus in the course of this study, I shall point out in detail the basic aesthetic differences that arise in the creative methodologies of verbatim theatre-makers according to the degree to which they engage with the tropes of documentary realism. To do so, this research project explores the theoretical approaches (epic, postdramatic, postmodernist, postructuralist and post-Brechtian theory, etc.) that one might use to analyse such a profusion of performances and divides the discussion into three sections that correspond to the identifiable recurrence of a common aesthetic trait, that is to say to three aesthetic formations.[44]

1.5.2 Detailed Overview of the Argument

> At one end of the spectrum are hyper-aestheticized productions that exploit and manipulate source material in the interests of spectacle, aesthetic appeal and audience engagement. At the other, are highly 'ethical' productions where practitioners inadvertently drain the drama from theatrical representation in attempting to preserve a perceived 'truth'. ("Reality")

44 In his *Representing Reality: Issues and Concepts in Documentary,* Bill Nichols identifies four modes of documentary film practice: "expository (the 'classic' mode of documentary), observational, interactive and reflexive" (23). Whilst these categories share some similarities with my own categories – in particular my understanding of "documentary theatre" can be aligned with Nichols' conception of "the observational mode" as marked by an apparent "nonintervention of the filmmaker" (38) – they also significantly depart from them as "reflexivity" is something that two of my own categories share.

Following Suzanne Little – who does not focus specifically on British verbatim theatre – three aesthetic tropes that display or prioritise certain characteristics within this spectrum will be mapped as follows:[45]

Chapter 2 forms a theoretical introduction laying out the key terms around which contemporary British verbatim theatre revolves, as it focuses on two main strands of documentary realist verbatim performances that "construct a discourse of authenticity in the performance, grounded in the reproduced testimony, which thereby makes its truth-claims inviolable" (Megson, "Testimony"). Documentary realism is understood here as the normative mode of presentation for verbatim performances that may indicate an unbroken tradition that links prior manifestations of documentary theatre to those of the present day. Often considered as a simpler form of openly oppositional theatre reflecting a relatively primitive stage of old-fashioned political drama of the socialist kind assimilated to more classical ideologies of representation – as opposed to the more sophisticated poststructuralist practices identified by Liz Tomlin in her *Acts and Apparition* (2013) – this part strongly questions the validity of these claims.[46] In order to fully comprehend the hypothesised shift in scenic practice over the period, it is necessary to define "documentary realism", the aesthetic concept that serves as a basis for all subsequent chapters. Chapter 2.1 will therefore provide such a grounding in order to offer a framework for the investigation of a significant number of verbatim performances in Britain. To flesh out the scope of this theorisation, two representative strands of verbatim theatre will subsequently be analysed: the "tribunal play" in Chapter 2.2 and the "documentary realist verbatim play" in Chapter 2.3. Before drawing some conclusions from this analysis, I turn to Chapter 2.4 and introduce more problematic case studies and complicating factors, serving as transitions to the next aesthetic terrain.[47]

Chapter 3 shifts attention to what I call "new realism", an aesthetic trope that simultaneously performs and criticises the processes of documentary realism, playing with its regulative mechanisms. In other words, it marks the first shift of the main parameters discussed in Chapter 2. More specifically, it considers the extent to which verbatim theatre's aesthetic trajectory has started to deviate from documentary realism in works I call "new realist." Chapter 3.1 will

45 I therefore do not perceive a simple dichotomy between "aesthetics" and "non-aesthetics", but rather a continuum of practice with three main points on the line.

46 For many critics, as will be explored later, the (re)emergence of realism as the central and structuring element in verbatim performances is no longer a satisfactory model of political theatre.

47 Clearly, such a neat compartmentalisation risks being more tidy than accurate, hence the addition of a chapter entitled "Transgressions" to Chapters 2 and 3.

look at what happens when fiction is openly included within the frame of a verbatim performance and Chapter 3.2 will address a new and intriguing type of verbatim theatre, the "headphone" practice, that seems to both sustain and break documentary realism. Chapter 3.3 will finally showcase more transgressive examples of verbatim theatre, fundamentally challenging my conclusions as well as offering a preview of the aesthetic contours of the next part.

The third part of this book is framed by the notion of "post-realism" – that is, the moment or situation when verbatim theatre wrestles with contradictory aesthetic paradigms and is transformed into an equivocal performance that challenges its own ontology. In other words, this is a seemingly postdramatic reinvention of verbatim theatre that radically transcends its own condition and profoundly destabilises its realist claims. Here, through this new mode of expression, the verbatim material is completely torn apart from its "original" context(s) and engaged in a free play of signifiers. This outright manipulation creates a new theatrical vision that wrestles with the very idea of realism. "Post-realism" thus names the strange state of affairs in which some recent British verbatim performances find themselves in the twenty-first century. In sum, this is verbatim theatre seeking to become post-realist without ceasing to be verbatim. This part is therefore concerned with the investigation of current "post-realist" practices within contemporary British verbatim theatre: Chapter 4.1 addresses physical verbatim theatre, Chapter 4.2, the verbatim musical and Chapter 4.3, site-specific verbatim theatre.

Undoubtedly, such theoretical moves possess a certain dose of artificiality. However, they do not constitute a cavalier dismissal of documentary realism but, rather, introduce the notions of degree, accent and emphasis. As the book progresses, each part is meant to gain force and precision from its relation to the other two, as verbatim theatre almost effortlessly flows from one aesthetic instantiation to another. In other words, the spectator may experience those modes simultaneously within a single verbatim performance and one must therefore be alert to anything that breaks a particular mode. Therefore, this book seeks to establish an aesthetic typology of contemporary verbatim theatre practice in Britain and attempts to find commonalities across richly diverse articulations and practices. Documentary realism can be understood as a common denominator, engine and aesthetic framework but, as one delves more deeply and carefully into these works, it becomes evident that the notion of documentary realism certainly does not apply exhaustively to today's verbatim output.

1.5.3 Case Studies

The book bases its analysis on seven main case studies and focuses on the performances themselves. It describes the experience of seeing them in a theatre or reconstructs what it may have been like for the spectators, thanks to a video recording and other archival sources. As such, each of the seven performance in my corpus was either seen first-hand or made available to me by way of a video recording and script, courtesy of the theatre-maker or various archives and libraries in London. The central case studies chosen are the following: **Richard Norton-Taylor's *Justifying War*** (directed in 2003 by Nicolas Kent at the Tricycle Theatre, London); **Robin Soans' *A State Affair*** (directed in 2000 by Max Stafford-Clark, at the Liverpool Everyman); **David Hare's *Stuff Happens*** (directed by Nicholas Hytner in 2004 at the National Theatre, London); **Alecky Blythe's *Come Out Eli*** (directed by Sara Powell in 2003 at the Arcola Theatre, London); **DV8 Physical Theatre's *To Be Straight with You*** (directed by Lloyd Newson in 2008 at the National Theatre, London), **Alecky Blythe and Adam Cork's *London Road*** (directed by Rufus Norris in 2011 at the National Theatre, London) and **Jericho House's *Fallujah*** (directed by Jonathan Holmes in 2007 at the Old Truman Brewery, London).[48] The prominent practitioners of this type of theatre draw on a rich series of historical precedents (tribunal theatre, documentary theatre, etc.), while some others seem to have broken new ground and invented whole new ways of making verbatim theatre. The above seven performances – not usually considered together in discussions of verbatim theatre – that make up the major portion of this book reflect how these artists have transformed the familiar documentary devices and strategies of previous decades in response to the growing chaos of global culture in the information age. Taken together, they exhibit the richness of the field at the present time and its promising future. I have chosen these particular performances in my corpus mostly because of their usefulness in foregrounding a set of aesthetic preoccupations that I see as recurring in the vast field of contemporary British verbatim theatre. That the work of Alecky Blythe has a central place in this study, for instance, has less to do with her unquestionable mainstream prominence in the period under discussion than with its manifestation of the aforementioned preoccupations. I readily recognise the London-centric bias of the case studies. Readers may also wonder about the paucity of case studies from Wales, Northern Ireland and Scotland. Given that – to my knowledge – most theatre reviewers and blog-

48 As I was unable to see the production of Robin Soans' *A State Affair* in 2000, I have relied on a performance recording from when it reached the Soho Theatre in London.

gers are still very much London theatre-goers and that video recordings of performances are rarely made outside London, I was unable to gain access to enough documentation that would have enabled me to focus on a more geographically-balanced corpus of performances. Due to my ignorance of the Welsh language, I could not include Bethan Marlow's *Sgint* (2012) – the first ever Welsh language verbatim play – for instance.

A few additional words on this book's procedures: each analysis is strictly based on a specific production in the UK – in order to include a consideration of casting, physical imagery, props, lighting design, stagecraft, acting, sets, soundscape, directorial decisions and any other element that makes up the unpredictable excess of performance. Here, I follow the conviction of Alan Filewod in *Collective Encounters: Documentary Theatre in English Canada*, who argued that "the documentary theatre must be considered a genre of performance rather than a form of literary drama" (ix). Indeed, the operative principles of verbatim theatre are to be found in the structures of the performance, rather than the playtext. Most tellingly, verbatim performances appear to be less rigidly defined than traditional plays and it is rather common to see the performance texts being updated, as was the case with David Hare's *Stuff Happens*, the case study in 3.1. The last chapter on "site-specific verbatim theatre" certainly accounts for its fluidity as the physically mutable works crucially changed with each spatial incarnation. This being said, this study also makes reference to the published playtext or unpublished typescript in order to draw on its possible actuality and variables. In my view, this procedure justifies the concentration on high-profile theatre-makers who still serve as recognised models of possible ways of conceiving a verbatim performance.

1.5.4 Transgressions

> Every text participates in one or several genres, there is no genreless text; there is always a genre and genres, yet such participation never amounts to belonging. (Derrida, "Genre" 212)

> A question that is frequently asked about verbatim theatre is whether it really has any place in the theatre at all. (Haydon, "Terrorists")

The attentive reader will have aptly noticed the omission of "transgressions" from the earlier survey of the key terms for this study.[49] In fact, it is because it

49 I also omitted a section on aesthetics because I am focusing on specific manifestations such

deserves its own separate account here. One of the most common questions raised by contemporary verbatim theatre (besides the question of authorship) is the following: "But is it theatre?" (*Economist*).[50] Indeed, as Sara Soncini has claimed, verbatim theatre's overlap with non-fictional domains, as well as its aesthetic project of "anti-theatricality and facticity" (*Conflict* 80), urgently raise this question. Interestingly, she then goes even further by suggesting that it prompts "a reconsideration [of] the very fundamentals which define the field of drama" (*Conflict* 80).

Fundamentally, the constraints of orthodoxy and audience expectation have not, it would seem, weighted so heavily on contemporary British verbatim theatre-makers and some performances even appear devoid of the semiotic codes of theatrical representation, tacitly breaking an unspoken contract and leading to a great richness of forms as well as a relentless bending of norms.[51] In a paper given during the "Verbatim Practices in Contemporary Theatre" Symposium at the Central School of Speech and Drama, London in July 2006, Jonathan Holmes stated the possibility within verbatim theatre to "sidestep certain structural expectations" (Jeffers 5) such as narrative closure and linear narrative, based on causes and effects. As a result, there has been a strong, widespread feeling that its place may, after all, not be in the theatre, as Andrew Haydon above, and countless others have suggested.

In this book, contemporary British verbatim theatre is therefore re-examined through a lens focused primarily upon aesthetic innovation and how it troubles and upsets conventional critical affiliations and categories in three overlapping domains of expression. The term transgression – that Hans-Thies Lehmann understands as "the infringement of prescriptions" (*Postdramatic* 178) – is crucial to my arguments about verbatim theatre because it is a chief characteristic of its aesthetics and its capacity for rebellious deviance. By "transgressions", I mean those breaches between the perceived mechanisms of verbatim theatre and the mainstream British stage that indicate new directions and strategies that depart from the accepted rules of realism. This book thus turns to the

as "documentary realism" within the sphere. I consider aesthetics here as a reframing of the verbatim material as art.

50 The article was written in response to the verbatim play *Guantanamo: Honor Bound to Defend Freedom* (2004).

51 For instance, verbatim works can be seen as a threat to the cultural significance, status and prestige of the British playwright or to the text-based tradition more broadly, thus provoking an acute sense of disquiet. One instance of this is the Royal Court Theatre's attribution of authorship to "writings of Rachel Corrie" in the case of the verbatim play *My Name is Rachel Corrie* by Alan Rickman and Katharine Viner.

new experiential verbatim performances that theatre-makers are now producing on the British stage at a time when audiences appear more and more trained in the new habits and practices of mainstream verbatim theatre.

In his "A Preface to Transgression", Michel Foucault argues that "[t]ransgression is an action which involves the limit, that narrow zone of a line where it displays the flash of its passage, but perhaps also its entire trajectory, even its origin; it is likely that transgression has its entire space in the line it crosses" (33–34). In other words, the relationship between transgression and limits is far more complex than it would seem at first. In the quotation from Jacques Derrida that opened this section, it is proposed that the notion of genre itself implies the notion of transgression and vice-versa. Indeed, the reigning question of genre and its taxonomic certainties is never long in raising its head and transgressions could be conceptualised in terms of degeneration, that is to say as performing minor deviations with regards to both production and reception. Transgressions therefore characterise here those performances that demarcate themselves and do not intend to remain within the existing limits drawn, situating themselves at the edge of such an established demarcation. Ceaselessly inventive and shifting stylistic registers, their gestures consistently express a certain tiredness with the conventions of mainstream British theatre and may not fail to strike one, in this context, as shocking, artistically and culturally. In other words, transgressions are anomalies that cannot be equated with accidents, lapses or mistakes. The verbatim performances analysed in this book represent a disruption or deviation and a challenge to received understandings of verbatim theatre and re-perform an aesthetic split (break with established and mainstream British theatre practices) by unapologetically breaking both the promise of documentary theatre and the rules set down by the British theatre establishment.[52]

It should now be noted that the term "transgression" is reputedly "quite elusive" (Cieślak and Rasmus 2012, 1), as well as heavily weighted by historical and political critical contexts. It might be seen as inadequate or naïve today since Philip Auslander was already writing in 1987 that "the postmodern condition" had "necessitated the transition from *transgressive* to *resistant* political art" (21). In particular, Hal Foster described this notion in terms of "avant-garde transgression" and by that he meant that "it is culturally specific and historically bound – to a productivist model (as in 'change the apparatus'), to a simplistic notion of ideology as encoded class beliefs (as in 'transgress the conventions'), to a now tenuous idea of art as instrument of revolutionary change – all this

52 Janelle Reinelt considers "the promise of documentary theatre" to be one in which "[t]he value of the document is predicated on a realist epistemology" (2009, 7).

apart from its complicity with capital" (149).[53] He then explained that the situation obtaining at the time of writing indicated "a later, greater moment of capital, which in turn suggests that the structural codes which the modern avant garde sought to transgress no longer exist as such or are no longer defended as such by the hegemonic culture. In this new, all-but-global reach of the capital, there may be no natural limit to transgress" (152). It is significant too that Foster also stresses the fact that he is not saying "that there is no structural outside or cultural others" (152). Instead, of a strategy of transgression, he advocates "a new critical strategy of resistance from within" (152). As Robert Radin argues via Paul Mann, "the current move toward neocriticism, with its emphasis on 'internal resistance' is just as suspect as any notion of critical autonomy; that is, it serves the same function within the discursive economy. It is another instance of the way power deploys the resistance that is necessary for power to exist in the first place" (49–50). Today, as Patrice Pavis argues, and this is a view I endorse, transgression has changed but is still a possibility within the social/cultural codes of the moment and "has a bright future ahead of it" (*Routledge* 266).

To perceive verbatim theatre as a site of aesthetic transgression is to endorse what Jon McKenzie in *Perform or Else* calls the "*liminal norm*" (his emphasis, 50) which "operates in any situation where the valorization of liminal transgression or resistance itself becomes normative" (50). Transgressions also indicate in this context an engagement with the underside of our global culture, politically disenfranchised and marginalised subjects, previously excluded groups, that opens a space of possibility beyond representations in the media and our overarching cultural narratives.[54]Verbatim theatre is finally transgressive in the way in which it interacts with the world and may in effect have the potential to disrupt the perpetuation of dominant discourses by blurring the distribution of roles and going against received ideas. As Timothy Youker argues, "[a]ny documentary play can potentially register dissent from institutionally endorsed assumptions about who or what deserves a place in public discourse" (219). Perhaps more contentious – albeit fascinating – is his claim of another kind of transgression: a kind of "documentary theatre that overly dissents from the very conventions that struc-

53 Transgression here implies "revolutionary transgression of social and cultural lines" (149) and it is proposed that "resistance", in the sense of "immanent struggle within or behind [these social or cultural lines]" (149), is a more suitable term at the present time – at least at the time of Foster's writing – to describe political art.

54 If they are used well through a collaborative act inspired by oral history methodologies, these transgressive voices can be capitalised upon strategically to instigate change. However, in some other circumstances in today's culture, their power may be diminished and recuperated as a useful marketing tool.

ture public discourse itself, particularly insofar as public discourse has been structured by the disciplinary conventions of journalism, historiography and the social sciences" (219). Together, these transgressions share an oppositional stance towards more "orthodox" approaches and suggest that in an age of global scepticism, it is all the more important to avoid dismissing the radical potential of contemporary performance in practice. Finally, Chapters 2 and 3 of this book feature themselves a "transgression" section to consider verbatim performances beyond the case studies, that step outside the theoretical framework I have attempted to draw each time.

Before embarking upon a discussion of the further implications of the terms put forward above, I must now turn to "documentary realism" in more detail to make somewhat clearer the unifying thread tying together contemporary British verbatim performances as well as the universe of aesthetic discourse within which the ideas in this book have been formulated.

2 Documentary Realism

2.1 Verbatim Theatre in its Aesthetic and Theoretical Context

At every turn, as already seen in the Introduction, theatre commentators would have one think of verbatim theatre as a unified body of works that mounted a successful assault on new writing and exuberantly took command of the established cultural field. And so say "new writing dramatists" themselves, with considerable confidence, as they perceive verbatim theatre and devised performances – however strange and various they may be – as partners in the same cultural project, presumably working towards the one rather tight single end of making themselves accessible to ordinary consumers.[1] So much the worse, it would seem, for multi-authored verbatim productions.[2] On this basis, that same reductive gesture has been repeated by several disgruntled critics many times over and in many ingenious and abundant forms. Against these, however, my investigation begins where most such inquiries end as I remain unconvinced that the carefully drawn line separating verbatim theatre from new writing entirely boils down to a one-way street.[3] Thus, I maintain that the renaissance of verbatim theatre in Britain since the mid-1990s has been by no means a comprehensive uniform trend and has arguably been a manifold and diffuse phenomenon, especially if one attempts to trace its recent developments within applied theatre settings (a fascinating trajectory beyond the remit of this volume).[4] Clear-

---

1 Devising is to be understood as "a method of performance development that starts from an idea or concept rather than a play text [that] is from the start significantly open-minded about what its end-product will be; and uses improvisation – by performers, but also creators, including writers, designers, directors and choreographers – as a key part of its process. Its composition often happens concurrently in a variety of creative areas, including live performance, mediation, and the development of props/objects, machinery, texts and images" (Harvie, "Introduction" 2). In other words, as Alison Oddey already wrote in 1994, this way of working "challenges the prevailing ideology of one person's text under another person's direction [...] the single vision of the playwright" (4).

2 See for instance, Buffini *et al* (2011) or Wilson *et al* (2006).

3 See (Garson "Kindred") for a more elaborate articulation of this issue.

4 Thompson defines "applied theatre" as projects that "always take place in communities, in institutions or with specific groups. They often include the practice of theatre where it is least expected; for example, in prisons, refugee camps, forgotten estates, hospitals, museums, centres for the disabled, old people's homes and under-served rural villages: sometimes in theatres. Applied theatre is a participatory theatre created by people who would not usually make theatre. It is, I would hope, a practice by, with and for the excluded and marginalised [...]. In circumstances where celebratory escapism is dominant, it can be the theatre of serious enquiry" (15–16).

https://doi.org/10.1515/9783110715767-004

ly, it needs to be understood as a complex and contradictory process influenced by a range of different contexts as verbatim theatre now largely operates within a mainstream globalised territory. The postdramatic turn[5] as well as the rise of the independent sector in the 2000s have indeed prompted a number of irreconcilable changes and these profound changes have inevitably transformed the field of verbatim practice.[6] In other words, verbatim theatre thus simultaneously shapes (Sierz, Schnierer and Middeke xiv) and is shaped by the British theatre landscape at this point in history. What emerges so strongly among the various aesthetic temptations currently exercising the British stage is exactly an unspoken project to rehabilitate realism best exemplified by the contemporary resurgence of verbatim theatre and its turn to documentary realism in the period under scrutiny.

Documentary realism – a historical aesthetic exponent that unproblematically draws the audience into the reality of a particular situation, topic, event or narrative being dramatised and authenticated through verbatim sources – has not become obsolete by the advent of more "self-reflexive" modes of performance (Tomlin, *Acts* 208) and is persistently treated as the default representational mode for verbatim performances.[7] As such, documentary realism can be understood as a paradigm, that is to say a set of aesthetic principles and assumptions which provides the framework within which all verbatim performances can take place, settling on this initial formulation. As indicated earlier, there are two broad categories of verbatim plays that could be termed documentary realist: the tribunal play or verbatim courtroom drama and a subdivision of the verbatim output in the past two decades that is hardly stylised and features

5 Mark Berninger *et al.* define this movement via Lehmann towards postdramatic theatre as "a disruptive, non-conventional theatre which relies on physicality instead of text-based drama and stresses the importance of the spectator, thus creating an aesthetic of 'response-ability'" (10).
6 Liz Tomlin uses the term "independent" in her *British Theatre Companies* to replace the term "alternative" in the contemporary period. The term applies to any UK theatre company that is not "building-based or, at least, [when] the company identity pre-dates any building that has been acquired" (xi). She draws this understanding from a 2004 book chapter by Baz Kershaw which describes a change that has been occurring since the 1990s: "claims for a cohesive alternative theatre movement in the 1970s were beginning to look like nostalgic myth-making [...] experimental practitioners increasingly embraced the pluralism of post-modern theory and culture, often under the banner of 'independent theatre' [...] [s]o in the early years of the 1990s there was 'a new spirit of innovation' in British Theatre" (371).
7 The notion of "documentary realism" is not solely confined to the world of theatre. See Sauerberg (1991) on how this term applies to the contemporary novel, Nichols (1991) in relation to film studies, Cooke (2015) for an insight into its usage within television drama, Wells (2015) as regards to photography and Priseman (2015) on how this intersects with contemporary painting practices.

a strong affinity with theatrical realism.[8] The plays in this last category encompass a very wide spectrum of activities ranging from Robin Soans' *Talking to Terrorists* (Royal Court Theatre, 2005), Gary Owen's *Big Hopes* (Royal National Theatre, 2008), Clare Summerskill's *Hearing Voices* (Cochrane, 2010), Bette Bourne and Mark Ravenhill's *A Life in Three Acts* (Traverse Theatre, 2009), Nell Dunn's *Cancer Tales* (Royal Academy of Dramatic Art ,2001), Steve Gilroy's *Motherland* (Live Theatre, 2007), Zoe Lafferty's *The Fear of Breathing* (Finborough Theatre, 2012) to Victoria Brittain and Gillian Slovo's *Guantanamo* (Tricycle Theatre, 2004) to name only eight of them.[9]

In a 21st-century British context, the feasibility and desirability of documentary realism as an aim for the aesthetic expression of verbatim material in performance needs to be addressed. One question that springs to mind here concerns what it means to have documentary realism in the present moment. From the outset and excluding any political and ideological oversimplification, the very term "verbatim theatre" itself has a strong association with certain assumptions and expectations that generally characterise theatrical realism. As described in the Introduction, this is what Australian playwright Christina Wilson has referred to as the "fetish of the verbatim" (122), what Janelle Reinelt concedes to be "needlessly up[ing] the ante on the promise of documentary" ("Promise" 13) and what Paola Botham has rightly signalled in the following manner: "[i]t is highly significant that the name 'verbatim' has become the metonymical denomination for all contemporary documentary theatre in Britain [...] it involves a risk of placing too much trust on the spoken word as a carrier of truth" ("Deconstruction" 316). Importantly, verbatim theatre always appears to exceed the lure of mere "theatrical realism" insofar as it categorically triggers an unwritten contract stating that the verbatim play has a potential referential quality and this is where it reaches the shore of documentary realism "conjoining [...] two concepts involved into one representational format" (Böger 12).[10] In other words, it can be heralded as a sort of heightened version of realism. In the literature, its conceptualisation remains ambiguous, new and old terminology endlessly seeking to capture its essence with more or less unease: "a 'real' theatre" (Harvie, "Introduction" 13), "theatre of the real" (Martin, "Introduction" 1), "reality theatre" (Wake, "Definition" 5), "theatre verité" (De Jongh, "Bloody" 473), "nat-

8 Other designations in the literature include (but are not restricted to) "documentary trial play" (O'Connor 6), "verbatim tribunal theatre" (Sierz, "1990s" 48), "tribunal verbatim theatre" (McCormick 61), "tribunal-style verbatim theatre" (Cox, *Migration* 28), "inquiry drama" (Lachman, "Journalism" 306) and even "tribunal documentary drama" (Mason 268).

9 Clare Finburgh also classifies *Guantanamo* as a an example of "documentary realism" (2014).

10 See Richard Rorty (1976).

ural performance" (Stucky) "non-fiction playwriting" (Sally Bailey qtd. in Landy and Montgomery 230), "theatre of fact" (Grellet 63), "[t]heater of [d]ocumentation" (Freshwater, *Censorship* 76), "word-for-word theatre" (Saldaña 14)... the list could go on ; the slightest variation in nomenclature crucially altering what can be expected of the performance.[11]

Does it make sense, then, to think of verbatim theatre as (or in relation to) realism? The answer is, I would argue, a resounding yes although the exact nature of that relationship is complex and intricate, straddling conflicting paradigms within its methodological developments.[12] As to be seen more clearly in what follows, one is to regard "realism" as a shifting set of co-ordinates contingent on "the frame, window, or perspective of its mise-en-scène" (Murray, "Introduction" 7). In this chapter, I want to chart some of the lines of connection, tension and disjunction that hold together this strand of verbatim works and their historical model(s). Far from advocating an outright rejection of realism in this context, what this overall part will also attempt is a reassessment of verbatim theatre's engagement with the tropes of realism. It will thus take issue with the central tenet of much critical writing on verbatim theatre through a detailed study of some representative examples of works operating in this mode, namely the Tricycle Theatre's *Justifying War: Scenes from the Hutton Inquiry* (2004) in Chapter 2.2 and Out of Joint's *A State Affair* (2000) in Chapter 2.3. The endpoint of this discussion will be the various ways in which theatrical realism has been modified and its rules transgressed through the use of verbatim strategies. Beneath this argument is another one which runs throughout the entire undertaking of the study, pursuing the following enquiry: do traditional dramatic and theatrical categories have any purchase on the presentation of verbatim material in performance and conversely, does the use of verbatim material impact on these established dramatic and theatrical conventions?

In the following, I will provide a succinct review of past prominent debates and theories of theatrical realism and naturalism (Emile Zola, Bertolt Brecht,

11 See Saldaña (13–14) for a more exhaustive list.

12 I therefore align my thinking with Caroline Wake who pointed out "how closely the practice of recording the voices of "ordinary" people resembles the early practices of realism and naturalism" ("Mimesis" 107) and that, as a consequence, "theatre scholars would do well to reconsider claims that verbatim theatre is a form of documentary or epic theatre and instead consider it in terms of realism or neo-realism" ("Mimesis" 107). Similarly, Michael Rothberg strongly argues that "the need for a rethinking of realism is signalled by the emergence in the last decades of various new forms of testimonial and documentary art and cultural production" (9). Conversely, Stephen Lacey writes that "the term documentary was sometimes used in the discussion of working-class realist theatre" (*Wave* 121).

Raymond Williams, etc.) as well as a contextualisation of the notion of "documentary realism."

2.1.1 Documentary Realism in Hostile Context: A Sceptical Age

In the first decades of the 21st century, perhaps as never before and with increasing insistency, one inhabits a reality overdetermined by mass-mediating culture telling one (to one's greater lassitude) what is real, one of the further refinements of that process being Periscope, a new live-streaming app for iOS that is now joined by several competitors including Facebook Live, YouNow, Youtube Live and Instagram Live among many others. As these lines are being written, there are clear indications that this new technological advance along with VR will be greeted – as was the onset of Twitter and YouTube over the past few years – by equally rapid theatre and performance developments[13] staging and conceptualising the real anew but I will not be dwelling on these here.[14] Most important to this project, however, is how these everyday ordinary "performances" are symptoms of the pressing "current demand to explore the frontier between mediated and 'real' experience" (Schlote and Voigts-Virchow 108) and to realise once and for all a perfect symbiosis between the "real" and its representation validated through the latest in-real-time global activity of reception.

It will become apparent that this desire to palliate the aforementioned imbalance, quandary or impasse has been solicited, consecrated and staged – to the horror of one's postmodern eyes – in numerous works that establish their truth and documentary stimuli through recourse to extra-performance elements such as non-actors in verbatim theatre and non-verbatim theatrical pieces alike.[15] The use of such markers and vibrations of the real in theatre performances keeps these works as close as possible to an erasure of the stage that forces one to rethink notions of theatrical presence over the actual practices of imitation, as one knows them. I will go even further as I want to stress the fact

---

13 See Megan Vaughan's 2015 article in the Work Cited section.

14 Technology has played a key part in verbatim theatre from its inception and this will specifically be the object of Chapter 3.2 as I delve into the headphone-verbatim strand. See also my own article (Garson "Remixing") for a more focused study.

15 British director Katie Michell experimented for instance with the presence of "real" scientists in lieu of actors in two Royal Court productions, namely *Ten Billion* with Stephen Emmott in 2012 and *2071* with Chris Rapley in 2014. In Philip Ralph's verbatim *Deep Cut*, the "real" Des and Doreen James came on stage at the very end of one performance in a riveting finale and "addressed the audience directly as all actors did" (Ralph qtd. in Radosavljević 212).

that my own research practice, for the most part, is not exempt from an implicit desire to appropriate such epistemological groundings or plain absolutes. As I have already suggested in referring to my methodological practice in the Introduction, there is a "corrosive" side to my own activity as a researcher for, "theatrical representation is finite, and leaves behind it, behind its actual presence, no trace, no object to carry off" (Derrida, "Cruelty" 15) and, as Peggy Phelan makes clear in *Unmarked*, "[t]o attempt to write about the undocumentable event of performance is to invoke the rules of the written document and thereby alter the event itself" (148).[16] For Caroline Wake, the challenge raised by the specificities of researching documentary theatre demands a "self-reflexivity within the discourse" ("Review" 8), which I will ceaselessly attempt to provide in the following pages as I identify what I take to be the most basic and enduring components of a documentary realist paradigm in verbatim theatre.

Despite the immense popularity of verbatim theatre in recent years, the turn to predominantly documentary realist strategies has tarnished – at least for now – its credibility as a viable form of political theatre. For its detractors, these "backward-looking" aesthetic strategies belong to "an old fashioned theatrical vocabulary"[17] (Monks 347) that invoke a nostalgic mood of some sort and are therefore deemed out of touch, not having kept up with the critique of representation in philosophy, critical theory and stage alike.[18] This was most famously expressed by David Edgar as he argued that contemporary verbatim theatre constituted "the form of a current renewal of political theatre [...] [and] in essence, we[re] back in the 1970s" ("Secretary" 112). I may also mention Elisabeth Angel-Perez's claim according to which "verbatim theatre remains [a] form of modernity, not of postmodernity" ("Nation" 68) as well as Michał Lachman's contention that "verbatim drama [...] innocently believes that the objective truth of facts can be reconstructed from the chaotic pool of rioting, subjective voices" ("Colours" 321). If one is to believe these critics, verbatim theatre (in its documentary realist format that is) would represent a comfort in a world of ever more insecure truths

16 A splendid "corrosive" side that Ruby Cohn has considered as a sort of "skyzophrenia" (2), as one can hardly avoid with such works that "the life of performance congeals into type" (2) and that "plot summar[ies] [...] impos[e] linearity on deliberately disjointed works" (2). However, I do not see this as necessarily giving a lethal blow to any scholarly endeavour, quite the contrary, as long as one manifests early on an awareness of these limitations.

17 This is not the position of Monks, in the quoted article she only summarises the dominant critical narrative around verbatim theatre.

18 Edward W. Said writes of the contemporary intellectual climate (regarding these contested matters) that it "has recently brought forth literally acres of disputatious prose among critics and philosophers" (xxiv).

and the last desperate and futile grab at an increasingly elusive referent. More surprisingly, even playwrights and makers of other stripes of theatre in British theatreland use the familiar and recognisable narratives of a pre-postmodern era, or appear to when describing verbatim theatre. Here are a few exemplary quotations:

> it is certainly as close as we are going to get in the theatre to the 'truth'. (Ralph 23)

> [g]iven that most art-forms, in the hands of metropolitan elites, tend to drift away from reality, what could be more bracing or healthy than occasionally to offer authentic news of overlooked thought and feelings. (Hare, "Factual" 112)

> [t]he role of the theatre in exposing the truth and reality, unvarnished, is making a welcome comeback. (Norton-Taylor "Spirit")

> [t]his is also about truth, about facts, about quotes. This is not about opinions or fiction. (Levy 3)

> [w]hat I value above all else in theatre is its immediacy, the directness with which it can communicate: the unmediated transfer of ideas, emotions and stories from one person – the actor – to another – the audience member. (McLean, "Introduction")

Taking these statements together, one gets a sense of how the verbatim enterprise might arouse the suspicion of some people. Frequently, then, on the grounds of this presumed attitude towards the use of verbatim material in a theatre setting, contemporary non-verbatim playwrights, such as Simon Stephens, began to express their accusations: "I am [..] nervous about current trends for verbatim theatre, because the writers of many verbatim plays seem to place too high a value on questionable notions of authenticity" (xii). Although there are excellent reasons as well as "a necessary impulse" (Botham, "Deconstruction" 315) for pursuing such debates, I find these ardent critiques useful only up to a point as it is my contention that proceeding exclusively in this manner does cut off certain lines of enquiry.

Intrinsically, the purposes of the documentary realist project are in some ways hard to read and far from being conclusive. It is also my claim that the critical denigration and outright rejection of these plays as theoretically compromised, untenable or "inadequate" (Upton 186) is quite problematic. Indeed, as noted in the Introduction, these plays typically make empirical claims that are always vulnerable to evidential checks (admittedly a kind of process of trial and error) according to one's effective cultural epistemological practices and not *a priori* claims.[19] The term itself invokes theoretical speculation, a complex

---

19 However a strict adherence to this narrowly prescriptive programme may have some disas-

and constantly shifting field in its own right. In fact, so pervasive are its heuristic documentary claims that one feels compelled to a process of critical verification that corresponds to the "inferentialist" or epistemically dependent picture of testimony identified by Robert Audi:

> it is apparently not as direct a source of belief as perception: it yields belief only through both the testimony itself *and* one or two premises that support the proposition attested to or at least the attester's credibility [...]. (406)

By the same token, Derek Paget even proposes that "documentary *theatre* has always been reflexive in that its means of persuasion are so obviously 'on display'" ("Broken" 228) while Noël Carroll maintains that "*not calling attention to* such construction processes [does not necessarily] amount to *denying* the existence of those processes" (293). What is more, Linda Ben-Zvi even contends that, on the contrary, documentary theatre can and quite often does challenge the master narratives:

> theatre is always an illusion, one that may represent reality but is not identical with that which it imitates—always 're,' as Elin Diamond puts it (2000:32)—it has the potential to function, like citation, in a double game, using materials of daily life while offering up its own performative, inexact duplication to provide a space for destabilizing from within the simulacra that posit themselves as totalizing reality. (44)

In suggesting this possibility, I acknowledge and depart from several established critical perspectives. Amusingly, a sound designer has belatedly glimpsed this alternative dimension in reference to the plays I am to analyse in the following chapter:

> [i]n the early days of what is now known as verbatim theatre, the director Nicholas Kent asked me to design the sound for one of his Tricycle Theatre productions. He said he wanted recordings of actual location sound to run under whole scenes. I declined the gig because I thought this approach was theatrically naïve. With hindsight I regret not taking the job – I think he was actually suggesting something daring and innovative which was consistent with the emergent premise of verbatim theatre. (Brown, *Sound* 35)

In this climate, the standard academic response has been to consider verbatim theatre as a fascinating theoretical anomaly and advocates of postmodern tech-

---

trous consequences for documentary realism as an aesthetic model as Jameson has posited in his *Signatures of the Visible* in the case of realism: "if realism validates its claim to being a correct or true representation of the world, it thereby ceases to be an *aesthetic* mode of representation and falls out of art altogether" (158).

niques have thus vividly expressed their concerns regarding its "manipulative and worryingly unreflexive [treatment of] the 'realities' [it] purport[s] to discuss" (Bottoms, "Document" 67).[20] In fact, one has been so critically encouraged to treat that type of realism with suspicion (if not disdain), that one is desperate for theatre-makers to acknowledge the garments in which they have clothed the verbatim material only so that they can eventually be torn. One longs for the subtler and seductive momentary interruption at a point of symbolic significance, when the verbatim veneer gleefully cracks open causing a destabilising effect to reveal the mechanisms under which the performance functions, a deeper, further significance that would irresistibly cater for what current global dominant ideologies bestow upon it. Under these conditions, the more recherché calculated disclosing arrangement only serves to legitimise the work by confirming the facile suspicion under which every verbatim performance is automatically subjected so as to perhaps, ultimately, fetishise its own failure.[21] Thus, one might say that the fetish I have already mentioned is only displaced from one position to another theoretically above reproach (under the terms I have been sketching here).[22] The proof of this hypothesis may be found in the elaborate and trenchant process by which Dennis Kelly famously penned in 2007 this new fetish in his fictional play *Taking Care of Baby* about a young mother, Donna McAuliffe, convicted of the murder of her two infant children. The play is essentially composed of "fictional" edited interviews conducted by the "fictional" playwright Mr Kelly with Donna and her surrounding world in this sinister story. In *Taking Care of Baby*, Kelly suddenly revises and dissolves the classic

20 This is of course not a position shared by all in the literature. For instance, Paola Botham argues that "the strength of this form lies precisely in its power to exceed postmodernism's infinite itch for deconstruction" ("Deconstruction" 316) eschewing the often-discussed political apathy and relativism of postmoden theatre.

21 Deirdre Heddon even argues that, on the contrary, "these rhetorical appeals to "fairness" serve to further mask the playwright's power" ("Ethics" 119) as does Ryan Claycomb when he writes that the foregrounding of mediating practice can have the opposite effect "to suggest authenticity, without necessarily affirming authenticity itself" (102).

22 A more vivid critique of the postmodernist and poststructuralist theories of the documentary is formulated by Carl Plantinga as regards non-fiction films. He professes that "such 'apparatus theory' empowers the critic and theorist as one who 'sees through' the duplicity of representation, while the critic implicitly characterizes the filmmaker as one who naively constructs psychological and ideological traps for the viewer" (309). However, while I partially agree with this position it is important to recall that so called postmodernist and poststructuralist theories are more varied and complex than the positions alluded to. Of equal complexity is the audience's response to these plays which cannot, as Paola Botham has argued, "be attributable to simple deception" ("Witnesses" 45) or a kind of *trompe l'oeil*.

framing announcement that typically generates documentary realism in verbatim plays:

> The following has been taken word for word from interviews and correspondence. Nothing has been added and everything is in the subjects' own words, though some editing has taken place. Names have not been changed. (15)
>
> The following has been word from taken word for interviews and correspondence. Everything is in the subjects' own words and place, nothing has been added though some has taken editing. All names have been changed. (41)
>
> The taken word for following word has been correspondence from and interviews. Nothing has been everything and words added in the subject's is own, though taken editing has pomle sace. Chamed nanges heeve ban all. (71)
>
> Te foling has beelown takhen wormed for wspoord frondrm intews and cughorrevieence. Nothything has been odded and evering is in the subjts' awn wongrds, tho sam editing hoes keplan tace. All nas havece been chaed. (97)

Of course, there is no mistaking Dennis Kelly's intentions and his readers and spectators would have gathered well before this point that the standards of documentary realism were only in place to be violated. Kelly, then, is not only changing the rules of documentary realism within the play, he also wanted readers to experience the failure of the overall verbatim project. Thus, on this basis, one can assign it to the category of the "fake" which – beyond its status as "a piece of art in its own right" (Voigts-Virchow 176) – incorporates "a critique of the artistic field" (Voigts-Virchow 176).

For Stephen Bottoms, what is crucially needed in verbatim theatre is:[23]

> [a] kind of theatrical self-referentiality [...] if they are to acknowledge their dual and thus ambiguous status as both 'document' and 'play'. Without a self-conscious emphasis on the vicissitudes of textuality and discourse, such plays can too easily become disingenuous exercises in the presentation of 'truth', failing (or refusing?) to acknowledge their own highly selective manipulation of opinion and rhetoric. ("Document" 57–58)

This yearning is also expressed and cuttingly denounced within Stuart Young's criticism of Jonathan Holmes' verbatim play *Fallujah* as Young points out, with focused vehemence, "Holmes's failure to register the possibilities and virtues of a more reflexive approach" ("Playing" 74). So perceived, it is thus incumbent on those irresponsible theatre-makers and verbatim mediators to recognise the true nature of their questionable mode of production. By the same token, one

23 Stephen Bottoms' contentions were eventually materialised in the site-specific verbatim performance *Counted* in 2010 that he co-created with Look Left Look Right and whose guiding principle was to "dramatise the interview process itself, rather than obscuring it" ("Staging" 193). This will be further developed in Chapter 4.3 in this volume.

might regard such contentions as just another totalising narrative succumbing to what they set out to debunk and therefore desperately in need of deconstruction themselves. Another strain of this argument would thus propose a critique of the critique of documentary realism. Paradoxically, this is not to suggest that some critics secretly longed for the other side of the coin, that being – if followed through consistently – (ineffective) aloof postmodern abstractness such as Liz Tomlin pertinently warns against in a recent publication: "[s]uch a development risks pushing verbatim practice fearfully away from an engagement with 'the real' and into a potentially totalising poststructuralist self-referentiality, where all attempts to interrogate how 'the real' might be understood, re-figured or represented in the age of scepticism are abandoned" (*Acts* 138). Little wonder, then, that verbatim theatre finds itself dubious about the extent to which postmodern speculation has a home within its walls. These relentless debates tend to get very polarised and the extreme position neither side are virtually positions that nobody holds at all. In a symposium report on verbatim theatre, Chris Megson writes that these very debates (in which there is much at stake) on "the extent to which processes of theatrical mediation should be acknowledged reflexively within the verbatim performance itself" ("Report" 531) constitute, he continues, "one of the more contentious issues identified" ("Report" 531).

On scrutiny it turns out, pace these critics, that the documentary realist project is seriously vulnerable and bruised almost to the point of collapse following the disintegrating authority of those concepts so dear to artists such as "truth",[24] "the real" or "authenticity": "since reality is increasingly revealing itself as a simulacrum, realism as an artistic mode of representing reality is now becoming more inadequate and obsolete than ever before" (Sakellaridou 47). In many ways, the very problematics raised by the documentary realist verbatim practice connect with what Richard Schechner has called "Believed-in-Theatre" in 1997, a type of work that happens when "real life has invaded theatre" (90).[25] In this article, Schechner is led to posit the following:

---

24 Charles Colson in the winter of 2002 is reported to have coined the phrase "post-truth society" in America (see Alterman 11), and more recently, we seem to have entered "the so-called era of 'post-truth politics'" (Hannan 94), an era in which the semblance of truth is no longer a requirement.

25 This type of theatre can be considered as documentary theatre but it is arguably extremely different from the one I am describing in this book on high-profile British verbatim productions. Schechner gives the following example of such practices in the context of the believed-in theatre of the AIDS epidemic: "AIDS sufferers perform documentaries of their lives, a touching theatrical realism – touching because spectators know that the actors are condemned persons, infected and dying" ("Excerpt" 9).

> The *believed-in theatre* I've discussed shows a strong, conscious relationship to 'history', 'truth', 'authenticity', 'origins', 'presence', 'experience', 'lived-through', and so on. These terms, and terms like them, have been 'problematized' by poststructuralists. The makers of *believed-in theatre* either are unaware of this vein of thought, ignore it, or reject it. I am fascinated by how far away from academic theoretical concerns the practitioners of *believed-in performances* are. The makers of *believed-in theatre* actually believe; and to believe one must entertain the possibility of truth, wherever it might be found. (90)

The documentary realist verbatim practitioners also assume a logic that appears to work in concert with some realist playwrights in the UK, that is if one is to follow Liz Tomlin's account below:[26]

> In the United Kingdom in particular, the continuing dominance of realist drama in mainstream theatres and new writing venues demonstrates a resolute refusal, on the part of many established and emerging playwrights and directors, to accept the theoretical framework that the varied applications of the poststructuralist critique of dramatic theatre, now established in the theatre departments of most academic institutions, would seek to impose on their practice. Unlike the ideological battlefield of literary theory, however, theatre scholars and artists who still adhere to a predominantly Marxist understanding of the political have tended to dismiss or simply ignore the radical claims of the poststructuralists, rather than engaging with the work defined as such or the discourse that underpins it. (*Acts* 11)

Nonetheless, ironically, and despite the vigorous arguments mounted against verbatim theatre in the critical literature, its impact on the wider culture has never been felt so strongly and there seems to be a striking consensual recognition of "the value-added potential of acting with facts" (Paget, "Editorial" 135). Even those who excoriate verbatim theatre still insist upon the remarkable "real world" effects of these works. This allows Janelle Reinelt to make, quite unequivocally, the following statement: "although it might seem that postmodernism would gradually empty documentary of its authority if not its appeal, this is not what has happened" ("Poetics" 83). In other words, verbatim theatre appears to speak more prominently for the times and "can even be understood as intervening in history – as changing, or trying to change, history itself" (Martin, *Real* 5). The discussion of its reception is thus often disturbingly moved beyond the

26 According to Stephen Bottoms this logic might even extend to British playwrights in general and to a significant proportion of the British public: "British dramatists seem to have retained a basic faith in their ready apprehension of 'the real': most Britons still believe (somewhat gullibly?) in the underlying truth/reality of the news as mediated by the BBC and by newspapers such as the *Guardian*" ("Document" 57).

category of theatre and reminds one that theatre can still act as a powerful medium for effecting change.

Consequently, this analysis suggests that there are certain forms of realism that have been lengthened well into the new historical moment without succumbing to postmodern doubt (Canton "Reality"; Botham "Deconstruction"). At any rate, it follows that I shall now turn my attention to the definitional ideas that coalesce when one speaks of "documentary realism" in this context.

2.1.2 Redefining Documentary Realism

The term "documentary realism" is preferred in this monograph as it clearly points towards a substantial aesthetic difference in kind when compared to conventional naturalistic or realist plays in Britain.[27] According to Katie Beswick, typically "the representations offered in naturalistic or socially realist forms reinforce the kind of normative assumptions that already exist in dominant representations" (2014, 109). Given that one of the fundamental lasting values of verbatim theatre is exactly to resist these dominant narratives through "a variety of voices" (Beswick 111), theorising these plays in terms of realism or naturalism would present some a priori inconsistencies. For Derek Paget, the standpoint of verbatim plays is crucially different "from drama located within a realistic/ naturalistic tradition" ("Lovely War" 67) in the light of its attitude towards "fact." In the same vein, Johnny Saldaña highlights this shift from the standard notion of dramatic realism by stating categorically that what is at stake in these plays is "not realism, but *reality*" (14).[28] He then explains this curious state of affairs as follows:

> when a participant/character in performance makes eye contact with and speaks to an audience, we are presumably brought closer into his or her world. The connection is more intimate, immediate, and an unspoken contract that naturalism is suspended (yet a sense of realism can still be maintained). (65)

27 Gary Fisher Dawson uses the term "documented realism" (12) to differentiate documentary theatre from historical drama. Crucially for Dawson, within this specific mode of realism, "[t]he shift is from the appearance of truth to a documented truth" (93).

28 This claim strangely and remarkably echoes Piscator's formulation in 1928: "It is not theater we want, but reality" ("Letter" 329) and, in a cinema context, Michael Renov's contention that the documentary "most actively promotes the illusion of immediacy insofar as it forswears 'realism' in favour of a direct, ontological claim to the 'real'" (71–72).

It is precisely this specific sense of realism that I have called "documentary realism" in this study. I therefore contend that "documentary realism" constitutes in itself a departure from the traditional conventions of both naturalism and realism. The revision of realism advocated by verbatim theatre-makers thus marks a mutation and, more significantly, an intensification of the classic realist model that Yvette Hutchison describes in terms of "extra-realism" (209), that Sara Soncini has termed "hardline realism" (*Conflict* 225), that Mireia Aragay and Enric Montforte have jointly called "documentary hyper-realism" (96) or "verbatim hyper-realism" (101) in relation to Richard Norton-Taylor's *The Colour of Justice* (1999) and, finally, considered by Susannah Clapp as "hyper-realistic meticulousness" (2007) in relation to *Called to Account* ("Double").[29] However, to avoid further confusion with the notion of the hyperreal and the simulacrum, arguably in direct contradiction with a contemporary British verbatim practice (Garson and Gonzalez), I have decided to adopt the phrase "documentary realism" in the sense of "the conservative and conventional dramaturgy of realism that was so much a part of documentary theatre in the late twentieth century [...] [and] continues today" (Martin, "Introduction" 1). Unsurprisingly, the critical literature is replete with confusion and the exact same phenomenon is described within the same book in terms of an "essentially naturalist impulse" (Watt 192) by David Watt and, by Deirdre Heddon, as "consciously nodding towards a mode of realism" ("Ethics" 115).

Interestingly, Janine Hauthal in her study of "Realisms in British Drama since the 1990s" identifies two dominant realist trends,[30] what she terms "absurdist meta-realism" in reference to postdramatic theatre texts and "documentary realism" to characterise stage docudrama and verbatim theatre (146).[31]Separating verbatim theatre from postdramatic theatre stubbornly forfeits the possibility of them working in tandem and, once again, one should remind oneself that other scholars, such as Ulrike Garde and Meg Mumford (2013) or Campbell (2015), have already written about verbatim theatre's new natural habitat in the postdramatic ecology.[32] Hauthal then goes on to describe the documentary realist British plays as "new forms of social realism in which theatre and journal-

29 The term "hyper-realism" is also used by Aoife Monks in her study of *Bloody Sunday* (347).

30 A third example of realism "experiential realism" in relation to In-yer-face plays is also considered in her article.

31 The term was originally coined by Richard Schechner in his *Performance Theory* (1988) but I allude here to Lehmann's conceptualisation of the term (2006). This theoretical framework will be discussed in Chapter 4 of this book in relation to what I have called "postdramatic verbatim theatre."

32 I will fully discuss this third-stage evolutionary mutation in Chapter 4 of this volume.

ism overlap [...] [d]rawing on a diverse literary, theatrical and performance heritage ranging from Erwin Piscator's living newspapers, through radical street performances since the 1960s, to Rolf Hochhuth or Peter Weiss's documentary plays, as well as Anna Deavere Smith's solo performances" (150).[33] She perceives the manifestation of realism within these plays as "inferred from a (paratextual) frame which authenticates these works as documentary by vouching for a minimized gap between 'reality' and theatrical representation" (152). In other words, realism is created through the turbulent movements of signs framing a typical verbatim production and its relationship with the world outside the theatre. Such relationships are more intense than in realist fictional plays and programme audiences for interesting new habits and practices.[34] Actress Bella Merlin offers an animated account of this curious phenomenon in the context of a discussion she had with playwright David Edgar (after a performance of David Hare's *The Permanent Way*) who remained tied to an earlier claim according to which "documentary drama relies on its facts being correct" (*Farce* 54):

> the playwright David Edgar who came to see it [...] said to me, 'Gosh, isn't it fascinating that a woman was involved in this privatisation of the railways?' And I went, 'Oh, no, it's not a woman in real life, it's a man.' And he went, 'You can't do that in a verbatim drama. If you're gonna put a woman on stage you gotta know that it was a woman who did it'. ("Bella" 8)

What this highlights is symptomatic of a category of theatrical reality that is expanded to the field of non-theatrical reality. Put more productively as regards to films, William Guynn argues that there is a fundamental difference between fictional films and non-fictional ones as the latter succeed "in producing in the spectator a specific sense of the real, which we must distinguish from the *impression of reality* experienced by the spectator of the fictional film" (40). Such a framework and reformulation of realism raises important issues and its relationship with the aesthetic ideologies of naturalism and realism needs to be urgently elucidated in the next section of this chapter.

33 My definition of "documentary realism" is slightly different in terms of its affiliations. See the next section for a fuller theoretical statement of this hypothesis.

34 Some will rightly argue that this recalls the Bourdieusian concept of "habitus" articulated in his *Outline of a Theory of Practice* (2013).

2.1.3 Theoretical Frameworks for the Continuation of Documentary Realist Verbatim Drama

The idea that a piece of theatre could accurately reflect the world around is hardly new and, as a consequence, for Attilio Favorini, documentary theatre "may sound superficially like a rehash of the project of Naturalism or Realism" ("Introduction" xviii). The term "superficially" is important here, as it will be recalled that the relationship between the "two camps" – that are "naturalism" and "realism" – is of a very intricate and fraught nature and that they "do not denote the same aesthetic" (Katsouraki 64).[35] The type of realism associated with British verbatim theatre in this period is heightened in performance and is thus generally understood to intersect very deftly with the concept of naturalism, rendering any attempt at theorising ever more complex as, even in its era, the concept meant different things to different people. In other words, what one tends to associate with the notion of realism corresponds to the way in which one perceives one's relation to the real at a given time and place.

Naturalism was one of the first two waves of modernist theatre and Emile Zola,[36] in his 1881 *Naturalism in the Theatre*, conceptualised this novel practice in reaction to the well-made play.[37] Therefore, this was a type of theatre more attuned to the development of new realities in a nineteenth-century context (scientific advancements, an encroaching pictorialism, the rise of the city, industrialisation, the dominance of the middle-classes, a growing focus on material existence and so forth):

> It seems impossible that the movement of inquiry and analysis, which is precisely the movement of the nineteenth century, can have revolutionized all the sciences and arts and left dramatic art to one side, as if isolated. The natural sciences date from the end of last century; chemistry and physics are less than a hundred years old; history and criticism have been renovated, virtually re-created since the Revolution; an entire world has arisen; it has sent us back to the study of *documents* [...]. An irresistible current carries our society towards the study of *reality*. In the novel Balzac has been the bold and mighty innovator who has replaced the observation of the scholar with the imagination of the poet.

35 The term "naturalism" was first coined by French art critic Jules-Antoine Castagnary in the context of his review of the Salon of 1863 to refer to an expression of contemporary life in contrast to realism.

36 In theatre history, modernism dates from the 1880s to the mid-twentieth century.

37 This corresponded "in the nineteenth century to a play characterized by the perfectly logical arrangement of its action [...] each act comprises a rise in the action punctuated by a period. The story culminates in a central scene in which the various threads of the action are brought together, revealing and resolving the central conflict" (Pavis *Dictionary* 438).

> But in the theatre the evolution seems slower. (my emphases, Zola qtd. in Bentley, *Theory* 356–357)

It was thus truly innovative and revolutionary, prompting a rash of new possibilities and techniques for the theatre – such as "the building of 'lifelike' stage sets" (Williams, "Theatre" 310) – that Dan Rebellato considers as "the most influential theatre movement of the last 300 years" ("Lecture" 5). It was also, one might be tempted, close to the idea of documentary theatre, as Chris Megson has pointedly remarked: "[n]aturalist art, whether in painting, literature or theatre, is characterised by the pursuit of documentary veracity in the portrayal of subject matter" ("Introduction" vii). Importantly, Zola formulated the existence of an external ontological "real" that could be captured in naturalistic drama: "the truth does not need clothing; it can walk naked" (qtd. in Bentley, *Theory* 362). For Arnold Hauser, naturalistic theatre also went hand in hand with a necessary radical practice based on an adherence to factual reality:

> artistic truth is valuable as a weapon in the social struggle [...] the faithful reproduction of facts leads automatically to the dissolution of social prejudice [...] those who fight for justice need not fear the truth in any of its forms [...] there is [...] a certain correspondence between the idea of artistic truth and that of social justice. (408)

The term "naturalism" before being reprised by Zola was a philosophical position "allied to science, natural history and materialism" (Williams, *Problems* 125). Within a British context, naturalism in drama as distinct from realism tends to be equated with "surface representation" (Patterson 16) or "a 'photographic' approach to theatre" (Lacey, *Wave* 99) that is "linked to a documentary and sociological impulse and generally accorded low status" (Lacey, *Wave* 99) but, again, this is far from being a view accepted by all.[38] One could allude for instance to Strindberg's understanding of the term which seems to make the exact opposite argument:[39]

38 For Lacey, this was mostly the case "among the protagonists of the New Wave [1956–1965]" (*Wave* 99). I would argue that today the situation has changed again and naturalism seems to be endowed with a new desirability among British practitioners who wish to avoid the still-dominant realism of the mainstream sector.

39 In fact, Williams argues that there are "two senses of 'naturalism'" (*Keywords* 85) and "[w] hat was later meant by 'naturalism', almost everywhere, was a kind of photography: what Strindberg had attacked as passive, uncritical reproduction of a surface reality" (*Keywords* 85). In this sense the first meaning of "naturalism" can be referred to as "critical naturalism" (*Keywords* 86).

> Naturalism [...] is not a dramatic method like that of Becque, a simple photography which includes everything, even the speck of dust on the lens of the camera. That is realism; a method, lately exalted to art, a tiny art which cannot see the wood for the trees. That is the false naturalism, which believes that art consists simply of sketching a piece of nature in a natural manner; but it is not the true naturalism, which seeks out those points in life where the great conflicts occur, which rejoices in seeing what cannot be seen everyday. (qtd. in Williams, *Keywords* 83–84)

Christopher Innes even suggests that a play can be both realistic and naturalistic if one follows this distinction:

> [i]t would be logical to use 'Naturalism' to refer to the theoretical basis shared by all the dramatists who formed the movement, and their approach to representing the world. 'Realism' could then apply to the intended effect, and the stage techniques associated with it. (*Sourcebook* 6).

In *Naturalism In Theatre: Its Development and Legacy*, Kenneth Pickering and Jayne Thompson tentatively incorporate verbatim theatre within the long-standing tradition of naturalistic theatre in their last chapter: "[i]f we return to some of the original ideas of Zola, we might consider the verbatim form as one which similarly is attempting to represent an authentic reality" (217). But these acknowledged traces of naturalism in verbatim works do not only emanate from Pickering and Thompson as Chris Megson's insightful proposal based on David Hare's reception of Richard Norton-Taylor's *The Colour of Justice* seems to exemplify, suggesting that it "resurrect[s] Zola's resounding plea for a Naturalist revolution in theatre" ("Frisson" 197). In fact they exceed the critical narrative of the strand when, for instance, Gary Fisher Dawson is discussing the aesthetic movements he deems to have relevance to verbatim theatre and includes "Emile Zola (1840–1903)" (xiii) and "naturalism" (xiii) as one of "the forms' antecedents" (xiii). In *A Sourcebook on Naturalism*, Christopher Innes also concludes that "Docu-drama is a derivative of Naturalism" (18), albeit with "a fundamental difference" (17) as does Gerhard Stilz when he claims that "above all, the 'documentary turn in contemporary drama and the return of the political' [...] ha[s] used and continue[s] to reactivate quite a number of the features, both structural and strategic, that had shaped and qualified naturalistic drama since the 19th century" (249). Derek Paget also comes close to such an understanding of verbatim theatre by alluding to the fact that "we expect to be closer, somehow, to 'real' life as we watch and listen. To this very considerable extent documentary partakes of naturalism's original premise that to be more 'real' was to be, in some way, more 'true'" (Paget, "Lovely War" 51). More narrowly, Anneka Esch-Van Kan concludes that "one strain of recent documentary plays bonds with naturalism and strives for the uncovering of a hidden truth" (416). More curious still, Duška Radosavljević asks the following question in a recent study: "Does verbatim theatre – which has reached other theatre-making cultures as a specifically British technique – represent a continuation of the British tradition of naturalism, and therefore a defeat of the avant-garde?" (140). Much more interesting for my purposes, however, is Lib Taylor's suggestion that in British verbatim theatre, since the 1990s, "emotion was recuperated, not in the old Naturalistic modes" ("Experience" 226) but perhaps in "a kind of realism that we can call revealing or unveiling" ("Experience" 227). This would seem to coincide with Louise D'Eristal's contention that "documentary realism is buried in the person of Zola" (409).[40]

It is noteworthy that the dramatic categories of "realism" and "naturalism" have been used interchangeably in contemporary critical writings on theatre[41] as well as throughout history (Williams 1976) and even by Zola himself (Pickering and Thompson 2013, 3; Lacey 1995, 99). In *Modern British Drama: The Twentieth Century*, Christopher Innes explains that "'Realism' is generally used as a synonym or derivative of 'Naturalism': a critical label describing the objective reproduction of ordinary contemporary life" (8). Similarly, Paul Allain and Jen Harvie note that "in the theatre they are almost interchangeable, with only nuances of difference between them [...] but these distinctions are too subtle and contentious to be of much use today and it is difficult to achieve consensus on this subject" (178). In the general wake of confusion, Dan Rebellato adds to the debate by explaining that "[t]he term 'naturalism', in particular, has become flattened out to refer to any theatrical production where the set and the acting attempt vaguely to resemble real life" ("Practice" 6). In quite another way so does "realism" as David Tucker puts it: "[p]art of the problem for realism [...] is one of definition" (11).

40 My translation from the original French "[l]e réalisme documentaire est enterré dans la personne de Zola" (409).

41 In a recent article, Elaine Aston acknowledges the current confusion between them and proposes instead, via Raymond Williams, to refer to "naturalism when invoking the nineteenth-century tradition" ("Room" 33) and realism "for the twentieth- and twenty-first century manifestations of the genre" ("Room" 33). Whilst I sympathise with this perspective, it seems that a new kind of naturalism drawing from its 19th-century predecessor is also visible today on the British stage. The term is reclaimed at the present time by practitioners such as David Eldridge and Katie Mitchell. An example of its key features lies within its rapport with time that ressembles Schechner's concept of "event time" whereby the activity on stage takes as long as is required to be completed (Schechner, *Domain* 87). By comparison, realism is more conventional and tidied up.

Bertolt Brecht is also an important figure in the debates over theatrical realism[42] and his influence on the British stage is often considered as "massive" (Worthen, *Modern* 148); an influence important enough for British director Stephen Unwin to recently assert that it "is visible everywhere, from the most politically explicit theatre to performance art, from new plays to revivals of the classics, from art house to community theatre" (*Toolkit* 136). Brecht's understanding of theatrical realism demonstrated a vision challenging the commonly held belief that it simply equated "the reproduction of observable reality in theatrical performance" (Gale and Deeney 293). For Brecht,

> [w]e shall not restrict ourselves to speaking of realism in cases where one can (e. g.) smell, look, feel whatever is depicted, where 'atmosphere' is created and stories develop in such a way that the characters are psychologically stripped down. Our conception of *realism* needs to be broad and political, free from aesthetic restrictions and independent of convention. *Realist* means: laying bare society's causal network/showing up the dominant viewpoint as the viewpoint of the dominators/writing from the standpoint of the class which has prepared the broadest solutions for the most pressing problems afflicting human society/emphasizing the dynamics of development/concrete and so as to encourage abstraction. (qtd. in Willett 109)

In this sense, to some extent, his realism expands Lukacsean realism (which failed to explicitly tackle the theatre ("Against" 70)), incorporating the possibility of formalistic experiments.[43] As Brecht conceived them, "the terms *popular art* and *realism* become natural allies" ("Against" 80). His Marxian realism, a type of realism that conveys a social function or purpose, was to be found within a significant amount of documentary plays in Britain in the 1960s, 1970s and is, I want to argue, still observable in the verbatim performances to be analysed in Chapter 3 of this study.

In the field of verbatim theatre, a further distinction must be made between what Caroline Wake describes in terms of "diegetic realism" offering "a realist narration of historical events" ("Mimesis" 106) and "mimetic realism" offering "a 'realist imitation' of historical events" ("Mimesis" 106). To make matters more concrete, in Jonathan Holmes' verbatim performance *Fallujah: Eyewitness Testimony from Iraq's Besieged City* (to which further attention is devoted in

42 It must be noted, however, that he has sometimes wrongly been seen as embodying "the antithesis of naturalism" (Unwin, *Toolkit* 27) and as "being directly anti-realist" (Gale and Deeney 296). Of course, the aforementioned authors do not endorse these positions and only mention in passing some problematic positions in the critical narrative of Brecht.

43 This neglect of the theatre is rather surprising considering that Lukács was a drama scholar in the first place.

Chapter 4.3), this is the difference between Rana, an Iraqi aid worker's onscreen telling of what happened in scene 7 (diegetic realism):

> The evidence I've seen about that; I mean I've seen it, not even someone told me about it. When I went looking for a woman, left with a blind old man in the house, we enter into the house, start calling her – no reply. We went out and said let's look in other houses, but I just noticed something in the garden and it was a body but I couldn't recognize it, and it looked really bad – it was a body with the colour green, and I have never seen this in all my life, and my work is dealing with dead bodies and all this stuff, but I never seen a body with a green colour. (170)

and Jo, a British witness' re-enactment of what happened in scene 14 (mimetic realism):

> MAN: Do not worry. We will not hurt you. You are safe. (*Beat.*) What is your name? [...]
> JO: Jo.
> MAN: And what do you do in Iraq, Jo?
> JO: I help to get medical supplies to hospitals. The Americans usually don't shoot if they see a white face. [...]
> MAN (*re-entering*): Come now. We will take you. It is time.
> Jo (*trying to be courageous*): It will not help if you kill us. You know that. (*Man moves towards her.*) A minute; do you have children? (*Man hesitates. Slowly JO pulls from her pocket a balloon. Expertly she blows it up and twists it into a balloon animal, before offering it to the man. They both stare at it for a moment.*)
> JO: No. Perhaps not. (178–180)

To complicate matters further, there is another distinction to be made between "dramatic realism" and "theatrical realism" which is important enough to quote at length here:

> Instead of claiming that there is something called 'realism', which effaces the theatrical framework in performance, and its antithesis, which one could call 'formalism' (although never called that by its adherents) or 'deconstruction,' perhaps it would be better to recognize that *each side* can be seen as a species of realism insofar as realism means the attempt to locate and present the truth of a situation, whether the situation refers to the dramatic material or to the theatrical frame. Thus the 'realism' that political-theatre theorists dismiss or complain about can be called 'dramatic realism,' insofar as it tries to conceal or efface the apparatus of presentation – that is, theatricality – in an attempt to enhance the illusion of reality of its subject-matter. And we can call the other 'theatrical realism.' It is applied in all uses of self-conscious, ideology-unmasking theatre – insofar as it either concentrates on showing the reality of the apparatus of illusion (or at least refusing to conceal it), but in the best instances attempting to create a self-conscious dialectical relation between the form of presentation and the content, the matter that is represented. I realize that calling both these dramatic and theatrical tendencies 'realism' might be confusing, so I will refer to 'theatrical realism' as 'theatricalism.' Theatricalism is the intentional foregrounding of theatricality, of

> theatrical form, as opposed to theatricality's conventionalized, self-effacing uses for the mediation of dramatic content. (Erickson 160)

To come back to verbatim theatre, the notion of "documentary realism" used in this volume is thus closer to what Erickson has termed "dramatic realism".[44]

Having critically reviewed the terms "realism" and "naturalism" in some of their complexity and examined the aesthetic debates revolving around these concepts as well their ideological foundations, one should be prepared to examine the ambivalence that colours the history of British verbatim theatre, ambivalence produced by a contemporary gesture of disavowal, which I will describe briefly below.

2.1.4 The "Broken Tradition" of British Verbatim Theatre

To characterise the fluid and complex shifting output of verbatim theatre in Britain that some have framed as part of a (satisfactory progressive) narrative of cohesive movement (Waters, "Political" 139; Inchley 45), Derek Paget suggests instead a "broken tradition"[45] of theatre practice (224). Indeed, the history of verbatim theatre is anything but linear and, for Paget, this type of theatre is "always condemned to being forgotten then 'rediscovered' in another conjuncture" ("Documentarism" 130).

Rehistoricising the field of British verbatim theatre in this century has been perhaps a more ideological endeavour than the putative political nature of these plays in performance and, in what follows, I will attempt to sketch the particular approach that has construed the interplay of significant theoretical and aesthetic concerns anew. Verbatim theatre, as I have argued, has experienced a new burst of energy since the mid-1990s which coincided with its gradual general move towards (and embrace of) the mainstream and its newfound canonical status.[46] Following Peter Weiss's seminal articulation of the strand in 1968 pointing out that "[w]hen the performance takes place in a commercial theatre, with high prices of admission, the Documentary Theatre is caught in the very system it wants

---

44 However, the verbatim performances to be studied in Chapter 3 engage with what Erickson calls "theatricalism."

45 Paget borrows the phrase from Richard Stourac and Kathleen McCreery (xiii).

46 In the UK two documentary theatre pieces suffered from censorship when they attempted a mainstream production. Laurence Olivier was heavily censured by his National Theatre board of directors in the context of Hochhuth's *Soldiers* in 1967 and the Royal Court Theatre cancelled at the last minute Jim Allen's *Perdition* in January 1987.

to attack" (43), one might regard this as an inversion or a momentary suspension of its historical priorities. This heightened contact and *rapprochement* with mainstream theatre perhaps also meant that verbatim theatre was more implicated in its operations and therefore sailing closer to the winds of literalism and aesthetic realism,[47] both disputably characteristic of the dominant theatrical traditions of post-war British theatre (Luckhurst, "Why" 76; Cohn 1). By extension, Elżbieta Baraniecka via Lyotard suggests that "[r]ealism [...] provides the means of sustaining [a] particular 'mainstream' image of reality and the 'mainstream' values" (57).[48] The extreme version of this argument would suggest that this noticeable change, that is to say, verbatim theatre's current presence in the mainstream, may indicate an attempt to reach the standards of "excellence" favoured and consumed daily by the mainstream's commercial imperatives – in the words of Lachman "the strategy of an attentive catering to the popular taste" ("Colours" 321) – and, as I have argued elsewhere (Garson "Changing"), perhaps even an attempt to silence its radical lineage.[49] Once one recognises this, it might be fair to say that this attempt to dislodge verbatim theatre from this radical heritage appears to endorse Raymond Williams' position on the hegemonic sense of tradition as "a deliberate selective and connective process which offers a historical and cultural ratification of a contemporary order [...] [and that] has in practice to discard whole areas of significance, or reinterpret or dilute them, or convert them into forms which support or at least do not contradict the really important elements of the current hegemony" (*Marxism* 116).

It is interesting to note that some commentators considered verbatim theatre as a brand new contemporary practice, inadvertently (or conveniently) contributing to a problematic mainstream narrative; one that ruthlessly harnesses the energies and radical gestures of past oppositional/upstream practices in order to refresh itself and neutralise any subversive potential.[50] Within the mainstream

---

47 Although not referring directly to verbatim theatre, Howard Barker talks about "*a flight into realism*" (his emphasis) in contemporary British theatre, exemplified by the presence of "a bottomless supply of the drama of topicality" (Barker, "Ethics" 87). It is also interesting to point out that Alecky Blythe's verbatim plays to be studied in Chapter 3.2 started to feature a more realist casting (i.e. actors looking more like the people they represent in terms of gender, age, etc.) as she gained commercial recognition.

48 On this point, Alison Shonkwiler and Leigh Claire La Berge remind us that "[r]ealism, after all, has long been considered the aesthetic mode most intimate to capitalism" (1).

49 This is in direct opposition to verbatim theatre's alleged history according to which it used to counterbalance "the frequent commodified plays that became London successes" (Anderson and Wilkinson 158).

50 See for instance Bradshaw who considers that verbatim theatre is "a new form of contemporary political drama" ("Arbor"), Derbyshire and Hodson who write that it is "a mode of theatrical

theatre language, verbatim productions can then pride themselves on their ability to perform extremely well in challenging financial times,[51] entertaining new and larger audiences.[52] In the case of the verbatim plays to be studied in the next chapter, Anneka Esch-Van Kan puts it bluntly: "[t]he commodification of political theatre becomes blatantly obvious when tribunal Theatre is turned into a trademark of the Tricycle Theatre and when audiences and media evoke more excitement about the true-to-life impersonation of public personae than about the core issues of the plays" (423);[53] as does Douglas Murray: "'[t]ribunal theatre' is simply filling a gap in the market for no-strings-attached, neatly packaged, moral tourism" (2005) and Mary Luckhurst in the following: "[t]he label implies that no words are invented, which suggests that some marketing strategies are perhaps more focused on appealing to fashion than they are interested in certain kinds of integrity to the work (and of course to the interviewees)" ("Ethics" 213). In this sense, David Watt aptly posits that this unacknowledged past constitutes in the light of today's output "another direction entirely [...] which gives the form a political bite of a different kind" (198) while Chris Megson argues that the plays to be studied in Chapter 2.2 demonstrate "differently-nuanced inflections of the documentary mode" ("State" 115). Another version of this argument can be found in Liz Tomlin's *British Theatre Companies 1995–2014* that sheds light on the bigger picture. Indeed for Tomlin, the contemporary main-

---

intervention developed since the 1990s" (198), Dominic Dromgoole attesting in 2004 that "Hare has invented a new form [...]. This new form is one future for political theatre," Susannah Clapp stating in 2005 that "Richard Norton-Taylor pioneered verbatim theatre" and, more recently, Mark Lawson claiming that "[t]he recently popular form of verbatim drama – staging speeches based on interviews with witnesses and participants – was pioneered by director Nicolas Kent and journalist Richard Norton-Taylor" (2015).

51 A particularly good example of this is Gillian Slovo and Victoria Brittain's *Guantánamo: 'Honor Bound to Defend Freedom'* (2004) that had an extremely successful run in the West End (after its first Tricycle Theatre outing) before eventually becoming a "hit on Broadway (with Archbishop Desmond Tutu making a couple of guest appearances)" (Edgar "Doc"), thus reaching "over five thousand outlets" (Colleran 141).

52 Strikingly, director Chris Honer who worked on many verbatim productions in the 1980s and more recently on a tribunal play, Dennis Woolf's *Beyond Belief: Scenes from the Shipman Inquiry* admitted in an interview with scholar Alison Jeffers that "a substantial section of the audience went on strike [as] [t]hey thought it was exploitative and that it might make money out of victims" (2005).

53 This is echoed by Michael Billington who writes "I'd be astonished if, right now, someone (ex-Tricycle artistic director Nicholas Kent, perhaps?) is not dreaming of a way of turning the Leveson inquiry into theatre" ("V") and even more openly by Alfred Hickling in the following: "[i]t's increasingly the case that no official inquiry can be considered complete until it has been transcribed, condensed and turned into a piece of verbatim theatre" (2004).

stream theatre industry has successfully absorbed the performance apparatus and aesthetic models of the independent theatre sector (verbatim and otherwise) of the past 25 years to create a "golden age of experimentation" (124):

> From the capacity to build diverse audiences demonstrated by companies such as Tamasha and Eclipse, to the site-specific experimentation adopted by all three National Theatres, to the innovations in multimedia technology and new forms of playwriting that have now permeated the establishment theatres, the UK has never, I would argue, enjoyed an aesthetically richer mainstream than it does at present (124).

Thus, following the logic of this line of argument I have tentatively constructed, it appears reasonable to assume that this deep investment in reframing verbatim theatre as a predominantly realist art form presents certain limitations that could be seen as contentious or wrong-headed in the light of the long history of verbatim theatre in the UK. A typical example of these framing and structuring activities is Alan Rickman and Katharine Viner's verbatim performance *My Name is Rachel Corrie* – "the fastest sell-out at the Royal Court since *Look Back in Anger* 50 years earlier" (Martin, *Real* 145) – that was first performed on 7 April 2005:

> Even in [...] verbatim theatre [...] source materials are often edited and performed so as to respect theatrical conventions of narrative. For example, *My Name Is Rachel Corrie* [...] excerpted from the journals of an American peace activist killed by an Israeli Defence Force bulldozer [...] is notable for the poignancy of its character development. (Rae 14)

This, at least, has affinities with Lehmann's question regarding "whether the much implored political claim of documentary theatre could be upheld by adapting it formally to the dramatic form" (*Postdramatic* 55).

Crucially, realism has not always been conversant with verbatim plays and the new characteristics I have described set it apart from the past current of British verbatim theatre which tended to avoid portraying "the real *realistically*" (Paget, *True* 83). A telling example of this "tradition opposed to realism" (Paget, *True* 42) is Peter Cheeseman's early experiments with the technique at the Victoria Theatre in Stoke-on-Trent, the only permanent theatre-in-the-round in Britain at the time, which became "a test-bed for new styles of performance" (Cheeseman, "documentary" 105). *The Knotty* (1967), one of his most successful performances among the "regional documentaries" or "Stoke documentaries",[54] was a musical documentary that, in the words of Peter Cheeseman,

54 The series started in 1964 with *The Jolly Potters* – on the bleak working conditions of the

"seemed to suit the description DOCUMENTARY rather than PLAY" (vi), thus doing away with the conventional British play format at the time. In this way, actors "used a non-naturalistic costuming convention which allowed for speed of scenic transformation" (Paget, *True* 72) as well as an acting style distinct from the tradition of psychological realism. Largely, in the 1960s and 1970s amongst "alternative" theatre groups,[55] the typical staging was Brechtian, it included an extensive use of music and experiments with popular and entertainment forms. The vocabularies of feminist companies' innovative devising theatre drew on punk, surrealism, contact improvisation and even Freud. In the playtext of Gay Sweatshop's *Care and Control* (1977) – a verbatim play on the custody struggles of lesbian mothers that helped fuel the campaign to change the law – the departure from kitchen-sink realism is made explicit as early as Act One, Scene One when we are told that "at the back is a large screen or a blank wall on which to project slides" (67).[56] For Michelene Wandor, the play benefited "from popular culture, television entertainment, and art-theatre" (11), creating a particular style of performance "which continually break[s] through the boundaries of naturalistic storytelling" (11). Recognising the essential presence of realism at work in the piece, Wandor argues that "the realistic elements have a didactic purpose, and this purpose is sharpened by the non-realistic elements, such as music, stylisation, and direct contact with the audience" (11–12).

Similarly, Theatre Workshop's *Oh What a Lovely War* used the Pierrot show (drawing also on music-hall, variety and ENSA-show forms) while the 7:84 Theatre Company drew on the *ceilidh* form, "an adaptation of the traditional form of entertainment of the Highlands" (McGrath 122). Openly or covertly, critics perceive the previous wave of documentary theatre in Britain embodied by Joan Littlewood's *Oh What a Lovely War* in 1963[57] as diametrically opposed to the historical first performance of *Look Back in Anger* on May 1956 at the Royal Court

---

19th-century potworkers as well as on the 1842 Chartist riots – and featured almost one documentary every year.

55 More than 700 alternative companies were formed in the UK between 1968 and 1988.

56 This is to be understood as "an indication that [this type of realism] is associated with the domestic and the everyday" (Lacey, "Staging" 64).

57 1963 was also significant as it saw the production in London of Rolf Hochhuth's *The Representative* which would be followed by a whole series of documentary plays by German playwrights staged in the UK in the 1960s. Ironically, the first verbatim play of the new wave to be staged at the Royal Court Theatre that reignited this tradition was also by a German playwright: Klaus Pohl's *Waiting Room Germany* (1995).

Theatre[58], an alternative theatrical tradition operating in the shadow of the established mainstream theatres. These plays showcased a different approach to the aesthetics of realism at the Court[59], something akin to the Piscatorian tradition of epic documentary. This tradition of local documentaries continues to this day mainly in the applied theatre sector of practice and with companies such as Banner Theatre (Tomlin 2015, 66) that started in the 1970s. Far from what we can observe today, then, verbatim theatre was mainly perceived as an antidote to the British realist tradition and the dominant mode of theatre-making:

> the actor is freed not only from some of the burdens of conventional playwriting within the naturalistic mode, but also from some of those attendant upon the characteristic economic determinations of theatre productions in this country [...] it offers that ensemble method of working. (Paget, "Oral History" 318)

From this perspective, then, the use of documentary realism in verbatim would appear to be the yoking together of opposites. As Jonathan Holmes remarks in his introduction to the verbatim play *Fallujah*, documentary realism is *per se* a self-defeated project:

> In comparison with prose, theatre cannot create a sense of pictorial realism with anything like the equivalent level of conviction. Theatre as compared to prose, cannot give the audience access to detailed description, cross-referenced sources and comprehensive evidence. It is not on the surface, the medium most suited to a documentary enterprise [...] All this is to say that theatre, more than most artforms, struggles to represent events in as realistic or objective a manner as a textbook or a documentary film (themselves not uncompromised in this arena). (141)

In short, the earlier traditions of verbatim theatre were modified and arguably institutionalised in the 21st century as they moved from a local small-scale alternative context to a growing concern for the medium-scale (McMillan, "Fiction") and also the global large-scale sector, a scale not previously envisaged. However,

58 In spite of the narrative that has pervaded to the present day – put forward by commentators and the Royal Court itself – according to which the Royal Court's output is uniform (new writing and realism), one might find on closer inspection a significant number of plays that do not fit the main thrust of this official narrative such as Keith Johnstone and William Gaskill's near verbatim *Eleven Men Dead at Hola Camp* (1959), Louise Page's documentary play *Falkland Sound/ Voces de Malvinas* (1983) and Stephen Daldry's own experiment with the verbatim method, *Body Talks* in July 1996.

59 This type of realism is curiously described by Katie Mitchell in a way that recalls "documentary realism": "if the genre is realism (as in John Osborne's *Look Back in Anger*) the characters will obey the logic of real-life behaviour with the accuracy of a film documentary" (50).

at the opposite end of what I have posited thus far, Derek Paget contends that looking back at the history of British verbatim theatre, one found another structural variant,[60] one that uses an aesthetic apparatus that has the ability to mask its own performance techniques in order to present the "transparent" illusion unencumbered by the reality of its "montage":

> [t]his tradition is a *recording* one; that is, it believes that the effacement of the subjective creator(s) of cultural production will produce an 'objective' account. This literal 'record' of the event/activity is placed at the service of a universalised 'public good'. Film and television, initially confident in their ability to capture and fix the living moment, have made most use of the 'recording' tradition. Cultural producers have claimed to use documentary material without imposing themselves on it directly, mere 'flies on the wall', incapable of disturbing what they record. (*True* 39–40)

In this last variant that artificially re-produces realism in conjunction with "a documentary discourse of factuality" (*True* 87), Derek Paget draws attention to its relation to the position of the "cultural tourist" as "problems must inevitably be perceived at a distance whatever immediacy is claimed by such mediations" (*True* 87). In some respects, one can easily see at work the same discourse that had earlier characterised my depiction of the contemporary documentary realist verbatim plays, a depiction that I intend to probe further through a close study of two representative case studies in the following chapters. In much the same way, Stuart Young writes the following:

> self-reflexivity has been largely absent from the wave of documentary theatre that has swept Britain in the last few years. These British plays, still locked in a recording tradition, take a variety of forms, including tribunal plays and most conspicuously verbatim theatre, in which a text is composed wholly, or almost entirely, from interviews recorded, transcribed, edited, and then re-presented in performance, usually by those who conducted the interviews. ("Playing" 73)

and Alice Saville considers that "[t]rue to stereotype, the current wave of verbatim theatre practitioners are more self-effacing – to the point of rubbing themselves off the page altogether" (2015). While I agree with these analyses, I also believe that there are some theoretical inconsistencies with this past framework

60 The other variant is "[t]he alternative tradition [...] a radical/revolutionary *reporting* one, which recognises that facts and information can never come value-free [...]. The second tradition assumes knowledge and allows the citizen access to the makers' own place in the mode of production [...]. The information it contains is "enabling" rather than compensatory" (Paget, *True* 40).

even though its markers of influence are still visible and linger on to the present time.

2.1.5 Conclusion

I had to start with the diagnosis that the possibility of verbatim theatre changed radically under the new conditions of contemporary Britain in the age of globalisation. As such, verbatim theatre has become an essentially contested art form within theatre's representational networks. Upon investigation, it emerged that some critics' discussions of verbatim theatre had oversimplified what is involved in its performance as a type of regression. One also saw that many competing definitions sat within the broad church of realism and that the question of theatrical realism had taken a number of different theoretical turns no longer referring to a simple (and primitive) correspondence between language and visual evidence as well as between sign and referent.[61] I then looked at Brecht's version of realism which constituted a more complex and sophisticated engagement which what is commonly understood as realism in performance. Importantly, the strategies of typically Brechtian *Verfremdungseffekte* were a known feature of earlier periods of documentary theatre (Peter Weiss, alternative theatre movement like Monstrous Regiment and Gay Sweatshop).

The existence of a plurality of realisms in contemporary discourse renders my attempts at theorising more challenging and it is now time to focus on the case studies to get a better grip on the problems at hand. Having discussed how a documentary realist theatrical apparatus could work in correlation with verbatim performances, the following chapters in this part will now address how this theoretical understanding applies in practice. In Chapter 2.2, tribunal plays, "the hallmark of the present wave of documentary" (Soncini, *Conflict* 80), constitute a primary site of investigation in my attempt to articulate the possibility of documentary realism in the 21st century.

61 This echoes Baudrillard's conceptualisation of the real in *Simulacra and Simulation*.

2.2 "Tribunal Plays": The Tricycle Theatre's *Justifying War* (2003)

Having made an argument for the contemporary endurance and relevance of the documentary realist mode, let us now turn to one of its most prominent (re)incarnations, the tribunal play that Alan Filewod has described as "the most easily identifiable form of documentary theatre" (10) but also as "perhaps the most deceptive" (10). The term itself is to be understood as a single-sourced, two-hour (or more) verbatim play scrupulously based on the transcripts from a high-profile legal case that can be "trials, tribunals [or] public inquiries" (Paget, "Tradition" 223), an unusual format that one of its makers calls "an inquisitorial tribunal of inquiry" (Norton-Taylor qtd. in Hammond and Steward 114). Within a contemporary British context, Chris Megson defines this strand of verbatim theatre as follows:

> the meticulous re-enactment of edited transcripts of state-sanctioned inquiries that address perceived miscarriages of justice and flaws in the operations and accountability of public institutions. Typically, tribunal productions take the form of a forensic simulation of the inquiry's disputations and setting, with actors playing the roles of the actual witnesses and judicial personnel. ("Frisson" 195)

In other words, since the 1990s, these productions have been reconstructing portions of public inquiries for a theatre audience.[62] The premise behind these UK productions is that inquiries constitute a matter of public interest that are denied an appropriate coverage by being unaccessible,[63] poorly reported in the traditional journalistic media outlets and, more crucially, by not being broadcast on television.[64] Hence, theatre compensates for and serves as a corrective, pro-

62 William Twining defines public inquiries as "normally set up by Ministers to investigate and pronounce on matters giving rise to serious public concern, such as major disasters, allegations of corruption or of serious impropriety of public life or of police malfeasance. The main objective is to restore public confidence by a formal, independent, and open investigation of the facts, and to make recommendations to prevent a recurrence of the matters causing public concern" (34). This is indeed very different from an adversarial trial.

63 In a 2005 article, Chris Megson reminds us that during each of these Inquiries "only around ten members of the public were allowed to observe each session" (370).

64 In the US such inquiries are broadcast live and, famously, Sir Richard Scott refused to allow cameras into the inquiry which resulted in the first tribunal play *Half the Picture* at the Tricycle Theatre, London. In an interview Nicolas Kent expressed his conviction of the necessity to have a wider coverage of these inquiries in Britain and declared that he was pleased in the fact "that more and more inquiries [we]re starting to be televised [...] hop[ing] [...] [to] be put out of business eventually" (Hoggard).

viding the British public with "a readily accessible overview" (Woolf, "Programme" 7).

Trials are intense dramatic spectacles and compelling public events that therefore lend themselves readily to theatrical reconstructions. For centuries – ever since Aeschylus' *The Oresteia*, a trilogy first performed in 458 BC and Plato's *Apology of Socrates* (399 BC) – they have exercised a powerful hold over the imagination of writers as their particular structure offers an ideal template for plays. In a 1971 journal article, distinguished California attorney Richard Harbinger unapologetically describes trials as theatre: "[a]s a dramatic thing, it has dramatic attributes, such as dramatic form, dramatic substance and, most significantly, dramatic effect. When one observes an adversary trial, he sees a play" (122). In other words, theatre and fiction are specters hovering over these real judicial and human activities, sometimes even strikingly colliding with them: "there are recorded instances of a criminal being cast in the role of Christ in a mystery play, so as to exploit his actual torture and death as part of the spectacle" (Favorini, "Introduction" xiv). Similarly, if much less harrowing, one of the lawyers of the Stephen Lawrence Inquiry, a full judicial inquiry headed by Sir William Macpherson, five years after the murder of a black teenager in South East London – an area notorious for racist attacks and stabbings – makes the following comment after having heard the evidence of William Mellish, formerly detective superintendent:

> MANSFIELD: Can I just ask, the officer we are calling XX. This is becoming like a Pinter play with surreal references. (Norton-Taylor, *Justice* 400)

At one point in the Shipman Inquiry – investigating the actions (including those of the statutory bodies, authorities and responsible individuals) that led to Dr. Harold Shipman's deeply shocking series of murders at Market Street, Hyde, near Manchester, England – Dame Janet Smith aggravatingly alerts us to the fact that "[t]he way that Shipman could kill and walk away unsuspected would be dismissed as fanciful if described in a work of fiction" (Woolf, *Belief* 50). Even more indicative for our purposes is that when the actual Hutton Inquiry was unfolding, Geoffrey Robertson QC on behalf of ITN and Sky used the tribunal plays at the Tricycle Theatre as a legal argument to make an application for the evidence of key witnesses to be broadcast on television:[65]

65 Along the same lines, Lord Justice Scott who is represented in *Half the Picture*, the first tribunal play in the series at the Tricycle Theatre, was reported in *The Times* on 1 July 1994 to have taken "time off from writing his report on the inquiry" (qtd. in Megson, "Frisson" 205) to see the theatrical version of said inquiry.

> My Lord [...] your Inquiry will appear on television anyway, just as Lord Scott’s Inquiry into the Matrix Churchill trial and Sir William McPherson’s Inquiry into the Stephen Lawrence killing appeared on television. It appears by way of what is called a dramatic re-enactment. It usually begins as a play put on at a theatre in North London, and then transfers to I think the BBC, maybe in this case to ITV, I do not know. But the participants and witnesses are played by actors; and there is nothing that one can do to stop this. Indeed, it shows, perhaps, the appetite, legitimate appetite of the public, for information beyond the press, beyond the bare written words of what happened. So in Lord Scott’s case a play called ‘Half the Picture’ was performed on television with, I think, Ms~Sylvia Simms playing Lady Thatcher. Of course it was played very well, no doubt; but there is an element of mimicry. There is, of course, a large element of selection. The value of television, of televising Inquiries, is that it puts an end to this kind of dramatic re-enactment which, however well intentioned and obviously catering to a public interest, cannot be the truth. The evidence of the Prime Minister, whoever plays him, is not seen for really it is, for what it was, and not for what a skilled actor with all sort of acting abilities embellishes. My Lord, this is not a light matter that I am canvassing. It is a matter that I would respectfully ask your Lordship to consider. It may appear to the Martian rather bizarre that we live in a country where these great constitutional issues are discussed by evidence given by public men and women and yet we are not allowed to see it, but when we do see it it is by actors mimicking the evidence that was given. So that we would say that it is time that television was allowed to show the public the real thing [...] and not have it re-enacted. (2003)

In an interview, Charlotte Westenra, co-director of *Justifying War*, staunchly describes the actual Hutton Inquiry she attended in theatrical terms:[66]

> the court case became like a piece of theatre [...] The first three days would be quite boring because you wouldn’t know who everyone was. But then after that you’d start to see the different themes, as if you were watching a play. (qtd. in Stoller, *Tales* 158)

Additionally, Nicolas Kent reveals the endemic dramatic structure of these official legal cases:

> [a]ny inquiry is rather like a good courtroom drama, because they always follow the same narrative form: you lay out the circumstances first, then what happened, and then the reaction of the authorities to what happened, and then the examination of whether those people reacted correctly or incorrectly; then they come to some form of conclusion within the inquiry before the inquiry report is written. (Hammond and Steward 139)

66 This was not confined to Court 73, as Janelle Reinelt reminds us in an article: “[i]n the summer of 2003, actors playing ‘real’ politicians reenacted nightly, just after the evening news, the appearances their characters had made that day at the judicial inquiry into the suicide of David Kelly” (“Poetics” 71).

The exact background behind the revival of the tribunal play in Britain is difficult to pin down and Derek Paget often refers to this tangle of historical complications as a "broken tradition" (as seen in the last chapter) due to its breadth of influences, but suffice it to say here that the current upsurge of verbatim theatre is very much the result of the collaboration between the then artistic director of the Tricycle Theatre, Nicolas Kent (1984–2012) and *The Guardian* security affairs editor, Richard Norton-Taylor (Favorini 2013, 104–105; Adiseshiah 2012, 104; Billington 2007, 385). Their collaboration gave rise to a total of seven tribunal plays from 1994 to 2011,[67] while Nicolas Kent self-mounted *Srebrenica* (1996) – another tribunal play based on the transcripts of the Rule 61 Hearings in The Hague regarding the UN inquiry into war crimes committed in Bosnia in 1995 – as well as two non-tribunal verbatim plays, *Guantanamo "Honour Bound to Defend Freedom"* (2004) in partnership with Gillian Slovo and Victoria Brittain and *The Riots* (2011) with Gillian Slovo.

Back in 1982, at a time when he was the artistic director of the Oxford Playhouse, Nicolas Kent produced his first tribunal play as he tried to reconstruct the 1979 *Romans in Britain* obscenity trial at the Old Bailey (with writer Guy Hibbert) as the actual trial took place. This was broadcast by *Newsnight* at 10.30pm and caused quite a stir at the time. Importantly, various court trials had been put into verbatim (mostly in London) before then and these certainly constitute the precursors that might have inspired Kent's foray into this strand of theatre. The nearest parallel that springs to mind is *Are You Now or Have You Ever Been* (1972) by British-born American playwright Eric Bentley on the investigation of show business by the Un-American Activities Committee (1947–1956) during the period of McCarthyism. The play first opened at Yale Repertory Theater in New Haven, Connecticut in 1972 and was followed by a short run at the Riverside Church in New York. In 1975, a production of the show was mounted in Los Angeles with great success (running for nearly a year) before being transferred to the Bush Theatre, London, in June 1977. Arguably, *Are You Now or Have You Ever Been* gives at least a glimmer of a documentary realism paradigm in verbatim practice. In one particularly eloquent passage in his preface, Bentley explains what he tried to achieve: "I have succeeded for you, the reader, if the reading of this script holds you from beginning to end and leaves with you an

67 These include *Half the Picture: The Scott "Arms to Iraq" Inquiry* (1994) with additional material by playwright John McGrath, *Nuremberg* (1996), *The Colour of Justice* (1999) based on the transcripts of the Stephen Lawrence Inquiry, *Justifying War: Scenes from the Hutton Inquiry* (2003), *Bloody Sunday: Scenes from the Saville Inquiry* (2005), *Called to Account: The Indictment of Anthony Charles Lynton Blair for the Crime of Agression against Iraq – a Hearing* (2007) and *Tactical Questioning : Scenes from the Baha Mousa Inquiry* (2011).

impression of wholeness, of a single tale told at the proper pace in proper sequence, without waste motion, without loose ends” (*Now* 4). Nearly fourteen years prior thereto, the London public had already experienced a series of tribunal plays by German dramatists including Rolf Hochhuth’s *The Representative* based on the trial of Adolf Eichmann in a Royal Shakespeare Company production in September 1963 at the Aldwych Theatre in London. Another production by the Royal Shakespeare Company, Jon Blair and Norman Fenton’s *The Biko Inquest* (under the title *A Miserable and Lonely Death*) at the Adwych Theatre, London in April 1978 on South African political leader and thinker Stephen Biko’s tragic murder, would also appear to have struck a deep chord.

At the core of the Tricycle approach is a fervent desire to disseminate the work of its playwrights beyond their traditional theatre audience and most of the plays discussed here have been adapted for television or radio by the BBC.[68] It was even reported in 2006 “that a global audience of 25 million people” had seen the tribunal plays (Megson, “Testimony” 2). Structurally, the tribunal play model lends itself well to a certain type of realism as “the procedure of the courtroom makes it possible for the author’s editorial cuts and emphases to pass without notice” (Filewod 10) and therefore effectively denies and undoes any sense of theatricality. In this sense the contemporary tribunal play represents an extension and a modification of a historical tradition of theatre-making.

2.2.1 Towards a Realist Reconception of the Tribunal Play Model

It has been said in the opening chapter that the contemporary verbatim tribunal play in Britain operates within what I have framed in terms of a documentary realist aesthetic, perhaps echoing Aoife Monks’ extrapolation that there has been “a long history [...] of a move to a ‘realist’ aesthetic in the courtroom” (348) itself.[69] However, it is crucial to note here that the pervasive documentary realist tropes that punctuate the reception of these edited inquiry transcripts in performance found themselves clearly woven with fiction in monologue form (written by playwright John McGrath) in the first Tricycle tribunal play ,[70] *Half*

---

68 The Tricycle Theatre is a small theatre (235 seats) in North London (Kilburn), an area notable for its Irish and Black communities.

69 Here, Aoife Monks means a repression of rhetoric and spectacle in the law’s operations.

70 There are five monologues in total spoken by five different characters (a Matrix Churchill employee, an Iraqi Kurd, a Palestinian, John Kenneth Galbraith – the US economist – and Paul Henderson, the former Matrix Churchill managing director), giving each time a different perspective.

the Picture (1994), which re-enacted a moment of recent British history at a time when it was revealed that John Major's Conservative government was covering up the sales of arms-making equipment to Iraq. For instance, in Act One, Scene Five, right after having heard the (edited) evidence of David Gore-Booth, the then British Ambassador for Saudi Arabia, who was called to account regarding his activities as a former Assistant Under Secretary at the Foreign Office, the audience encounters "Mustapha, a Kurd, in Kurdish costume" (59) who makes an intervention by speaking directly to the audience:

> On 16 March, 1988, Iraqi planes flew over Halabja. They dropped cyanide, mustard, nerve gas. Five thousand people died this horrible death. Nearly all of them were civilians. They died because they were born Kurdish. Your Government knew about this. Yet your Government allowed your countrymen to help these unspeakable people to make the shells they fired, even the bombs they dropped the chemicals in. They found out, as we found out, what a friend they had in Saddam after he defeated Ayatollah's people. He turned on us with even more cruelty. Then he turned on you: he took Kuwait. And you may think you have won. But he is still there. We know he is still there. And we are still there. *(He goes*). (60 – 61)

As a result, *Half the Picture* opens up a space between the heightened realism of the court (or "dramatic realism" [Erickson 160]) and the theatricality of fictional characters (or "theatricalism" [Erickson 160]) that continuously troubles the "illusionistic frame" (Megson, "Frisson" 196). What is intriguing is that the tribunal play takes us through a different aesthetic in that instance without entirely giving up a certain idea of realism. More poignantly, the final scene in Act II leaves any ounce of realism at the door by repeating the words we have just heard in a new context where "Leithead, Higson, Thatcher, Clark, Heseltine, Gore-Booth, Garel-Jones, Major and Waldegrave reappear" (110) together:

> THATCH: They are exactly what they say – guidelines – they are not law. They are guidelines.
> BAX: Do they have to be followed?
> THATCH: I beg your pardon?
> CLARK: There was an implied invitation to them to participate in a fiction. [...]
> THATCH: Remember, they are guidelines. It's not like interpreting law...They are guidelines. (110)

Remarkably, these earlier experiments with theatricality and "non-naturalistic devices" (Megson, "Iraq" 370) akin to "a European tradition of documentary performance-making" (Megson, "Frisson" 198) were to be chased off in favour of a "strand of British verbatim work, associated principally with the Tricycle Theatre, which consciously seeks to exclude any hint of fiction" (Rebellato, "Exit"

14). This new documentary realist strand would thus focus its attention on a more tacit (political) intervention deemed more suitable at this point in history. Nevertheless, it stands to reason that *Half the Picture* laid the ground for the documentary realist turn and this moment arguably marked "[a]n alternative narrative of British theatre in the 1990s" (Megson, "Report" 530), this period being generally perceived as the one of "In-Yer-Face theatre" (Sierz 2001).

The idea of realism in these performances arises from an aesthetic project that takes it upon itself to realise the utopian impulse towards the real, essentially through the repudiation of anything that can be construed as theatrical. This position can be better grasped as an ongoing obsession with the rectitude of its own intentions within a theatre setting, promising to give the audience access to a world on the other side of mediation and doing so by offering certain kinds of verbatim information. In practice, such a pursuit is achieved through the absence of aesthetic embellishments and a muted theatricality (e.g. the house lights are never turned off). The way in which the tribunal theatre-makers strive to minimise theatricality is visible in their compelling reconstruction of the minutiae of the everyday clutter of today's courtroom. We witness an almost replica of actual events that happened in the courtroom: "the production locates performers within a closely observed mimetic version of the environment in which they gave their testimony" (White, "Compelled" 75). To that end, and spurred on by a belief that documentary realism can boost political consciousness, Hannibal House at Elephant and Castle was accurately re-created by designer Bunny Christie for *The Colour of Justice* and, as Terry Stoller puts it, this was extended to the mise-en-scène: "the computers, the rows of desks, the ring binders, lawyers whispering to each other, and people walking in and out" (*Tales* 153). The exceptional level of accuracy of these productions may be judged by the fact that the desks used in the Tricycle's *Srebrenica* were requested by an administrator of the court at the time of the trial of Slobodan Milosevic:

> he asked if he could send a United Nations lorry to pick them up [...] It turned out that the trial was going to be bigger than the previous hearings and they didn't have enough desks. Although we hadn't actually used mahogany, we had reconstructed the furniture very carefully and we did still have them. So the lorry came and as far as I know the desks are still in the Hague being used for the trial of Slobodan Milosevic. (Kent qtd. in Wroe)

Importantly, one of the keys to this documentary realist staging lies in the use of plasma screens that project and enlarge the documents that the lawyers refer to in the inquiries. In the case of Dennis Woolf's *Beyond Belief: Scenes from the Shipman Inquiry*, the screens were also used to show "the face of whoever is speaking" (1).

By this transformation, the aesthetic horizon of the contemporary tribunal play decisively moved away from Peter Weiss' use of Dante's *Inferno* as a structural template for *The Investigation* in which nine nameless characters represent the testimony of hundreds of Auschwitz witnesses in the context of the Frankfurt trials.[71] Strictly speaking, the German tribunal plays that reached London in the latter half of the sixties – generally framed as the legitimate historical antecedents of the British strand – were more theatrical and poetic in their presentation:

> Hochhuth's dramatic accusation about Pope Pius XII's apparent complicity with the Holocaust in *The Representative* may have been based on fact and is certainly incendiary, but the play is written in heroic couplets. And this verse, together with its five-act form, shapes the horrific events into conventional theatrical aesthetics, with documentary factuality being lost in Schillerian tragedy. (Innes, "Mainstream" 449)

But if anything distinguishes the tribunal plays of the early 1960s/1970s from those of today, it is their relative absence of a stated thesis or rather their most careful and astute dissimulation of one as their makers skilfully distance themselves from any particular ideological stance, with all standpoints being supposedly given equal legitimacy. They also seem most stubbornly attached to "irrelevant details" and the edited version of the transcript that ends up in the published playtexts is peppered with so many unremarkable *sur le vif* moments that they strategically prevent the reader or spectator from reaching *ipso facto* conclusions. Indeed, what has been characteristic of the past two or three decades is a continuing sense of loss, that theatre-makers are no longer operating within a shared purpose or political consensus. While these past political commitments have become somewhat diluted, one can still behold some vestige of the erstwhile engagements. In other words, there are always the materials for an argument within the verbatim transcripts but the spectators have to assemble it by themselves: "[g]rounding the documentary idea in reception rather than in representation is exactly the way to preserve its validity. It allows for the audience to make the truth claim for the documentary rather than the documentary implicitly making the claim itself" (Winston 253). So one may argue that this creates an onstage ethical-decision situation involving the audience more intensively, emotionally as well as intellectually.

71 The play was subtitled "an Oratorio" in the German edition and was given a public reading at midnight, directed by Peter Brook with the Royal Shakespeare Company at the Aldwych Theatre, London, on 19 October 1965. For Michael Kustow, this particular production "pioneered, in this country at least, the stage as courtroom" (134).

Importantly, there is a kind of compelling reverent realism that inheres within the situation,[72] that the whole apparatus of that type of theatre event is designed to uphold. It makes a demand of total complicity and assent upon the audience, along the lines of a constructed "overall coherence" (Lane 69), so coherent that it becomes, in the words of playwright David Hare, "a clarity no other medium can achieve" (*Diary* 118). As Charlotte Westenra writes as regards the first night of *Bloody Sunday*, "you could have heard a pin drop" (Stoller, *Tales* 174). Indeed, always fundamental to this style is an over-arching frame that typically consists in artificially keeping the verbatim material orderly, neat and tidy within a coherent dramatic narrative with a satisfying well-marked beginning, a middle and an end where an audience can get angry about political issues together and get inspired to do something.[73] Within a British realist theatrical context, this fulfils a number of expectations according to which the play's argument(s) spring(s) from the playtext and end(s) with words: "a certain British puritan attitude towards the theatre that [...] prefers theatre just to get on with providing a series of 'pungently instructional points'" (Rebellato, "review" 530), strangely echoed by journalist Richard Norton-Taylor who admitted his intentions to simply "present the argument in a different medium" (qtd. in Megson, "Frisson" 204). Therefore these arguments can be assessed in a critical manner and what happens in performance squarely fits within these parameters towards a synthesis. Such a tidy-minded and questionable reasoning then culminates in Michael Billington's contentious statement which would have one believe that "[v]erbatim Theatre [...] provides us with the data that enables us to make up our own minds on a particular subject" ("Introduction" 2).

Interestingly enough, Michael Billington recognises in Richard Norton-Taylor's *The Colour of Justice* something that seems far removed from the hitherto mentioned desire to narrowly think of these plays in terms of theses or arguments:

> equally impressive was the structure which revealed, in a form reminiscent of Sophocles or Ibsen, the way racial prejudice was embedded in the Metropolitan Police from the lowest level to the highest. I'm not equating the play with *Oedipus Rex* or *Ghosts:* I'm simply saying it used the same tactic of peeling off the layers of deception. ("Introduction" 3)

---

72 In an interview for TheatreVoice (2009) conducted by Aleks Sierz, Paul Unwin and Sarah Beck, the authors of the verbatim play *This Much is True* (Theatre 503) describe this in terms of "a very respectful earnestness": http://www.theatrevoice.com/audio/verbatim-drama-jean-charles-de-menezes/. Similarly, Paola Botham argues that these plays display a certain "austerity on stage" ("Deconstruction" 315).

73 This is generally combined with an artifical sense of progression so that the audience feels that they are getting close to the truth so to speak.

The comparison with Ibsen's classic Naturalist play *Ghosts* is interesting as regards to the points made in the preceding chapter. In the same way, Terry Stoller argues that the success of *The Colour of Justice* is partly due to the fact that "it exposed the layers of deception within the Metropolitan Police with a Sophoclean irony" (*Tales* x). At another discussion, Richard Norton-Taylor describes the entire series of tribunal plays as "what theatre can do and should do, and [...] did do way back in ancient Greece" ("Process" 3). The affinities between ancient Greek theatre and the tribunal play model were also picked up by David Edgar very early on in his discussion of Rolf Hochhuth's *The Representative:* "[h]is act of documentary revelation performed, in his play, an equivalent function to that of the peripeteia of Greek tragedy: the sudden, unexpected, and shocking reversal of fortune that captures and freezes the themes of the play as if caught in a sudden shaft of bright light" (*Farce* 54). Taken together, these remarks would seem to indicate that verbatim theatre clings to a well-trodden dramatic realism at this juncture in contemporary theatre. In other words, if there is a refrain here it is that the contemporary tribunal play creates what one might call a documentary realist verbatim performance through a theatrical reconceptualisation of the tribunal from the point of view of structure.

2.2.2 An Overview of Justifying War

Justifying War: Scenes from the Hutton Inquiry opened at the Tricycle Theatre on 30 October 2003 and marked Nicolas Kent's fifth collaboration with the security affairs editor of *The Guardian* newspaper Richard Norton-Taylor. It dramatised Lord Hutton's independent inquiry into the events surrounding the sudden death in July 2003 of a key government Iraqi weapons expert, Dr David Kelly, that took place in Court 73 of the Royal Courts of Justice, offering insight into the workings and mindset of the BBC, the Ministry of Defence and Downing Street. The 25-day inquiry was condensed into two and a half hours and the production was subsequently adapted for television by BBC 4 to be broadcast in January 2004, only a few weeks before Lord Hutton published his report on 28 January 2004 so as to give "another chance to hear the evidence." Amongst the "official" documents consulted, featured – for the first time in history in the context of a public inquiry – a certain number of emails.

Dr Kelly had been caught up in a row between the BBC and Whitehall about the use of intelligence reports in the run-up to the war with Iraq. On 29 May 2004, the then BBC Radio 4 journalist Andrew Gilligan made an allegation according to which the Government had "sexed up" the dossier. The following day, the soft-spoken Dr. David Kelly was named as being the source of this

BBC news story regarding the September Dossier (2002) which comprised false claims about Iraq's Weapons of Mass Destruction, stating that they could be deployed within 45 minutes. More specifically, it turned out that even the claim that Iraq had such weapons relied entirely on one source. When Lord Hutton's final 740-page report was made public, it put more blame on the BBC than on the Labour Government to the extent that – besides journalist Andrew Gilligan – Gavyn Davies, its chairman of governors and Greg Dyke, its director general, had to resign, causing a significant crisis.[74] This particular moment in international history will also be discussed in Chapter 3.1, albeit with recourse to the different verbatim lens that is David Hare's *Stuff Happens* (2004).

Like Peter Cheeseman before them at the Victoria Theatre in Stoke-on-Trent in the 60s,[75] Richard Norton-Taylor and Nicolas Kent often stress the importance of being as verbatim as possible and ensuring that "answers to one question are not put against another" (Wroe). As such, the playtext of *Justifying War* is "strictly governed by 'pure' verbatim protocols" (Bignell *et al.* 31), thus featuring square brackets when some words have been changed in the context of the dramatisation process of the official transcripts. These serve several purposes. First, they tend to bring clarification to an audience not accustomed to legal and governmental proceedings and help identify individuals as the following quotes exemplify: "HUTTON: In the report I think the F[oreign] A[ffairs] C[ommittee] said that Mr Gilligan's sources should be investigated" (448); "CAMPBELL: John Scarlett [chairman of the Joint Intelligence Committee.]" (452). Second, they make the exchanges between the different "characters" more "theatre-like":[76]

> GILLIGAN: [...] This is me speaking live and unscripted.[77]
> DINGEMANS: Can I take you to this: [You say] '...and what we've been told by one of the senior officials in charge of drawing up that dossier was that actually the Government probably, erm, knew that that forty-five minute figure was wrong, even before it decided to put it in' (433)
> DINGEMANS: [...] Dr Kelly [says]: 'I cannot recall that. I find it very difficult to think back to

74 Alastair Campbell, then Downing Street's director of communications and strategy, also resigned from his position after the Inquiry.

75 Peter Cheeseman's methodology for verbatim theatre was based on one strict rule: "Everything used on stage, all words and actions, must be based on primary source material. If there is no primary material available, then no scene is created" ("Casebook" 87).

76 This is my own interpretation and does not reflect the official statements of the authors in the producer's note where they only mention that these are "made for clarification purposes" (421) to describe a character's title or insert a date, which is arguably linked to my third point.

77 These square brackets are not in the original playtext and only serve here to reduce the length of the quotations.

> a conversation I had six weeks ago.' (434)
> DINGEMANS: [...] 'Mr Kelly [said]: It was a statement that was made and it just got out of all proportion. [...] [You say] 'Okay, just back up momentarily on the 45 minute issue [...]' (441)

Third, they ultimately reinforce, to a considerable extent, a relationship of trust with an audience through a strict adherence to the verbatim methodology and its standards of objectivity so that they appear to circumvent the baleful operation of bias. In addition, the producer's note lays bare how the verbatim playtext was constructed, notably in terms of condensation as the 75 witnesses heard over 25 days in the actual inquiry are only represented by edited transcripts "taken from the evidence of 12 witnesses in the first part of the Inquiry" (421) and purposely omitting Prime Minister Tony Blair's testimony.

The performance started with minor court officials entering the Inquiry room, including an usher and a typist, before the arrival of Lord Hutton and his team of lawyers. The set was mainly composed of lawyers (on the left) and witnesses (on the right) facing each other, chaired at one remove by Lord Hutton. This was followed by Lord Hutton's opening statement before the audience in attendance was to hear the first witness, Patrick Lamb, who "worked very closely with David Kelly" (424), in his role as deputy head of the Counter Proliferation Department in the Foreign and Commonwealth Office.

2.2.3 Time

The experience of a tribunal play is in general linear – following with exactitude the chronology of events in the actual proceedings – from an opening statement to a closing one pronounced by the Chair of the Inquiry. One can see at work the organising principles of realism which artificially create a real-time feel as "we are to imagine that the time of the performance and the time of the original inquiry are identical" (Rebellato, "Kelly" 605). Sometimes realism meets reality as it were, involving far less strain on the imagination as when Sir William Macpherson of Cluny, the Chairman to the inquiry in *The Colour of Justice*, announces that "we will break off now for twenty minutes" (366) which is made to coincide with a 20-minute interval for the audience at the Tricycle Theatre. The same strategy can be seen in the opening statement of Lord Hutton in *Justifying War* which recreates "a minute's silence in memory of Dr Kelly" (423) as well as in the closing statement of Macpherson in *The Colour of Justice* which demands "a minute of silence to remember Stephen Lawrence" (415).[78] Evidence of the

78 This strategy may be seen as an example of what John Langshaw Austin has termed "per-

success of this strategy may be found in the fact that “the whole audience at the Tricycle st[ood] up” (Butler “Trial”).

Very much like realism, the majority of the lines are in the present dramatic tense. This may appear self-evident as verbatim words are known to be previously uttered words. However, a close study of verbatim playtexts reveals that these plays tend to display a disproportionate use of the past tenses, which may seem counterintuitive given that the verbatim playwrights arguably possess “a passion for the present moment” (Waters, *Secret* 193).[79] To make matters more concrete, let us have a look at a scene from a recent verbatim play. In one scene entitled “first meeting and falling in love” (6) in Sudha Bhuchar’s *My Name is...* , Suzy tells us about her first encounter with Farhan:

> Suzy The day Farhan met me, I was on the roller skates, and my wee ghetto blaster on my shoulder. I’d just come back from my Christian camp, which was run by Americans and I was confused. I knew his friend Naseem who was here from Pakistan [...] Naseem drove past and stopped [...] Naseem told me to come down to the flat. I’d sat on the sofa with my photos and Naseem came and sat, and Farah came over and sat beside me on the other side and I was like ‘AAHHH!!’ It creeped me out because he was still messing around in that accent. (6)

While past tenses are also used in more conventional realist plays, they are generally kept to a minimum in favour of generating a feeling of being in the moment and experiencing everything for the first time. Importantly, for Steve Waters, they constitute “the enemies of language’s potential energy, the deadening forces [...] [i]f the past tense is allowed to predominate, urgency is lost” (*Secret* 126). One way to explain this unusual abundance of past tenses in dramatic speech is to reflect on the verbatim methodology itself. Real-life people are interviewed regarding something that happened to them, something they did or witnessed, and a look at any verbatim transcript, under these circumstances, would disclose that the sentences are mainly, if not exclusively, in the past tense. As a kind of compromise within a verbatim play, the standard playwright’s gesture would be to opt for a change of tense in order to create a stronger sense of immediacy, action and confrontation with an audience.

However, the persistence of past tenses in these works – beyond a desire for absolute verbatim purity – means that they play their own literariness off against

formatives” in his seminal *How to Do Things with Words*, described in his first lecture as verbal constructions that indicate “that the issuing of the utterance is the performing of an action” (6).

79 This view is shared by Bernhard Klein who has invoked the term “the presentist play” in reference to verbatim theatre’s use of recent history “to exploit its potential for political interventions in the present” (208).

the idolisation of theatricality. For Gillette A. Elvgren, Jr., "by frequent reversion to past-tense descriptive remarks the sense of reportage is never lost" (95). Therefore this reliance on the exact verbatim words can be read as a strategy to elicit a different experience that can be tentatively described for my purposes as anti-mimetic, casting the audience as listeners within a theatre setting. To put it another way, the agencies of constructing a verbatim play are the milieu in which the original quarrel between verbal diegesis and theatrical mimesis can be resisted and reengineered, for verbatim theatre transforms the standard terms of these strategies. One could then argue that such plays, through their inclusion of direct quotations, are essentially diegetic in the sense that the present tense of the performance chimes in with the past tense of indirect narrative representations. As Tom Maguire has argued in *Performing Story on the Contemporary Stage*,

> this process [...] manifests an incongruity between what is represented and the mode of representation itself [...]: the telling is a different event from the events of the story. [...] Storytelling is a fundamentally anachronistic and displaced form: there is a distinction between the time and places of the storyworld, and the time and places in which the telling is taking place. (13)

Arguably, the significant importance of diegetic mechanisms in documentary realist verbatim performances reaches a climax in the tribunal play. The textual diegesis of these plays constantly competes with visual performative mimetic elements. For my argument, the overabundance of diegetic speech further undermines the theatrical apparatus in the service of a documentary realist aesthetic project.

In the Producer's Note, Richard Norton-Taylor contended that "[t]he evidence is presented chronologically, with the exception of that of Dr Jones. He gave his evidence two days after Mrs Kelly. It is presented here before her evidence" (421). *Justifying War*'s deviation from the original transcripts is striking and deserves to be further investigated. I argued in the previous chapter that the rules under which these plays operate, their formal properties, were subordinated to a documentary realist signifier that infiltrates, to a particularly heightened level, every aspect of the productions under analysis in Chapter 2 of this volume. In other words, the entire verbatim operation was brought under the aesthetic terms of documentary realism, making openly theatrical interventions inconceivable. Its significance is less to do with the complexity of the editing tasks in hand (of such a voluminous transcript, in search of the most noteworthy and representative sessions of the Inquiry) than the impression generated in a theatre setting. This change in the chronology then suggests an intensification of an audience's ordinary awareness of *Justifying War*'s own status as theatre performance and perhaps this also inadvertently ripped apart those vital unities of meaning and expression arbitrarily wielded by the documentary-realist theatre-makers. Structurally speaking, placing her arguably moving testimony at the very end of the play, reveals perhaps the underside of documentary realism, the paradox of an aesthetic dependent on that which it negates.

It is clear that this gesture was, as Tom Cantrell claims, responding very sharply to dramatic demands: "[t]here is little doubt that Norton-Taylor positioned Mrs Kelly's evidence at the end of the play to give the piece an emotional punch, as it is the play's most heartfelt and shocking interview" (*Acting* 172). It is worth noting that, although Norton-Taylor is a journalist and, as he often claims not "a playwright" (Boll 87) or "artist" (qtd. in Hammond and Steward 130), he is perfectly aware of the potency and exigencies of the theatrical medium: "[t]he experience of watching leads to an understanding that goes beyond the mere intake of information; it involves empathy for the victims" (qtd in Hammond and Steward 124). But more than that, one begins to suspect that the deliberateness of this gesture merely foregrounds a tacit strategy, arguably shared by the documentary-realist verbatim playwrights:

> Despite a surface level at which many of these plays offer a sober, even clinical approach to the events they depict, they aim to engage the audience emotionally. What looks like a forensic analysis is calculated to stir up indignation, public protest and sympathy. (Bignell *et al.* 33)

Janice Kelly's audio testimony, transmitted remotely, was accompanied by a photographic image of herself displayed on screens (496). What is more remarkable, perhaps, than this is that the testimony of David Kelly's wife was performed each night as if for the first time with the actress giving the evidence from an upstairs dressing room via a live link:

> instead of just having...the testimony on minidisc with the answers recorded, which would be absolutely foolproof [...] Nick [...] [insisted that] [i]t's got to be live every night. So we ran cable [...] from the control room, under the stage, up the back stairway into the dressing room. And Nick wanted two of everything, in case one system failed. (Shaz McGee qtd. in Stoller, *Tales* 161)

The audience was thus subjected to a particular dissonance between the visual (the immutable expression of Mrs Kelly on the screen) and aural dimensions (the faltering voice of the performer reproducing her testimony). For Paul Taylor, this documentary realist apparatus was disarmingly upping the emotional demands of conventional dramatic realism:

> The painstaking verisimilitude of the staging, with its plasma screens flashing up all manner of documents, is matched by the documentary accuracy that even reproduces the fluctuating drift in the voice link-up with Mrs Kelly. A virtue of dramatic reconstruction is that it can alert you to details that fall through the net in day-to-day newspaper coverage. I shall never forget the brief, harrowing silence at the other end of the line before Mrs Kelly, hitherto steady and stoic, confirms that the painkiller her husband used was the medication that she takes for arthritis. (Taylor "Justifying")

Another and quite difficult aspect of documentary realism needs pointing out here; difficult, for acting is often considered as the worm in the apple of documentary realism as Chris Megson has correctly stressed: "the figure of the actor continues to trouble today's documentarians in multiple ways" ("Act" 148). It is therefore hoped that the next section will illuminate some of these ways and critically examine how the tribunal playmakers have dealt with this disturbing "semiotic surplus" (Megson, "Act" 149).

2.2.4 Acting

Whilst reviewing *Bloody Sunday: Scenes from the Saville Inquiry* – one of the Tricycle Theatre's tribunal plays – for *The Financial Times*, Alastair Macaulay writes that "[f]ormally speaking [...] this is not true acting [...] [b]ut the formal questions pale beside the inescapable seriousness of the theatrical experience" (469). This would imply that the style of acting adopted for the production sits uneasily with the realist tenets of "the Stanislavski-based, stage-focused approach" (Paget, "Facts" 169) of the British actor. As regards to her experience of playing Clare Short in the tribunal play *Called to Account*, Diane Fletcher expresses this position rather vividly in terms of a "different skill" (Cantrell, *Acting* 136) and "not acting"[80] (Cantrell, *Acting* 136). The same pattern informs actor Terrence Hardiman's processes: "[y]ou can't use Stanislavski or method as you are not there to emote, you are there to present an argument as honestly as you can and fairly to the script you have" (Cantrell, *Acting* 100).

For Susannah Clapp, this type of acting conveys something genuinely unique, "a new transparency" ("Sunday" 469) that goes beyond conventional understanding of (stylised) acting practice. This distinction signifies that these particular acting processes produce another kind of theatrical communication, different stimuli arguably closer to those that happen in the "real world". Inter-

80 For Tom Cantrell this emphasis on non-acting is in fact the result of a lack of vocabulary to express "not Stanislavskian acting" (*Acting* 136) in a British context.

estingly, for increased effect, this "documentary realist" acting style is practised rigorously from start to finish in the Tricycle Theatre pieces as the audience is bereft of the traditional curtain call.[81] In addition, in *Justifying War* for instance, "the actors playing the legal counsels had their backs turned to the audience as they speak" (Megson, "Iraq" 370). This degree of realism, then, typically reduces the aesthetic gap between the world of the theatre and that of the audience. In other words, it decisively departs from the repertoire of acting conventions that a British theatre audience is accustomed to see in a playhouse, tentatively restoring a certain level of authenticity despite the fact that the testimony has been removed "from its human source" (Holmes, *Fallujah* 141) to be performed through "an actor's voice" (Holmes, *Fallujah* 141). This distinction from traditional realism is expressly drawn by actor David Michaels who acted in the last Tricycle Theatre tribunal play to date, *Tactical Questioning: Scenes from the Baha Mousa Inquiry* (2011): "I probably could get applause if I decided to make that last speech kind of impassioned. People don't do that in an inquiry. It's that fine line. You have to give it the weight it would get in an inquiry" (qtd. in Stoller, *Tales* 185). More specifically, the process involves mi-cing the actors so that "no one had to project, everyone could mumble" (Kent, "Verbatim" 23).

Given that the performances were intended to replace the broadcast of the inquiries on public television, it is tempting to draw parallels with non-theatre acting. This is a position very briefly touched upon by Derek Paget who argues that "the actors' task in Tribunal Theatre is to achieve an almost filmic naturalism – an acting- that-conceals-acting" (Paget, "Documentarism" 137). In a book chapter, Anneka Esch-Van Kan makes a similar point: "actors speak in an accelerated naturalistic TV-like fashion – striving for a high degree of verisimilitude and avoiding caricature" (422) while Chris Megson concludes that "Norton-Taylor's approach has been more intensely attuned to creating theatre that approximates the experience of live television coverage" ("Frisson" 207). Interestingly, actor James Woolley also implies an uncanny similitude with screen acting: "[i]t is more like a 'take' in a film. You have to print exactly the same one each time. In other plays, things develop and change, but in documentary theatre it is more like printing something – it has to be the same each time" (Cantrell, *Acting* 131). Ironically, the documentary realist acting style described as such oddly resonates with the one generally associated with the mega-musical whereby the actors have no creative input after the first performance in the

81 This was a common feature of the German tribunal plays in the 60s that were, as Christopher Innes explains, "in sharp contrast to normal theatre" as "applause was not wanted" (*German* 172).

sense that it is "a faithful reproduction of the original, all the way from the choreography to [...] the costumes [...] these musicals are usually miked, so as to allow the singer's voice to be modulated and cleaned up, and therefore sound like the record" (Rebellato, "Boom"). This perhaps reached its apotheosis in the BBC television adaptations as well as in the way actor Thomas Wheatley oddly claims that there is no distinction to be made between the stage and television versions: "[i]n the case of the Tricycle pieces, I don't think it makes very much difference at all because the pieces are...were in a small theatre. And because it's all miked, and, anyway, it's all quite...televisual already" (12). Importantly, I am keenly aware of using a term, "documentary realism", that invites one to consider further this relationship.

There is also a fundamental difference for actors playing real people as they are under more pressure than ever to adapt themselves to an illusion of reality, especially in the context of the tribunal plays that have portrayed extremely high-profile individuals (Margaret Thatcher, John Major, Michael Heseltine, Albert Speer, Alastair Campbell, etc.) – in a manner not dissimilar to Piscator who used "leading political figures as characters onstage" (Irmer 18).[82] Actor Thomas Wheatley, who appeared in all eight Tricycle tribunal plays, explains that he was cast for the role of William Waldegrave, a former minister at the Foreign Office in *Half the Picture*, as "a more-or-less credible look-alike or seem-alike" (qtd. in Stoller, *Tales* 138). This was further supported by his acting preparation which consisted in using a tape in order to examine some of his body language: "I watched that and I looked for the hand mannerisms and to some extent the way they speak" (Wheatley qtd. in Stoller, *Tales* 138). The same acting routine is voiced by actor Bill Hoyland who was playing John Major in *Half the Picture* at the time:

> What I mostly did was work from video tapes [...] I watched him, as I say, like an hour in close up and his top lip never moves [...] Then the Tricycle found a very similar pair of glasses, and we just put plain glass in them, obviously. I parted my hair – as it happens I do part my hair that side, me and John Major are some of the few. (3)

It should be said that this process was given full force in *Called to Account* (2007) as Tom Cantrell writes: "[e]ach actor was [...] given a DVD of their subject's interview. In comparison to the previous tribunal plays [...] in *Called to Account* the actors had unprecedented access to the specifics of the original interview [...] and [this] prompted a particularly focused concern with restraint and precision

---

82 Very often the people they portray (and their close circles) are in the audience on a performance evening.

among the cast" (*Acting* 89–90). Here the tribunal play overlaps with another kind of documentary realist verbatim strand pioneered in New Zealand that is beyond the remit of this volume and which characteristically combines the verbatim methodology to be discussed in Chapter 3.2 with the intense study of videos:

> In re-presenting the interviews, the actors used MP3 players (or iPods). With the testimonies of interviewees playing in their ears, the actors repeated not only the words of their subjects, but, as closely as possible, replicated their accents, inflexions, and hesitations. [...] However, we took this method a step further [...] [b]ecause we film all our interviews, in rehearsal the actors study intently the edits they are to re-enact, and therefore, in performance, they replicate as accurately as possible not only the words and the verbal delivery, but also each gesture and involuntary movement of their subjects. (Halba and Young 175–176)

Such a fidelity to the "real" perhaps reached a level of fetishism rarely encountered when actor Thomas Wheatley confessed that he had met Philippe Sands before attempting his portrayal in *Called to Account*, leading him to borrow some of his belongings: "I'm using the pen he uses...I'm wearing one of his own ties, which he has lent me" (Cantrell, *Acting* 122).

It is arguably the actor (contrary to what one might expect) who is, in many respects, the most intense site for these documentary realist practices.[83] As seen at the beginning of this study, this was reflected clearly in the unanimous critical reception which found these plays authentic and eerily real.[84] The actor in these works began to precisely inhabit this liminal space – a space that provocatively troubles the easy division of fact and fiction – as his role was unabashedly extended to the public sphere beyond the performance:

> it is a measure of the extraordinary verisimilitude, the reality of what we managed to do there...That...there was sort of two camps in the Tricycle. I mean, those of us playing dodgy policemen...I, I never had an exchange with either Doreen or Neville Lawrence. And...and...I think that was something to do with...I was Groves, I was also Youngerwood. I was...I was them, and not an actor...I think. You know, it all gets a bit blurred. (Wheatley 9)

83 Even more so in the plays to be studied in the next chapter as actors possess "a certain secondary 'authenticity' as witnesses of witnesses" (Kalb, "Solo" 19). In the tribunal plays, actors do not systematically attend the real Inquiries or meet the real people they seek to portray.
84 In the context of one evening performance of Dennis Woolf's *Beyond Belief: Scenes from the Shipman Inquiry*, Alfred Hickling amusingly reports that there was "a slightly comedic sense of confusion reign[ing] as to whether events [were] happening for real. When a court official announce[d] Dame Janet's arrival, several members of the audience stood up."

> Sometimes the boy who was playing Jamie Acourt, one of the accused, would feel threatened. People wouldn't speak to him in the bar. People would think he was the actual guy. His name was Christopher Fox. He had a tough time. And I said [...] Your portrayal of this horrible, evil young man is so powerful, is so real, so genuine that the audience believe you're him. (Jenny Jules qtd. in Stoller, *Tales* 151).

These paradoxes that can be said to prolong the event are undoubtedly the paradoxes of the documentary realist frame that creates a kind of hypertrophy of the realist apparatus, conferring "a semiotic frisson" (Megson, "Act" 147) which may lead to a stunning reversal of original and copy as the body of the actor is exposed to new socially and politically articulated forces and claims. These very forces and claims rupture the balance between the real and the representational and afford the actor a documentary importance as an almost "human document" (Scott, *Expressions* 6) he or she might not have gained otherwise. As Liz Tomlin has indicated as regards another verbatim play, this "seamless slippage from the actor to the real person [...] highlights the potential dangers in representing the 'real' person through [...] naturalistic techniques" (*Acts* 132–133). Curiously and conversely, this particular involvement, as Terry Stoller suggests, is internalised and taken on by the actors themselves, creating a structure of association and meaning which goes far beyond the performance: "[a] core group of actors regularly appear in the tribunals, and those I've spoken to express pride about their involvement in this important body of political theater work" ("PM"). One of them, Jeremy Clyde, describes how extensively emblematic of this, the backstage discussions had become:

> Because we were all reading the papers and everybody is up to date with the issues and has done the research, some of the most interesting debates and discussions I've ever had took place backstage at the Tricycle. Enlighting conversations. Often talked about from a position of great knowledge, from the position of your character. [...] Really arguing out the politics was another great element of the process. (Cantrell, *Acting* 111–112)

Yet, considering the change in actors' practices towards what I have termed documentary realism is only the first step. Before one can say decisively that these verbatim plays are infused, redefined and guided by documentary realist principles from production to reception, one needs to understand the curious dependence on these aesthetic strategies to achieve "political theatre", which has been observed in this period.

2.2.5 Political Theatre

As tribunal theatre re-emerged in the past two decades, the question of its relationship to politics often followed close behind. Certainly, the contemporary verbatim theatre I am discussing here cannot be equated with the overly "political documentary theatre" of Piscator or Weiss "tak[ing] sides" (Weiss 42),[85] a revolutionary political theatre seeking to intervene and exert a direct political influence whereby all aesthetic means are subordinated to an essentially political end so as to incite an audience to come to the "right" conclusions: "the presentation of solid proof that our philosophy and all that can be deduced from it is the one and only valid approach for our time" (Piscator, *Political* 93). For one thing, as Steve Waters reminds us, "the new wave of fact-based theatre inhabited a much more uncertain political context" ("Political" 139) and consequently, contemporary political verbatim theatre cannot hope or desire to achieve "the status of an irrefutable document" (Kipphardt 307). Michael Kirby, in his illuminating essay "on political theatre", identifies two main characteristics that distinguish it from other types of theatre: "it is a performance that is intentionally concerned with government, that is intentionally engaged in or consciously takes side in politics" (129). In this case, it stands to reason to argue that the tribunal play is political according to the first characteristic. As I remarked in the Introduction, the recourse to documentary realism in this period generally pertains to the fulfilment of social and democratic functions such as curbing governmental and media excess, providing a diverse forum for public debate and voicing minority or unpopular views in British society: "[i]n general, functional theatre-with-a-purpose, like Verbatim and Tribunal Theatre, has served political purposes broadly opposed to *status quos*" (Paget, "Documentarism" 138). In the case of the tribunal plays, more specifically, they fulfil one of the functions of documentary theatre identified by Carol Martin: "[t]*o reopen trials* in order to critique justice" ("Bodies" 12). Moreover, the fact that Lord Hutton came to a different verdict from the one "implicitly" advocated by the Tricycle Theatre production confers the piece an overt oppositional political status as Reinelt has forcibly argued: "*Justifying War* now constitutes a reminder of the material and critical connections overlooked by Lord Hutton; it is a kind of counter-discourse" ("Promise" 17).

85 For Bérénice Hamidi-Kim, this type of theatre has not completely disappeared from the contemporary stage and she calls these practices "historical documentary theatre." See Hamidi-Kim (*Cités* 461–473).

More surprisingly, this type of theatre possesses a radical potential that is extended into the political actual. In other words, the documentary realist tendencies of these aesthetic productions create a mode of representation that is more authoritative, convincing and therefore more likely to have tangible effects in the physical world beyond the act of performance. Many instances of verbatim theatre in the past two decades have tapped into this category of realism in a startling way and this leads one to question whether verbatim theatre might have distinct affordances that would allow the difference between art and politics to be no longer discernible, to cease being separate modalities. Such considerations take on a particularly compelling turn with the knowledge that the Arms to Iraq Inquiry by Lord Justice Scott was the first play ever to be performed in the Houses of Parliament and that the video recording of one of the performances of *The Colour of Justice* has been used "for training in Hendon, and the Leicestershire police have used it for racial awareness for police" (Kent "Empathy").[86] To sum up, this pertains to a conception of realism that Elin Diamond in her *Unmaking Mimesis* articulates in the following manner: "[r]ealism is more than an interpretation of reality passing as reality; it *produces* 'reality' by positioning its spectator to recognize and verify its truths" (4).

The impact of these plays was felt even more forcibly by riding on a momentum as most of them were systematically and strategically staged before the final verdict of the inquiries. In the case of *Justifying War*, the Hutton Inquiry was particularly fresh in the mind of the general British public as the witnesses had given their evidence a mere two months beforehand and "Lord Hutton [was] expected to present his report to the Government in the New Year" (421).[87] What is perhaps most striking is that *The Colour of Justice* "was part of the breaking news all the time [...] and there were extracts being played on news programmes" (Wheatley 9). Even the traditional post-show discussions were deliberately trying to discuss the issues raised by the Inquiries, intriguingly and firmly shifting the emphasis away from any theatre-related concerns. The first post-show discussion after *Half the Picture* in June 1994 is described in these terms by journalist Malcolm Rutherford: "[t]his takes place in the theatre with a formal chairman, a panel of lawyers, journalists, politicians and, as with Question Time on televi-

86 *Half the Picture* was performed in the Grand Committee Room of the Houses of Commons, hosted by MPs Menzies Campbell, Michael Meacher and Richard Shepherd on 11 July 1994. It also won a Freedom of Information Campaign Award. Another Tricycle Theatre verbatim production was presented at the House of Parliament, much later in 2006: Gillian Slovo and Victoria Brittain's *Guantanamo* which also played on Capitol Hill.

87 At the time of the Inquiry, BSkyB was transmitting nightly reports using edited verbatim court transcripts.

sion, the audience is encouraged to join in." This is reminiscent, on a less harrowing level, of the way in which these discussions were orchestrated in the context of the upsurge of German tribunal plays in Britain in the 1960s. On the occasion of the production of Rolf Hochhuth's *The Representative* by the Royal Shakespeare Company at the Aldwych Theatre in 1963 that I have already mentioned, Michael Kustow, who ran these post-performance discussions at the time, tellingly explains that "the auditorium became a pressure-chamber for the rage and short-lived relief of concentration camp survivors at seeing some part of their experience brought into shared light. English people who had lived a long way in space or time from what happened saw their 'positions' overwhelmed by this release of pain prompted by minimal theatre based on events" (Kustow 133–134). At this point it is pertinent to recall that the procedure I have just described was extended to the theatre programme itself as Chris Megson revealingly suggests in the context of *Justifying War:* "[t]he programme for the production is so brimful of copied documents and transcripts that it resembles a ministerial briefing" ("State" 116). One might also recall the ample and substantial supply of editorial apparatus in the published playtexts themselves. A good example of this is *Called to Account* which comprises a 27-page appendix.

Additionally, theatre critics seemed unwilling to review these pieces as theatre and the terms of their praise are fascinating, to say the least. Quentin Letts of *The Daily Mail* speaks of "a valuable service" (470), *The Telegraph's* Charles Spencer of "a signal public service", *The Jewish Chronicle*'s John Nathan of "public-spirited value" (471) and finally, *Guardian* journalist Michael Billington refers to these pieces in terms of a "source of information [...] and instrument for social change" (*State* 385) as well as "event" ("Sunday" 470) in their own right (in fact an event of the event).[88] Through this process, each of these discourses prioritised a certain vision of political theatre to the detriment of aesthetic questions. However, I cannot stress too strongly that broad aesthetic questions are the very ones that critics must pose and it is therefore precisely this lacuna that the present monograph is seeking to redress. Put differently, the fact that some critics are almost reduced to making such claims is also a sign of documentary realism's strategic efficacy in mediating verbatim material in performance and that one is, arguably, knee-deep within the elusive and obscure domain of documentary realism proper. Such an enchainment led Will Hammond to phrase the following hypothetical question: "Does it make sense for a verbatim play to be reviewed by a theatre critic, rather than, say, a political journalist?" (163). In a

88 Derrida reminds us in *Archive Fever* that "the archivization produces as much as it records the event" (17).

particularly telling instance, Catherine Love – in the context of Chris Goode's verbatim piece *Hippo World Guest Book* (2007) – also writes about the difficulty of critically engaging with the performance's aesthetic:[89] "There's a very low-fi aesthetic to the piece – so low-fi, in fact, that it would be easy not to think of it as an aesthetic at all, just to accept it as a man reading from a stand."

With even greater irony, the plot thickens as tribunal plays seem to operate in surprisingly similar ways to the very ideologies they are attacking, as Chris Megson has rightly observed: "tribunal theatre attempts at once to resist and reify the discursive technologies of contemporary media" ("Frisson" 207). Michał Lachman makes a similar point, perhaps more vehemently: "[b]y communicating with audiences accustomed to the 'totemic somnambulism' of the popular press, verbatim drama itself conforms to, and becomes part of, the cultural industry" ("Colours" 321). It would thus appear that these performances do not provide a critique of reification and, instead, their documentary realist structures might be said to enact precisely what they ostensibly attempt to define themselves against. The tribunal play's aesthetic strategy, then, perfectly articulates the paradox of documentary realism, as it reveals how the desire to get beyond the current hegemonic structuring discourses hopelessly intensifies and increases the dependency of its rhetorical seduction as it undergoes the precise processes it claims to combat. As Kaplan aptly puts it, it might be the case that "[r]ealism as a style is unable to change consciousness because it does not depart from the forms that embody the old consciousness" (80).

2.2.6 Conclusion

In this chapter, we found that it is not the use of verbatim techniques or aspects of the real alone as autonomous objects that generates realism within the tribunal pieces but their organised contextualization in performance, what Alan Filewod has termed "the aesthetic illusion of reality" (13) and Janelle Reinelt "an aesthetics that sometimes functions as an epistemology" ("Poetics" 72), that is an artistic supplement that actively works in this context towards its own disso-

89 *Hippo World Guest Book* tells the story of an American weapons manufacturing executive, Ramon Valencia, who, out of a particular fondness for hippos, decided to launch a website for like-minded people in 2000. The online guest book was not, however, only the recipient of love-filled messages (as his creator intended) and, soon enough, mocking and inappropriate messages started to fuse. Ultimately, trolls and spambots became the only vocal visitors of the website and Chris Goode is reciting here, in a solo performance, an hour-long edited version of the messages left over a span of six years.

lution so as to catalyse the Inquiry as real again. In other words, it implies that there is no verbatim theatre without a relation to some aesthetic frames – even in a minimalist form – that shape and encourage a specific audience's perception. As such, these performances exemplify many of the central concerns of a contemporary verbatim practice more generally, representing a sharp turn in theatre-making and a documentary-realist reappraisal of the British stage. As Carol Martin puts it, these practices are provocative through the way "in which [they] strategically deplo[y] the appearance of truth, while inventing [their] own particular truth through elaborate aesthetic devices" ("Bodies" 10). It is thus precisely the documentary realist frame that engenders a particular engagement, encouraging one to see not only these theatrical reconstructions as real but, more profoundly, the "original" as perhaps fully real for the first time as it eventually triggers its deep emotional force, potency and charge. This documentary realist process thus arguably fills its audience with a sense that it is imperative to reconsider the "original." Judith Butler's thoughts on photography in the context of contemporary war can arguably be extended to the tribunal plays discussed here as they also constitute "alternative frames" that "permit another kind of content [that] would perhaps communicate a suffering that might lead to an alteration of our political assessment of the current wars" (*Frames* 77). As I use the term, then, documentary realism is only the result of the collaboration in performance between theatre and the verbatim material. In the case of the plays I have just discussed, much of their aesthetics strive to emulate the tribunal template, scraping away the tropes of theatricality that have become associated with a certain breed of documentary theatre and thus redefining its boundaries.

21 years after *Half the Picture* – a dramatisation based on the edited transcripts of the Scott Inquiry, also known as "the arms-to-Iraq" inquiry – opened at the Tricycle Theatre, the verbatim tribunal play seems to have completely disappeared from the British theatre landscape. In a special interview to mark the 20th anniversary of the Tricycle's first tribunal play I have just mentioned, playwright David Edgar solemnly pronounces its demise in the presence of Victoria Brittain, Nicolas Kent, Richard Norton-Taylor and Gillian Slovo: "I get the feeling that people are now moving away from fact-based drama, that fact-based drama has, as it were, done its job. And it may be that part of that job has been to put theatre back at the centre of political life" ("Verbatim" 40). Other commentators routinely appealed to this idea that the "era" of verbatim theatre was coming to an end and that "the shine on verbatim theatre [was] starting to tarnish" (Gard-

ner "Talk") as its "popularity has [...] waned somewhat" (Radosavljević 119).[90] More specifically, they claimed that a new creative energy would be discovered, perhaps something nostalgically reminiscent of "British Theatre during its golden ages of creativity" (Sierz, *Today* xi), "[a]n avant-garde that explored theatrical possibility, [...] pioneered a new aesthetic [...] suggesting daring new experiments" (Sierz, *Today* xii). In many ways, of course, this is rather truistic and may be explained in part by the departure of Nicolas Kent in May 2012 due to new cuts in Government funding.

Despite these claims, by all accounts, it may be impossible to assert with any certainty whether or not the tribunal play is to reappear again in the mainstream theatre landscape. However, it is clear that the contemporary tribunal plays have been an important force in the establishment of a documentary realist tradition of verbatim performances and their effects are found in every "nontribunal" (see Dawson 18) play mentioned in this research (see Luckhurst, "Ethics" 209).

---

90 In Chapter 3 and 4 of this volume, I argue that verbatim theatre was in fact just coming into a new phase at the very moment critics started to claim it had begun its downward spiral.

2.3 “Documentary Realist Verbatim Theatre”: Robin Soans’ *A State Affair* (2000)

As I have already argued at the beginning of Chapter 2.1, against the presumed exhaustion of the documentary realist position in theatrical terms and concomitant with the popularity of the tribunal plays, there has been a remarkable persistence of this strand in Britain until this half of the decade. Indeed, the enduring relevance of these verbatim works would seem to contradict, or at least complicate, Dan Rebellato’s contention that, in response to globalisation, “the rules of political theatre have been transformed [...] Where realism seemed essential, now a kind of non-realism seems so [...], now aesthetic experiment may be the right means to achieve an effective political response to the challenges of a consumer culture and a marketized world” (“Nation” 259). As regards the plays studied in Chapter 2.2, this hackneyed critique has been neatly summarised by Aoife Monks in the following: “a distrust of the realist form used to present transcript material which appeared as ‘fact’ onstage, but which had been carefully crafted by the show’s ‘writer’, the journalist Richard Norton-Taylor” (347). These theatrical works would also dispute a certain prevalent discourse that broadly conceptualises “antirealist methods [...] [as] forms of political resistance” (Erickson 164). Arguably, these plays present certain limitations but the automatic dismissal of them on the grounds that they constitute the least favourable aesthetic strategies for a presentation of verbatim material on the contemporary stage needs to be critically re-examined. If I am to draw the logical conclusions from these aesthetic proposals “out of post-modern anxiety” (Hur 204), they may reveal an unspoken and parallel project aiming to rehabilitate realism[91] in a “post-postmodern culture” (Nealon xii).

Describing the presence of this strand of plays in “the English and British theatrical mainstream” (134), Duška Radosavljević states that the first wave of the revival of verbatim theatre inevitably “evolved in the context of ‘English realism’” (134). For Radosavljević, David Greig’s definition of “English realism” can easily be applied to contemporary verbatim theatre, but I would contend that this is especially true of the documentary realist category identified in Chapter 2

91 This will be further developed in the Conclusion Chapter, drawing on Sheila Stowell’s 1992 article “Rehabilitating Realism”. One must also remember that “realism” can be compatible with a postmodern project depending on what one ascribes to the concept. Therefore, it is perfectly possible for Stephen Chinna to write that “postmodern performance presents a ‘realism’ aware of its own entrapment within cultural and therefore aesthetic/political discourses, and thus it can be interpreted as the principle paradigm for the practices of a deconstructive postmodernism and its politics” (16).

of this book. David Greig's assertion regarding what he terms "English realism" corresponds to an important body of works that:[92]

> seeks out and exposes issues for the public gaze. It voices 'debates' rather like columnists in the broadsheets. Its practitioners are praised for their 'ear' for dialogue as though they were tape recorders or archivists recording the funny way people talk in particular sections of society and editing it into a plausibly illustrative story. English realism prides itself on having no 'style' or 'aesthetic' that might get in the way of the truth. It works with a kind of shorthand naturalism which says, 'this is basically the way I see it'. Distrustful of metaphor, it is a theatre founded on mimicry. In English realism, the real world is brought in to the theatre and plonked on the stage like a familiar old sofa.

The unfortunate irony is that verbatim plays have very often been compared to journalism (Norton-Taylor "Truth"; Preston) and the very use of tape recorders was deemed to be at the origin of the method (Paget "Oral History"). Furthermore, the problematic truth claims associated with verbatim practice are often invoked in conjunction with a desire to recreate on a stage real-life events as authentically as possible, and this literally happens on stools, cardboard boxes, chairs or even sofas following the interview process in the research phase.

In 2006, as the resurgence of verbatim theatre in Britain was starting to be felt more acutely, the symposium "Verbatim Practices in Contemporary Theatre" was organised at the Central School of Speech and Drama by the Centre for Excellence in Training for Theatre, describing these practices as "a new realism" (CSSD). This would imply that verbatim theatre aims to revitalise the category of realism and therefore constitutes an aesthetic project different in kind from what is traditionally understood as theatrical realism. Having looked at how the documentary realist representational system functioned within the tribunal play, I will now venture into the second perceived dominant narrative in British verbatim's centre ground. In this chapter, I will thus address this practice specifically through a detailed study of Robin Soans' *A State Affair* (2000), which seeks to affect the spectator's perspective (and own responsibility) on the ideological basis of the emotionally charged reality represented. The decision to dedicate a whole chapter to this strand (separate from the tribunal play model) was dictated by the crucial theoretical difference encountered when theatre-makers base

---

92 The term is used in this context against different traditions of theatre-making, notably in France and Scotland, that David Greig considers as non-realist. British playwright Simon Stephens uses a different phrase, "Anglo-Saxon naturalism" (Finney 9), to designate this perceived dominance of "English realism" in comparison to contemporary German theatre. In a similar fashion, Aleks Sierz opposes "English naturalism" ("What" 20) to what he calls "Continental modernism" ("What" 20).

their performance text on "aural testimony" (Paget, "Documentarism" 138) rather than on documents available in the public domain.[93] The challenges raised by the resurgence of the documentary realist mode of performance are far reaching in their implications, and the ensuing discussion will attempt to encompass, not only responses to this particular case study, but reflections on diverse verbatim plays in this alleged aesthetic category.

2.3.1 An Overview of *The State Affair*

Robin Soans's *A State Affair* was directed by Max Stafford-Clark in an Out of Joint[94] double-bill production with a revival of Andrea Dunbar's semi-autobiographical play *Rita, Sue and Bob Too* (1982) and opened at the Liverpool Everyman on 19 October 2000. This production was then first performed in London at the Soho Theatre on 5 December 2000, eighteen years after Max Stafford-Clark had originally commissioned *Rita, Sue and Bob Too* for the Royal Court. More significantly, the production had two national tours, a New Zealand outing in the context of the Wellington International Festival and even ended up being performed in the River Room in the House of Lords in 2002, in front of a select audience of politicians, businessmen and representatives from Addaction (a drug and alcohol charity) who were trying to raise money for a new centre in London at the time.

It also marked the first collaboration between actor and playwright Robin Soans[95] and "one of the pioneers of verbatim theatre in the United Kingdom" (Radosavljević 135),[96] distinguished director Max Stafford-Clark, on a series of relatively high-profile verbatim plays:[97] *Talking to Terrorists* (2005),[98] *Mixed Up*

---

93 For Caroline Wake, "the most significant difference between verbatim and tribunal plays is that tribunal playwrights rarely conduct interviews" ("Spectrum" 8).

94 Out of Joint was founded by Max Stafford-Clark and Sonia Friedman in 1993.

95 Robin Soans first encountered verbatim theatre in 1995 in the context of Klaus Pohl's *Waiting-room Germany* at the Royal Court Theatre in which he had a role.

96 Since 1974, Max Stafford-Clark has produced a certain number of verbatim pieces. Notably with Joint Stock Theatre (co-founded in 1973 with David Hare and David Aukin) he created *The Speakers* (adapted by William Gaskill and himself from the book by Heathcote Williams) in 1974, *Yesterday's News* in 1976 and later in 1983 he used the verbatim technique again, this time at the Royal Court with *Falkland Sound*.

97 Robin Soans also wrote three other verbatim plays as a solo playwright: *Across the Divide* (1997), *The Arab-Israeli Cookbook* (2004) and *Life After Scandal* (2007).

98 *Talking to Terrorists* was in the middle of its summer run when the 7/7 attacks in London happened, forever changing its public reception.

North (2008) and *Crouch, Touch, Pause, Engage* (2015), coinciding with the revival of verbatim theatre in Britain. Typically, their particular process consists in working with a group of actors helping them "carry [...] out the interviews and [...] the editing process, particularly trying to find a way to bring all of the separate interviews together in a coherent story" (Quinzaños 10). They tend to start from a position of ignorance and then use a specific approach whereby interview and feedback techniques are crucial to the process of making a verbatim play. Once an interview has been successfully conducted and some notes taken with extreme attention to any characteristic gesture or phrase, the actor then relates the interview back to the entire creative team. This is what Max Stafford-Clark has coined as "hot seating", a type of role-play where

> the interviewers 'become' the interviewees [...] the rest of the group have the opportunity to ask the same, or similar, questions to those asked at the original interview [...] they also gain a sense of the personality of the individual interviewed, and of the atmosphere of that interview. (Soans, *Affair* 149)

They recreate in the rehearsal room the exact location of the interview, trying to adopt the manner, gestures, tone of voice of the interviewee, but fidelity to the exact words is not the main focus. In this sense, Radosavljević points out that this process equally "involves an empathetic/neurological engagement with the speaker" (138). For instance, in the context of David Hare's *The Permanent Way*, another Out of Joint verbatim production, actress Bella Merlin was extremely attentive to one of the bereaved mothers and her every physical action as well as her "body language slips":

> I'd witnessed [...] how her voice faltered, how long she paused before talking again, how her vocal pitch cracked, how she boldly endeavoured to regain her composure as quickly as possible by brushing away her tears, straightening her back and munching on a chocolate digestive. (*Toolkit* 154)

A State Affair revisits the community living on the bleak Buttershaw Estate, one of the poorest estates in Bradford, West Yorkshire, and focuses on the recent devastating effect of cheap heroin: "[t]he magnitude of the problem was such that practically every other home on estates like Buttershaw had at least one family member who was an addict" (Soans, *Affair* 156). More specifically, the play concerns a broken working-class community plagued by many growing social ills such as mass unemployment, poor housing conditions, homelessness, sexual abuse, domestic violence, high incidence of youth crime, low educational attainment rates and self-esteem and a general lack of governmental responsiveness and support.[99] What fills the vacuum, however, is "the drive and the commitment of [...] various care workers" (Stafford-Clark qtd. in Soans, *Affair* 144). In the play, Sue and her husband Jim started to build "The Church on the Way" (107) with "a drugs and alcoholics meeting every Tuesday" (107) and "a [monthly] prayer night" (120) before investing in a second project "Agape House" (109) "for lads from eighteen to twenty-five roughly [...] lads who are wanting to come off drugs" (110) six years ago. The success rate of these initiatives has been noticeable in the area and "some people [were] getting their lives together with a lot of courage and determination" (134) but sadly "some others [were] going down a big steep hill, into a big black hole" (134) away from the public eye in the general indifference.

2.3.2 A New Documentary Realist Contract

A significant factor in the discussion of this particular mode of representation lies in the place accorded to the conventional agreement between actors and audiences that underpins most forms of stage realism within the established structures of British theatre. As in the case of the tribunal play, realism emerges from a "diegetising" technique in the form of a framing announcement stating that: "All the words in this play are taken from conversations with the people in and around Bradford in July 2000. Some of the stories have been condensed and conflated, but the words are theirs" (Soans, *Affair* 84). Yet, the verbatim artists also retain a strong idea of theatrical realism – despite the fact that they are working from fragments – as it reflects a certain consensual idea of the representation of reality on stage in Britain. This is particularly clear in David Hare's *The Permanent Way* (2003), one of the plays in this tentative category which uses a "metaphorical realism" to ground the different interwoven verbatim narratives:

> This term may be applied to plays where the location is non-specific: although the details of each moment may seem realistic (coffee cups, biscuit tins, newspapers, etc.), the fluidity of the changing stage space allows each 'mini' location to take on a theatrical resonance beyond its own reality. For example, the Scottish Literary Editor is able to move between other characters' individual spaces without breaking the 'reality', say of the Vicar of Hatfield's sitting room or the train carriage in which the Squadron Leader and the Bereaved Widow are sitting. (Merlin, "NT" 20)

99 In 2000, according to Joe White, "government authorities [were] beginning to recognise the extent of these problems and making resources available to deal with them" (qtd. in Soans, *Affair* 162).

Portrayed as such, as John Fiske has argued, "realism is not a matter of any fidelity to an empirical reality, but the discursive conventions by which and for which a sense of reality is constructed" (21) and this is to be understood as belonging to what Roland Barthes has expressed in terms of "the cultural rules of representation" (145). In this sense, Max Stafford-Clark in the same breath proudly disavowed any Brechtian potential as it would appear "intrusive" (qtd. in Merlin, "NT" 6). Obliterating these Brechtian devices was designed to supress questions which would undermine the documentary realist strain which was meant to be preserved in all its glory. A good example of this appears in Part Two of *The Permanent Way* which re-enacts part of the Cullen Inquiry into the Ladbroke Grove rail accident in 1999, but the style significantly departs from what has been seen in Chapter 2.2 as "*Everyone* [on stage] *becomes the Cullen Inquiry*" (48):

> Kika Markham doesn't make any costume changes at all for the tribunal. Now on the page, when you consider it beforehand, you think, 'Well, how can that work, because we've established a convention, she comes on in a different costume to be the Bereaved Widow at the end, so how can she now – without any costume change – be an attendee at the tribunal? Isn't that going to be confusing for the audience?' And then you try it in rehearsal, and you think, 'No, it's not confusing, it's perfectly followable' (Stafford-Clark qtd. in Merlin, "NT" 6).

The extent to which this is perceived as such by an audience is debatable, but these plays unfold as if this were true.

However, these documentary-realist conventions equally pursue realism to such an extreme point at times that it feels almost strange, as when a gap is just a bit longer than expected. In Chris Goode's *Stand* (2014), a verbatim piece about ordinary people changing the world, one of the six characters abruptly leaves the stage mid-sentence to check on some baking in the oven: "I need to go and check my puddings".[100] What this highlights, apart from a tenacious fidelity to the interview process in front of an amused audience, is a type of realism whose intentions go a step further than traditional realism, perhaps even further than one has been willing to go. This is what Chris Megson describes as "hardcore illusionism" ("Iraq" 370; "State" 116) in the context of the plays studied in Chapter 2.2. It is in this sense that Elaine Aston and Janelle Reinelt formulate a hypothetical proposition according to which *A State Affair* would seem "to be trying to go beyond the constructed realism of the first play" (290). From this perspective, it is evident that the pairing of Robin Soan's *A*

100 Notes taken on Saturday 2 May 2015 during one of the performances of *Stand* at the Battersea Arts Centre, London. The playtext has not been published.

State Affair with the social realist *Rita, Sue and Bob Too* by Andrea Dunbar moves the comparison well beyond thematic parallels. Crucially, this was also accompanied by the use of the exact same set design in both plays which reinforced a sense of "gritty realism" through its picture of "urban decay and decline" (Aston and Reinelt 286), establishing their aesthetic parameters immediately.

Contrary to the tribunal plays that cast the audience as a distant tribunal audience,[101] perhaps even as a televisual audience remote from the action, audiences in these plays are more openly solicited, essentially by means of direct addresses. In this way, the audience becomes the interviewer figure, "a key, if silent, character in the performance" (Soans in Hammond and Steward 21), and is intermittently invited inside the fourth wall. A good example of this can be found at the very end of *A State Affair* when Lorraine, Andrea Dunbar's daughter, enters the stage to express the difference between the world described in *Rita, Sue and Bob Too* and the everyday experience of people living on the estate in 2000:

> Then? Then there was a sense of community, but it was really rough. [...] Now? [...] the estate's been fractured. The drugs wrecked it really. If my mum wrote the play now, Rita and Sue would be smackheads ... on crack as well [...] Bob would probably be injecting heroin ... taking loads of tablets as well. (133)

Even when several characters on stage are talking across each other, they do not really engage or interact with one another and are again exclusively addressing the audience "on a purely personal and confidential level" (Soans in Hammond and Steward 23). In the following excerpt, for example, Paul, Marie and Dean explain the day-to-day struggle of heroin addicts, delivering their speech directly to the audience:

> Marie A typical day in the life of a shoplifter [...] You don't want to get in the shower...
> Dean No...
> Paul Not if you're rattling.
> Marie As it's the morning, me and Benny'll [...] target a particular shop, and a particular article [...] I'd go in with a detagger ...
> Paul Just a strong magnet really.
> Marie I'm good at it.
> Dean I'm not. (118)

---

101 Christopher Innes writes of the German tribunal plays of the 60s that the audience was cast "in the role of jury" (*German* 171). This is not as much the case in the tribunal plays discussed here apart from *Called to Account* (2007).

In short, these plays formulate direct address as the primary mode of communication on stage and very seldom feature dialogues or complex interactions between the characters.[102] The terms of the new documentary realist contract would thus seem to differ from the ones proposed in the tribunal play model. If the previous model was concerned with realistically reproducing the experience of attending a public inquiry, these plays would appear to attempt to restore the experience of hearing the voices first-hand: "in response to the impact that our interviewee had initially had on us – and therefore how we might now impact in a similar fashion upon our own listeners" (Merlin, "Hare" 125). Therefore it is not surprising to find a general absence of dramatisation of the content of the interviews and, instead, a dramatisation of the interviews themselves. For example, in the following, Peter, a youth worker in Bradford, tells us about his diary for the day "Eight-o-clock: Leave Fortress Tyrrell. Eight twenty: Meeting with the fire officer to discuss the introduction of the new weekly fire drill [...] Six thirty: Back to the office to do the accounts" (114–115). The actor playing Peter does not try to enact any of the activities on the list and certainly recreates the way the original individual conveyed the diary's entry instead of having the transcript extract divided between the other characters on stage. Accordingly, Peter makes some comments about his diary to the audience: "Two o' clock: Playscheme discussion with Mandy and Lorna ... publicity ... excellent publicity ... I devised that myself" (114). In this sense, the audience is never taken to the places, events, time they hear about and the emphasis is put on the aural dimension rather than on the visual. No doubt, this articulates a different involvement of the audience in the course of the performance.

An investigation of contemporary British verbatim plays also reveals that the concept of a realist theatrical scene is partially undermined. As Radosavljević acknowledges, these plays "forego dialogue, narrative structure and conventional dramatic climax" (137). Consequently, *Rita, Bob and Sue Too* keeps the conventional scene division to structure the play which enables the audience to accept the change of locale from one scene to the next. This is what Andrea Dunbar clearly indicated in the stage directions where Scene Nine takes place in Rita's house between Rita, Sue and Bob and the following scene unproblematically features "**Sue** [...] *at her local pub*" (77). Whereas Robin Soans does away with the conventional idea of a theatrical scene in *A State Affair*, giving the impression of a unity of time and place, as if everything were happening now and the spectators casually happened to be there. The fact that Es Devlin's stage design incor-

102 This is very clear in Robin Soans' *Talking To Terrorists* (2005) which features only one scene in Act II when two actors really talk to each other.

porated two swing doors enabled the audience to experience the role of accidental eavesdroppers at times:

> it was in the nature of the interviews that we’d done that sometimes you’d just overhear a conversation and actually in the background you’d hear a row between two people about funding that just passed over the horizons, so those two orange swing doors, where people could just come in, say something, and have a conversation on the move, were very much part of it. (Stafford-Clark in Ginman 24)

This impression of being “unselfconsciously involved” (Soans qtd. in Hammond and Steward 23) was truly maximised by the editing process which included an overabundance of such “impromptu” interruptions.

One of the most common devices used by verbatim playwrights to achieve this particular effect is to insert a section of the transcript that was interrupted by an actual phone call. In *Talking to Terrorists*, another play by Robin Soans, Faiser, one of the Luton Muslims, unexpectedly receives a phone call and decides to take it:

> FAISER: ‘I’m a celebrity, get me out of here’. You’re only gonna match that by having a guy with a hook in his arm preachin’ rubbish.
> *His mobile phone rings.*
> Sorry, I gotta take this. The bumper’s in the alleyway… The headlights are in the bedroom/ actually they’re in my brother’s bedroom. (34–35)

Similarly, in Act Two of *Called to Account*, one of the Tricycle Theatre’s tribunal plays, MP Michael Mates’ mobile phone starts ringing as he is answering Defence Lawyer Julian Knowles’ question regarding how the Intelligence and Security Committee carried out its work to support the Government’s case for war in Iraq. Obviously, due to the extreme seriousness of the situation, MP Michael Mates has no other choice but to apologetically turn his phone off:

> *MATES’s mobile phone rings.*
> MATES: Sorry I should have switched that off. I will switch it off.
> *He switches it off.*
> Where were we? (Richard Norton-Taylor and Nicolas Kent 705)

Admittedly, the inclusion of such an unremarkable episode does not serve the legal narrative’s progression and can thus be attributed to a different purpose. On one level, this alludes to dramatic functions, giving a better sense of character, affording a much-needed comic relief to relax the audience’s heightened emotions in the context of narratives of terrorism in *Talking to Terrorists* and those of the complex inner workings of the British Government in *Called to Ac-*

count. On another level, this also breaks down the allegedly fixed boundaries that separate MPs and terrorists from members of the general public, presenting someone with whom we are more likely to relate. For my purpose here, this consolidates the documentary realist project, reminding the audience that these are real-life events that the general realist aesthetic frame can never quite contain (these episodes are detached from the initial realist's semiotic structure). If the aesthetic of realism would generally avoid disturbing the unfolding of the narrative, the aesthetic of documentary realism encapsulates and capitalises on these moments to assert its claims with ever more force. In other words, the realist contract extends to incorporate a relationship with the audience within the fourth wall.

For purposes of clarity, let us have a look at how this device works within the category of (text-based) new writing. In Philip Ridley's *Radiant Vermin* (2015) these moments of interruption break the fictional play's engagement with realism. In the following extract, Jill and Ollie have just moved to their dream house but it appears that a lot needs doing before they can call it a home:

> **Jill** We haven't put a curtain up.
> **Ollie** There's no neighbours, Jill.
> **Jill** I know that but ... oh, it don't feel right.
> **Ollie** Okay, okay. I'll hang a sheet over it.
> **Jill** Thank you.
> **Ollie** Can't have you refusing to take your clothes off, can we. Not when I'm in dire need of some frivolous and hopefully indecent fondling.
> **Jill** Ollie! Don't!
> **Ollie** I'm trying to be erotic.
> **Jill** I know, sweetheart, I know. But ... not in *front* of everyone.
> **Ollie** Oh! Sorry – Sorry! (24–25)

This verbal contact with the audience disturbs the flow of the supposedly realist narrative. The audience suddenly enters their world, their presence is acknowledged, the fourth wall collapses. This is even more apparent at the end of the play when Miss Dee, the head of a Department of Social Regeneration through the Creation of Dream Homes, an offshoot of the government's housing programme, speaks directly to the audience in these words: "What do you think brought you all here? Eh? Seeing a flyer for a play? A friend saying, 'Oh, we must really "check this out"!' Oh no. You're here because *I* summoned you" (115). Going back to *A State Affair*, every interruption of the narrative is located within the fourth wall. For instance, after a rather animated phone conversation, Sue explains to the audience the following: "That was Sam ... he's been in here three or four times" (113). Indeed, as the contract between actors and audience is different in a verbatim play, these elements work together to support and strengthen the documentary realist picture in the same way as the incorporation of trivial details seen earlier, in the context of the tribunal plays.

This is where a fundamental difference between the kind of realism one tends to find in British new writing and the one operating in verbatim theatre arises, a difference that lies in the attitude towards the mediating role of a playwright in the process of constructing a play. For Elaine Aston and Janelle Reinelt, it is therefore "a matter of a different mode of theatrical discourse" (290). Contrasting the tribunal play with this type of verbatim performance illuminates new effective strategies penetrating into the kernel of documentary realism that this chapter will now examine.

2.3.3 Language

Writing on the representation of the council estate in contemporary British theatre, Katie Beswick argues that in many ways the sector has been increasingly engaged with "a commercial focus on the 'authentic voice'" (97). Arguably, this general climate in Britain can chiefly explain the sudden popularity and rejuvenation of verbatim theatre in this period, the strand holding some "transcendental pretensions" due to its privileged position as "the ultimate in lifelike-ese because it is based on real speech" (Dorney 223). Unlike new writing and its customary imitative language, and similar to oral history, then, these plays emphasise the authenticity of speech as the ordinary language tends to be less filtered out or aestheticised and the general reader of drama may find it surprising to encounter in the published playtext numerous remnants of hesitations, repetitions, speech-fillers, false starts, overlappings, speech mannerisms, bad grammar, etc. More broadly, such a use of the vernacular and its empirical disorder deviates from and overturns what is expected to be heard in the theatre and, by extension, in a public arena. In the 1950s, the radio ballads – that Derek Paget includes in his genealogy of verbatim theatre ("Oral History" 319) – produced by Charles Parker for the BBC with folk singer Ewan MacColl were extremely influential in diversifying the range of accents and classes that could be heard in mainstream broadcasts:[103] "[t]hese ballads really represent one of the most original artistic assimilations of the life and working experiences of

103 The voices heard then were restricted to the "BBC," "Oxford" or "RP" voice. If the voices of "ordinary" people were heard, they were always mediated by a "BBC" narrator as they assumed a middle-class audience. MacColl thus cut out the narrator to let these voices be heard with everything that they carried. Songs were found to have a narrative quality and came to replace the narrator figure.

simple people: railwaymen, miners, fishermen, street builders, polio victims, boxers, youngsters, gypsies" (Glinga 89).

In this way, these works greatly differ from the tribunal plays and their elevated "language of officialdom" (Lane 68) and language here acts as "the chief authentifying sign system" (Dawson 106). For instance, when Sue – one of the founders of "The Church on the Way" whose main purpose is to help the Bradford community with alcohol and drug-related issues – starts talking about how drugs affected her own son in *A State Affair*, the emotional content is clearly visible, increasing the level of realism:

> **Sue** And then it was my thirteen-year old son ... he was down on Lumb Lane ... Only on the other side, you understand ... well into crime ... in with the dealers ... at thirteen ... on heroin and crack ... and he came to me, desperate to find ...
> *Knock ... door opens.* (108)

In fact, this has even become a signature of the verbatim methodology and non-verbatim dramatists have drawn on this dramatic device for comic effect. Notably, Dennis Kelly in his mock-verbatim piece *Taking Care of Baby* (2007) recreated realistically within a fictional play the feel of unedited ordinary speech as the first words of Donna McAuliffe, a young mother convicted of the murder of her two infant children, can testify:

> DONNA: And so I'm, erm, just erm, standing there and this girl, there's this girl lying on the top bunk and she's not saying – she had thinning hair on one side, alopecia, I think it was – and she's not saying a word, she's just staring at the ceiling and I was too scared to move really, so I just stood there, I mean, and the door's erm, closed, just closed behind me and I'm, erm, you know, in this cell with this girl and she's not – do you want all this?
> Beat.
> and she's not looking, she's not looking at me. (15)

It would seem in such plays that verbatim becomes a stylistic device. For example, in much the same way, James Fritz's mock verbatim play *Lines* occasionally showcases this particular feature, albeit in a less extreme form:

> Terrence: No, no, no...more like a kid...you know, like a kid...like that feeling when you're little and everyone is laughing at you and you just, you just blow. (*Pause*) Look, I wasn't bullied mate, that's not what I'm saying. Just the odd thing you know. (43)

The fact that British playwright Mark Ravenhill exploited this startling approach to dramatic language in his non-verbatim epic cycle of short plays *Shoot/Get/Treasure/Repeat*, for instance, through the words of fictional soldiers delivered in direct address to the audience:

> Sometimes you look at me with dead eyes and I think I read, I, I, I ... maybe ... project, I project hate in your eyes. I used to see an analyst when I was ... back home. I would drive across the city very early before the day had begun and I'd see a little Jew and he told me I had a tendency to project so maybe – do you hate me? Do you hate me? Please don't ... it would very ... hurtful if you hated me. (181)

led Jenny Hughes to automatically assume that perhaps he attempted to "parody verbatim theatre's vocal conventions [...] within entirely imagined dialogue" (119–120).

Clearly, verbatim plays such as Soans' *A State Affair* rely heavily and perhaps disproportionately on language to build a documentary realist picture on stage.[104] As a corollary to this, it is worth noting that they exceed certain conditions of realism that are activated, as Raymond Williams reminds us, when "the written speech moves much closer towards the imitation of everyday ordinary life" ("Lecture"). As seen previously, language is not the only part of the story in a vigorous documentary realist strand and one must now dwell on the question of acting.

2.3.4 Acting

For Tom Cantrell in *Acting in Documentary Theatre*, the nature, scope and development of the acting processes in documentary theatre present new and different sorts of challenges which are hard to pin down theoretically.[105] If a "minimalist realism" (Cantrell, *Acting* 136) was *de rigueur* in the tribunal plays in order to undermine any acting or theatrical excess,[106] the impetus behind *A State Affair* may presuppose another conception of realism. As actress Emily Aston explains in an interview, the fact that the actors had met the real individuals affected their acting in significant and substantial ways: "It can also be very surprising – someone may laugh whilst telling you something really quite horrific. I mean if you received a script, without having met the person, you would probably never think to play it that way" (Soans, *Affair* 146). Far from the cutting edge

104 In an interview with Will Hammond and Dan Steward, Nicolas Kent reveals that later on (thanks to an access to the audio recordings) in the most recent tribunal plays they also decided to "include 'ers' and 'ums' and stutters" (qtd. in Hammond and Steward 150).

105 On this particular topic see also Cantrell "Playing" and Cantrell and Luckhurst.

106 Thomas Wheatley refers to his contribution in the Tricycle tribunal series in terms of being a "minimalist actor" (11). In addition, these productions feature very simple settings as well as a sparse use of props and costume.

of formal innovation in British theatre, these plays invoke the familiar tropes of realism essentially through acting conventions. In David Hare's *The Permanent Way*, an Out of Joint verbatim play also directed by Max Stafford-Clark, the adopted acting style is described in terms of Stanislavskian realism by one of the actors involved: "we [...] laughed full-belliedly or wept real tears: everything had to have a genuine base and a committed manifestation" (Merlin, *Toolkit* 255). Duška Radosavljević plainly argues that "the influence of Stanislavskian psychological realism" (147) is crucial to the actor's processes in verbatim theatre. Consequently, actors in verbatim performances deploy broadly realistic modes of delivery, or at least techniques related to them.

This documentary realist acting style, if I may call it so, is played closer to the audience to further grab their attention and increase the intensity of their listening. An important point in this consideration lies in the way in which this impacts on the reception, following Konstantin Stanislavski's envisaging technique that can "be so powerful that the listener is compelled to see the world from your PERSPECTIVE" (Merlin, *Toolkit* 240). As Merlin explains concerning her role of the Second Bereaved Mother in *The Permanent Way*, she "endeavoured to plant in [her] audience's head a conception – however fleeting – of the 'total literal' destruction caused by the terrifying train crash [and] felt that psycho-physically [she] was executing an ACTION on the audience, even though all [she] was doing was sitting and recounting [her] tale" (*Toolkit* 240). Convincingly, this statement reveals how the documentary realist project is tightly connected to a conception of political theatre whose principal aim is to make space for unheard, marginalised or overlooked voices. In addition, the connection established between actors and the real people on whom their characters are based offers a different channel whereby actors are somehow in a place of consonance with what they are doing onstage.

2.3.5 Political Theatre

> I believe there are four ind ispensable rules when writing a verbatim p lay. The first is: *don't* try to write a political play. If you do, it'll end up as agit -prop – worthy, one-dimensional and boring. If on the other hand, you write a humanitarian play, it has every chance of being funny, moving *and* political. (*Cookbook* 8)

In this, Soans clashes with a certain conception of political theatre and commitment whose innermost logic arguably repeats some common charges marking a perceived repudiation as well as a growing incompatibility between two theatrical strands. For Derek Paget, there are two "similar-but-different models of the idea of *commitment* that contrast past and present times" ("Acts" 173) that could be of help here. Apropos of the late 1960s through the 1970s, Paget tells us the following under the sub-heading "commitment then":

> For actors, writers, companies, venues, and audiences alike, the very concept of something 'alternative' denoted a fundamental rejection of the orthodox, the mainstream, and the conservative in politics and theatre. [...] Theatre events were part of a wider movement dedicated to changing the world [...] The 1960s was pre-eminently a time for commitment to a politics of the Left. ("Acts" 173–174)

In his second category "commitment now", accounting for the 1990s-2000s period, Paget brings to the light of day new principles that are of special interest for this analysis of Soans' work in terms of its departure from none other than Piscator and the tradition that ensued.

For Paget, there is an abyss between "the fierce but generalized political commitment of alternative theatre [and] the specificity inherent in theatre of the single issue" ("Acts" 175). He then proceeds to argue that this new mode of intervention is "part of a postmodern, 'post-politics' period, in which the absence of credible political narratives mirrors a wider intellectual distrust of theoretical 'grand narratives'" ("Acts" 176).[107]

Perhaps most importantly, one must be wary of Soans' self-declared humanitarian position that recalls (and updates) the age-old objectivity discourse (from which we seem further and further removed) under a different exponent, a reigning "default" non-ideological position uncritically accepted by journalists such as Paul Taylor who writes,[108] following the performance of Robin Soans' *Talking To Terrorists*, that it "allows you to see from both points of view in the humane and untendentious way it shuffles and shapes the material, creating telling juxtapositions and stage pictures." Arguably, this documentary realist framing only masks ideologies through a reassuring normalising discourse tied to some tendentious ideological assumptions about so-called humanitarian reality, as Žižek suggests in another context: "*the stepping out of (what we experience as) ideology is the very form of our enslavement to it*" ("Spectre" 6).

---

107 Bérénice Hamidi-Kim makes a similar division between what she calls "post-political theatre" and "the theatre of political struggle." However, she perceives in France a return of the "theatre of political struggle" which coincides with the "revival of documentary theatre" ("Struggle" 46).

108 Skip Shand and Ric Knowles call this contemporary position "the obvious and supposedly ahistorical quality of universalist and essentialist humanism: 'good' theatre is 'obviously' individualistic, universalist and politically 'neutral'" (3).

Significantly (these unavowed processes aside), from Soans' standpoint and following the deliberate pun in its title, *A State Affair* attempted to make a decisive intervention through the public representation of politically nonconformist perspectives such as the ones from addicts and the homeless who possess "a penetrative wisdom which the rest of us would do well to heed" (2011). In this sense, the play fulfils the second function of documentary theatre identified by Carol Martin: "*[t]o create additional historical accounts*" ("Bodies" 12). Katie Beswick notes that the airing of these perspectives can "meaningfully intervene in wider political and social inequalities that exist between estate residents and other types of groups at and beyond the theatre" (111) in a way that is perhaps more radical than "single authentic voices" (111). Crucially, the play was also adapted for radio in a BBC 3 production in the hope that these words would intrude on the listening space of more people in the UK:

> And what good can possibly come from exposing well-to-do Radio 3 listeners to this ghastly horridness as they turned off and turned into their nice, safe, cosy beds on Sunday night?
>
> 'I have no problem with doing plays for a middle-class audience. After all, it's them who can change things.' (Kershaw 2001)

Soans in a book of verbatim monologues including some extracts from *A State Affair* clearly stresses the transformative potential of placing their stories in the spotlight, making these part of conversations and public debates: "Well Paul, at least your voice got heard at the epicentre of power ... your bed may have been two blankets and a pile of leaves in the corner of a disused garage off Kirkstall Lane, and you may no longer be in this world, but your words live on" (*Heat*). Stated another way, the documentary realist aesthetics – through its deliberately tempered and repressed theatricality – is chosen as a more effective vehicle to participate actively in today's struggles.

However, the recourse to this mode has perhaps – despite the best intentions – rightly re-opened one of the most long-standing and contested areas of verbatim practice: the issue of authorship and construction. A familiar charge levelled at verbatim theatre-makers is that their use of personal testimonies is somewhat naïve, more so when it attempts to make truth claims for political effect. Many of these issues have been thrown into question recently by a wave of theatre-makers whose work has been transforming the field of British verbatim theatre and these will be fully explored in Chapters 3 and 4 of this volume. What the documentary realist framework implies is that the material presented to a live audience is more truthful, less mediated than in conventional plays and, as a consequence, the audience is more inclined to feel saddened and/or angered by it and thus take action. The skilful theatrical and editing work is obscured in such a way that the audience is led from the status of supposedly “passive spectators into the role of active witnesses” (Albright 20) of a raw experience.[109]

Interestingly, Jonathan Holmes suggests that in verbatim theatre, “[s]pectatorship [...] takes place on two levels: the reception of the performance on the level of art, and that of the script on the level of immediate testimony” (*Fallujah* 142). These works, then, call upon a very different engagement with the audience as they cannot easily be dismissed as fiction or, as Judith Butler puts it, in terms of “this is just an act” (“Gender” 527). In this respect, as Will Hammond and Dan Steward have suggested, “this claim to veracity on the part of the theatre maker, however hazy or implicit, changes everything” (10), the audience being more implicated and thus tempted to consider the performance as *the* story of what happened, rather than a (provocative) constructed story of what happened among many other competing narratives. Interestingly enough, a quick survey of the flurry of verbatim play titles reveals this position more clearly. Performances like Nicolas Kent’s *Srebrenica*, Tess Berry-Hart’s *Sochi 2014*, Kieron Barry’s *Stockwell: The Inquest into the Death of Jean Charles de Menezes*, Richard Norton-Taylor’s *Nuremberg* and more openly Paul Unwin and Sarah Beck’s *This Much is True* are *stricto sensu* exemplary in this regard.

Another problem resides in the recourse to testimonies from partial individuals that are not, unlike those of the tribunal plays, required by law to tell what they believe to be the truth. The verisimilitude of the entire performance is contingent upon the reliability of their testimonies although they are less “accessible to verification processes” (Tomlin, *Acts* 116). In addition, most verbatim practitioners, and certainly those I have discussed here, do claim in a ripple effect that the people interviewed are a representative sample in relation to one particular issue so it follows that something must be done. This may look like an unreasonable demand in the context of a theatre performance and it is not so surprising, therefore, that with the benefit of hindsight, Alex Chisholm reports some of *A State Affair*’s caveats concerning the use of fallible individual testimonies as evidence:

> I know one of the people interviewed for [*A State Affair*], who was working with young people in Bradford at the time, when Robin and the cast went up to interview them and create a story about the state of the city at that time, the nature of the drug scene, and so on. And I happen to know that many of those young people interviewed for that show lied their tits off. There are lots of incidents in the play shown in good faith by the company to say, this is

---

109 In the case of the tribunal plays, Michael Billington surprisingly used the phrase “raw information” (“V”).

> what it's like in Bradford and isn't it terri- ble . . . and it isn't true; those things didn't happen. (601)

One can find similar statements in the case of Robin Soans' *Talking to Terrorists* as Christopher Innes has rightly pointed out:

> in the interviews, the one-time Palestinian Martyrs Brigade leader laments being separated from his wife and children and from a son, born after his forced departure, whom he has never seen. Onstage, it forms a convincing plea for humanity and restitution. However, subsequently, Stafford-Clark found out that the man now had an Irish girlfriend and – particularly after the death of his primary patron, Arafat – was in no hurry to return to Palestine. Thus, despite the premise that a documentary is authentic, what, onstage (and in the script), seems a powerful political plea is, in fact, a lie. The political and the personal are severed. (His sexual betrayal of his wife was felt to be potentially dangerous for him if he were to go back to Palestine). And, at the same time, the man's situation clearly contradicts a central motif of the script, in which both terrorists and some of their victims rue their impotence or inability to form intimate relationships. It would also have undercut powerful statements like the speaker's '[e]verything good inside me is dead.' So the text was never altered to reflect this insight. ("Mainstream" 440)

It is true, of course, that such occasions arose and they certainly betray a problematic "conflation of first-hand experience and authenticity" (Beswick 99). Such a conflation would seem to account for the reaction of some audience members during the course of performances of *The Permanent Way:* "the audience would shout back, like at a pantomime, they'd answer the rhetorical questions, so that was really exciting" (Merlin, "Bella" 9). Furthermore as David Lane has observed "[i]n verbatim (rather than tribunal) plays, interviews are not objective" (67), they are even, in the words of Stephen Bottoms, "a form of dramatic encounter conducted according to certain established (though not always acknowledged) behavioural conventions" ("Staging" 193). In the case of *A State Affair,* following the Out of Joint method, the process is even further mediated through the actor-researcher in the context of the "hot seating" phase before being edited by a named author into a performance text.

Notwithstanding these weighty issues, theatre critics moved past the theatrical discussion of the production and exclusively focused their attention on the points raised. At the same time, these plays were commented upon well beyond the theatre columns: "Certainly *The Permanent Way* caused a lot of flurry and it hit the front pages of the newspapers several times which is rare for a play. Y'know, the arts pages, sure, but not the front pages" (Merlin, "Bella" 9). In this sense, documentary realism reduced the theatrical act to a mere direct reference to the "original" events, to neatly communicable points to be made. This suspect narrative that complacently rules out the possibility of aesthetic achievements is curiously reinforced by the theatre-makers themselves to a degree that relegates the dramatic interest to second or third place. It is therefore not an exaggeration to say that this constitutes one of the unresolved contradictions pertaining to the documentary realist aesthetic project.

2.3.6 Conclusion

The repeated message across the case studies is that dominant theories of realism do not provide clear ways forward for apprehending recent verbatim performances. At the same time, the different aspects of a documentary realist aesthetic strategy used in combination with verbatim material have proven to adhere to some conventions paradigmatic of theatrical realism such as the notion of cause and effect, a surface accuracy (setting, costume and props),[110] suspension of disbelief, gestures and movements in the service of portraying a character, psychologically explicable situations and motivations, conflict and resolution.[111] Mention should be made here of Soans' creative approach when turning verbatim materials into a public performance:

> a verbatim play should still be built around a narrative, and it must still set up dramatic conflicts and attempt to resolve them. Characters should be shown to undertake journeys of discovery of some kind, even if these journeys take place while the character is sat in a chair, talking. (qtd. in Hammond and Steward 26)

In *A State Affair*, resolution takes the form of t'ai chi movements performed by the ensemble of actors in character before the dramatic blackout, giving a fleeting semblance of community and hope: "*Music.* **Paul** *starts to remember the t'ai chi movements he learnt in Doncaster prison. One by one the others join in* [...] [Lorraine] *joins the others who move in perfect harmony as the music swells, and the lights slowly fade to black*" (134). Politically, the play's ending suggests some positive undertones that are traditionally excluded from popular "realist" representation of the estates, typically leading their residents "to feel they have

110 It would thus seem that documentary realism in theatre operates in a similar fashion to its counterpart in the novel, as defined by Lars Ole Sauerberg, who has so eloquently argued: "a narrative mode which, while adhering in principle to the time-honoured narrative conventions of realistic narrative, draws on verifiable reality to various extents, but invariably in such a way as to call attention explicitly or implicitly to the difference between the fictional and the factual" (6).

111 In the tribunal plays seen in the last chapter, the cross-examination process caters to this dramatic function.

little control in effecting change in their own lives" (Beswick 100). In an interview with Jon Bradfield, Max Stafford-Clark expressed this contrast with the traditional approach in the newspapers exhibiting "a readiness to blame and to find scapegoats" (125). Richard Norton-Taylor and Nicolas Kent, as discussed in Chapter 2.2, decided to end the tribunal play with the potent testimony of Janice Kelly. Her words, then, fulfil a realist dramatic project insofar as, to one's relief, they give a sense of temporary closure at a time of great uncertainty in Britain:

> Lord Hutton, on behalf of my family I would like to thank you and your counsel for the dignified way in which you are carrying out this Inquiry into my husband's death. We would also like to acknowledge the support our family have received from so many people all over the country and elsewhere [...] Thank you. (509)

More decisively, Dennis Woolf in his tribunal play *Beyond Belief: Scenes from the Shipman Inquiry* (which was staged after the trial had finished taking evidence) chose to end the performance in a way that fulfils the expectation of a realist closure:

> DAME JANET
>
> If you will bear with me, I just want to express my thanks to a few members of the Inquiry staff, in particular to begin with the boys at the back who have dealt with the sound technology. It is now two and a half years since we first sat in the hall and I am very grateful indeed. Also, the IT boys who have provided you with laptops and assistance with our difficult CD's. The transcribers, there is Georgina still typing away. There is also the witness support team. Those of you who have attended and given evidence will know how very well you were looked after and I am very much grateful to them. Finally, the Town Hall staff who fed and watered us day in and day out, very well, very courteously and very efficiently. I am grateful to them as well. Last of all, bon voyage. (98–99)

This type of onstage realism arguably corresponds to Lehmann's controversial use of the word "drama" in his *Postdramatic Theatre:* "[w]holeness, illusion and world representation are inherent in the model 'drama'; conversely, through its very form, dramatic theatre proclaims wholeness as the *model* of the real" (22). This search for a determinate unity, harmony, stability, coherence, linearity and progress-oriented universalist wholeness *par excellence* is particularly prominent among the documentary realist verbatim playwrights studied in this chapter who do not base their performances on an available ready-made continuous structure (e. g. a public inquiry) and artificially create instead a disciplined documentary realist aesthetic template along the lines of what I have called a "continuist perspective" elsewhere (Garson "Continuité"). Throughout this analysis, I have also raised the question of how the documentary realist structure is at odds with theatrical realism, intentionally resisting its conventional definition(s). What this type of realism seems to disclose is an ongoing quest for the real, a continual reorientation in performance to collapse the distinction between aesthetic perception and cognition. Structural analysis also revealed a pattern behind the standard sequence of the classic realist British play.

In this new format, documentary realism does not draw its authority from a scrupulous dramatization of some sections in the public archive but, instead, constitutes what one might call a shadow archive. As such, these pieces arrange and authorise by aesthetic means new narratives in relation to the official narrative(s), pointing to marginalised voices as if they were part of the same "reality." Documentary realism functions here as a radical restructuring force that creates new images and intermittently makes one feel again (in the original sense of aesthetics, "*aesthesis*") connected to "othered" identities and, as Janelle Reinelt has argued, it "calls the public sphere into being by presupposing it exists, [...] construct[ing] its audience to be part of a temporary sociability" ("Promise" 11). However, one must confront the possibility that the reverse may be true as well: the same sovereign talent for selecting the verbatim narratives, I believe, "*give[s] face*" or "*efface[s]*" (Butler, *Frames* 77), authorises another classification system endowing certain exemplary members of a group with uniquely complex and rich identities while silencing others. More saliently, beyond the general issue of representation, Linda Alcoff identifies an additional "complexity and specificity of the problem of speaking for others" (10) and this would seem to take a rather "dramatic turn" when it comes to the particular discursive context of a theatre event. Simply put, it can be argued that verbatim theatre is objectionable when its makers articulate its practices as belonging to a more ethical arena by virtue of its use of the exact (edited) words of consenting "real" people.[112] If so, then one faces a more perplexing question, one of a more problematic nature, as such theatre-makers attempt, in the same breath, to persuade one to accept their world of "good" mediation as if it were any more real than the familiar hegemonic "realist" regime of "bad" mediation:

> [h]ow is this creative editing any different from the spin that governments put on their material? I can only say that, unlike governments that edit in order to obfuscate and deflect attention, and therefore to limit our knowledge, I attempt to edit in order to enlighten and intrigue, and therefore to broaden our knowledge. (Soans qtd. in Hammond and Steward 35)

112 Soans is particularly guilty of this when he says that "[t]his is literally what verbatim theatre does: it allows people to speak for themselves" (qtd. in Hammond and Steward 33).

> We don't have the bad record journalism has for misrepresenting people [...]. I believe print is much more manipulative than performance. (Hare qtd. in Hammond and Steward 71)

Hare and Soans' statements are of course not immune from criticism focusing as they do on a supposedly privileged position of legitimacy regarding verbatim theatre's own process of enunciation. However, the fact that two of the verbatim speeches from *A State Affair*, the ones belonging to Marie, "a young drug addict" (Bluhm, *Women* 32) and to the "young criminal and drug addict" (Bluhm, *Men* 32), Paul, were used in a training book for actors touches the very nerve-centre of the issue and would seem to proceed in exactly the opposite direction.

Despite its radical potential, therefore, documentary realism only adds to one's accepted idea of culture without ever changing the "real" world. Indeed, one might even say that documentary realism in these verbatim performances gives value to certain narratives within the logic of global capitalism insofar as these narratives measure up to a fixed standard of representation. In Caroline Wake's words, this poses a particularly thorny problem for theatre-makers that must attempt "to intervene in this public sphere without reproducing its structures" ("Listening" 83).

2.4 Transgressions

It has been suggested thus far that the documentary realist model was perhaps the least prone to experimentation and motivated by different concerns from those conventionally proposed in a contemporary British theatre landscape. A general concern over the hazards that a perceived theatricality can set for a verbatim performance would thus seem to underpin every production discussed in Chapter 2 and it will be recalled that the result of all this was then to both exacerbate a certain sense of realism and evade making a direct political case: "this is a way of talking politics without seeming to come from a particular point" (Slovo, "Verbatim" 39). This conception particularly animates the tribunal plays studied in Chapter 2.2.

Against this, however, tribunal theatre has also been inching towards reassessment, helped by their extraordinary popularity in the last decade. Jacqueline O' Connor goes even further to argue that, even though tribunal plays are "linked by law's structures and themes" (171), they have also been the object of an "ongoing diversification" (171). Indeed, in the last few years, for instance, there have been Gillian Slovo and Victoria Brittain's *Guantanamo* (2004) that Chris Megson described as "a sort of Tribunal play without the tribunal" ("Testimony" 2) at a public symposium held at the Central School of Speech and Drama in July 2006, Tess Berry-Hart's campaigning verbatim piece *Someone to Blame* (2012) on the live legal case of Sam Hallam, a wrongly-convicted-for-murder teenager and Tanika Gupta's *Gladiator Games* (2005) on the events surrounding the death in custody of Zahid Mubarek, a nineteen-year-old Asian in March 2000 in which Gupta unproblematically combined verbatim quotes from the official Zahid Mubarek Inquiry with interviews she had conducted with the family as well as with her own writing. As Ursula Canton has justly remarked regarding the latter, however scrupulous the published text may have been, "spectators [were] thus unlikely to distinguish between the two, assuming that the whole play possesse[d] a similar form of reference to the source materials" ("Reality" 326).

In one scene pertinent to this study, in Act II of *Playing with Fire*, David Edgar tells us within a fictional public inquiry into a local riot that "the feel, structure and ground-rules of the dramatization follow those of the tribunal plays at the Tricycle Theatre in London" (83) until "suddenly, the Inquiry disappears and images of riot swirl across the stage" (107). However, Edgar initially indicates some minor alterations as the inquiry he describes is not judicial and therefore we are told that "in this case the physical set-up is less formal" (83). Faithful though this moment may be to the atmosphere of the tribunal plays discussed in Chapter 2.2, it is at the same time a profoundly non-realist sequence in its re-enactment of the inquiry proceedings:

> The presentation of the Inquiry is continuous in dramatic action but is clearly compressed – so while some WITNESS presentations are topped and tailed, others (we imagine) begin in the middle, or continue after the section we've chosen to dramatise. Similarly, we assume that the different COUNSEL introduce themselves to each WITNESS, but after the first time, this introduction is only represented when dramatically significant. (83)

As seen earlier, the necessary compression of the Inquiry to create an evening of theatre rested on a documentary realist frame which ensured that the audience's experience would be in no way compromised by editing and theatrical fripperies.

2.4.1 The Tricycle Theatre's *Called to Account* (2007)

Most notably, the Tricycle mounted in 2007 an imaginary trial of Tony Blair for war crimes – partly related to the September Dossier controversy explored in the context of *Justifying War* – entitled *Called to Account: The Indictment of Anthony Charles Lynton Blair for the Crime of Agression against Iraq – a Hearing* drawn from the testimonies of voluntary sources,[113] which predated by two years the actual Chilcot Inquiry into the circumstances of the UK's invasion of Iraq (24 November 2009–2 February 2011).[114] Strikingly, the production initially planned to directly involve the audience "by holding a vote at the end of the play [...] to decide the fate of Tony Blair" (Cantrell, *Acting* 101). It is also interesting to observe how this play draws on a documentary realist frame similar to the one described in Chapter 2.2 despite the fact that it was not based on a real legal case. In fact, given the notorious status of the Tricycle Theatre, *Called to Account*'s documentary realist project relied entirely on the audience's "horizon of expectations" (Jauss) and the play was misleadingly incorporated into the series of tribunal plays. The marketing apparatus at the time suggested an underlying continuity with the renowned series and therefore mobilised a certain force of authority. Yet, more perniciously perhaps, the 2014 Oberon volume seems to suggest, by including Gillian Slovo and Victoria Brittain's *Guantanamo* (2004) and Gillian Slovo's *The Riots* (2011) in their *The Tricycle: Collected Tribunal*

113 Coincidentally, Tony Blair announced his resignation during the run of the production. Nicolas Kent reported that, as a consequence of this, "the box office went down, just slipped away" (Hammond and Steward 144).

114 These included Clare Short, American security advisor Richard Perle, Chile's ambassador to the United Nations, Juan Gabriel Valdes and Edward Mortimer who served as the director of communications for UN secretary general Kofi Annan.

Plays 1994–2012, that they also have the same status as the other tribunal plays proper.

In other words, to come back to *Called to Account*, documentary realism is animated here essentially by means of the discourses that throng around the production and, in this sense, perhaps (if one may put it like that) this exemplifies a true instance of "hyper-realism" in reference to Baudrillard.[115] To put it yet another way, the documentary realist equation holds true even when a piece of verbatim theatre does not offer a replication of an actual event and it becomes enough to simply invoke the verbatim label. Furthermore, most of the features which had become familiar to a tribunal play audience were persuasively reprised. The two lawyers, Philippe Sands QC and Julian Knowles that cross-examine the 11 witnesses,[116] approvingly wrote fictional opening statements to help create the impression that the following verbatim accounts were heard in a real court: "SANDS: In accordance with Article 15 of the [International Criminal Court] statute, the prosecutor seeks the authorisation of the Court to investigate the facts, to ascertain whether they provide a basis for indicting Anthony Charles Lynton Blair for the crime of aggression" (667). Meanwhile, the attentive tribunal play reader can rejoice in finding precisely the use of the same codes to annotate the edited verbatim transcripts and the spectator can be comforted in the knowledge that designer Polly Sullivan recreated the "original" atmosphere "[i]n a style of design familiar from previous tribunal plays" (Cantrell, *Acting* 115). Other aspects of the legal process seen in Chapter 2.2 were meticulously reconstructed such as the final impression of the closing statement, while tempering the perceived parasitism of the theatrical signifiers (e.g. absence of a curtain call).

However, *Called to Account* also makes the familiar unfamiliar through the establishment of a different documentary realist audience-actor contract which cast the audience members "throughout in the role of jurors" (Derbyshire and Hodson 201) and showed less concern for chronological inaccuracies. Unsurprisingly, few critics have subscribed to this new breed of verbatim tribunal plays, arguing that it throws into doubt the very ground on which it stands and perhaps even throws a pall of suspicion over the integrity of the prior wave of tribunal

---

115 For Baudrillard, in his *Simulacra and Simulation*, the hyperreal indicates "the generation by models of a real without origin or reality" (1).

116 Nicolas Kent and Richard Norton-Taylor also invoke the usual curious involvement of the creative team at the time, blurring reality and fiction: "Kent: [Julian Knowles and Philippe Sands QC] took it terribly seriously...NORTON-TAYLOR: They made a real trial of it" ("Verbatim" 27).

plays, or, to be more precise, makes a fictional documentary realist tribunal play upset the delicate balance on which these plays depend.

2.4.2 Owen McCafferty's *Titanic*

Owen McCafferty's *Titanic* is also an interesting case in point. The tribunal play was directed by Charlotte Westenra who coincidentally had worked with Nicolas Kent on two tribunal productions at the Tricycle Theatre, as co-director for *Justifying War* (2003) and assistant director of *Bloody Sunday* (2005). The play received its premiere performance on 22 April 2012 as the inaugural production of the MAC, Belfast and on the occasion of the Titanic centenary events in Northern Ireland. Following the tribunal play model described in Chapter 2.2, the text was drawn from the transcript of an inquiry – the British Wreck Commissioner's Inquiry 1912 – and "[a]s is the nature of these plays [the author] ha[s] left out far more than [he] included" (McCafferty 3). After the Titanic sank, two inquiries were ordered a few weeks later. One of them was held in New York and this play is entirely based on the one that sat in London from 2 May to 3 July 1912 under the commissionership of Lord Mersey. However, more curiously, the author's note reveals that the play was in fact not completely governed by the usual rules of verbatim and documentary realist composition:

> While I have at all times tried not to interfere with the content of what a character is saying, I have on occasion, for the sake of clarity, changed how they might say it. I have also used the same style of punctuation I always use when writing any play. And for practical and artistic purposes I added a fictional character. (3)

While it is understandable that the punctuation of prose was changed to fulfil the oral quality of the theatre (sentences being turned into cues, dashes being added as an aid to actors), the introduction of a fictional character is all the more surprising.[117] More so, when at the very beginning of Act One, the fictional clerk of the court in a self-reflexive gesture unexpectedly says the following:

> i am a fictional character by the way – of the now – the present – i do not sit in judgement – functional not emotional – a made-up clerk of the court – the rest is real – or as real as can be made – or wanted. (9)

---

117 Punctuation in verbatim theatre is more problematic than in traditional plays as it constitutes an "ethical choice" according to a tweet once composed by theatre-maker Chris Goode.

By any stretch of the imagination, this very quote can hardly not suggest a postmodern twist within the apparent documentary realist project. The reader of contemporary fiction would have easily spotted – with a slightly smug smile betraying a momentary feeling of ontological superiority – the vast tracts of common ground with these novels, the familiar self-conscious strategies overwhelmingly challenging the authority of realism and establishing a more dynamic involvement with the audience. That expectation would be disappointed and it would be quite an erroneous conclusion for a number of reasons. First, as Baz Kershaw and Graham Ley warn us in the context of theatre practice, "postmodernism as performance has not produced an artistic movement, least of all one with a manifesto" (289). Similarly, Liz Tomlin argues in *Acts and Apparitions* that

> [u]nlike the ideological battlefield of literary theory, however, theatre scholars and artists who still adhere to a predominantly Marxist understanding of the political have tended to dismiss or simply ignore the radical claims of the poststructuralists, rather than engaging with the work defined as such or the discourse that underpins it. (11)

This would seem to imply that there are some potential dangers in importing such a framework wholesale to the theatre. Second, the fictional character in *Titanic* serves a different purpose and is used throughout to introduce people, make the transitions, give the play context, relate facts about what was on the Titanic, the number of dead, etc. He also reads out at the end of the play what William Alden Smith, an American Senator, had said about the disaster:

> Clerk [...] (*Reading from another page.*) senator william alden smith said during his address to the senate in delivering the titanic panel's report on may 28th 1912 – in our imagination we can see again the proud ship instinct with life and energy with active figures again swarming upon its decks [...] only a vestige remains of the men and women that but a moment before quickened her spacious apartments with human hopes and passions – sorrows and joys – [...] in such a heritage we must feel ourselves more intimately related to the sea than ever before – and henceforth it will send back to us on its rising tide the cheering salutations from those we have lost ... (117–118)

Third, he stays outside the play and never enters their world. Thus on one level the documentary realist project is not disturbed unlike the contract one might encounter in other verbatim performances that use the same device for different ends such as Cora Bissett and Yusra Warsama's *Rites* (2015) in the following example:

> FARA: I am real. I am a West African woman living in Scotland. I am being played by an actress for you. I gave them my words and they gave theirs. Every character you meet today is a real person. I am one of many stories, many different stories, which you may

> be surprised by. You may think you know a little of what this is about. You may know a lot. It's more complicated than it seems. We do not all agree. Let us start by listening. (9)

Yet, clearly, the performance text of *Titanic* bears the marks of a certain tension between the past verbatim words and the clerk's intervention in the present, between the vivid and playful story-telling form and the cold hard facts the play is meant to convey, perhaps also between oral and written forms of communication. The effect of these tensions is to paradoxically create a stronger documentary realist framework and here is a case in which our suspicions are aroused whilst avoiding taking the postmodern turn towards a complete displacement of its own verbatim-induced meaning(s). Instead, I want to argue that they constitute an engagement with epic theatre in the vein of Piscator whereby the use of a narrator to address the audience directly was one of the essential features of his political documentary theatre. It is precisely in responding to this impulse that these plays participate in the documentary realist trope I have been describing thus far. In other words, the clerk is part of the documentary realist paratextual frame that constructs the verbatim play in terms of a realist aesthetic experience. Without question, McCafferty was more interested in creating a coherent realist narrative than he was in creating a "postmodern" disruption:

> I felt I needed to top and tail the play with a fictional character [...] No matter whether it is verbatim or not it is still a play and you have to make it feel as such [...] The fictional character didn't disrupt the realism [...] so it looked like exactly what it was – a character outside the play.[118]

In short, these plays do not endanger the notion of truth as correspondence with reality or allude to the Baudrillardian "*simulacrum*" (1994) holding the possibility of non-referential simulation, that is to say of a sign without a referent. This very conception of verbatim theatre also comes to the fore in Tanika Gupta's exhaustive annotative gesture in the playtext of *Gladiator Games:*

> 1 Verbatim evidence from the Zahid Mubarek Inquiry
> -witness statements, oral and written [V1]
> The Inquiry includes evidence from:
> a MPS – Metropolitan Police Service [V1a]
> b Butt Inquiry/Feltham Murder Report [V1b]
> c Closing statements [V1c] [...]
> 8 Justice Hooper's judgement [V8]
> 9 Justice Bingham's statement [V9]

---

118 Email from the author received on 27.04.2015.

> Verbatim text reproduced from these sources is marked with an asterisk in square brackets [*]; the references are given at the end of the reproduced text in the form shown above [...] Anything unmarked is a dramatisation based on events/hearsay. (31–32)[119]

Having briefly surveyed how the tribunal play itself could potentially be tipped out of the axis of documentary realism, let us now explore a similar impulse in a different kind of verbatim works, those plays I had temporarily consigned to Chapter 2.3 of this volume.

2.4.3 Robin Soans' *Crouch Touch Pause Engage* (2015)

It is into the very debate of verbatim rules and conventions in Britain that Robin Soans enters with his latest play, *Crouch Touch Pause Engage*, that will serve as a temporary closure for this investigation of documentary realism. Like *A State Affair*, Soans teamed up with Max Stafford-Clark and Out of Joint and the research method was very similar to the one I have described in Chapter 2.3. However, the production departed from previous documentary realist verbatim performances in several respects. First, the main character, Welsh rugby legend Gareth Thomas, was played by the whole company of actors (regardless of age, physical appearance and gender) at different points in the performance. Second, the documentary realism of the piece does not attempt to recreate the conditions of the interviews. For instance, in a particularly moving scene at the end of Part 1 (the climatic moment of the performance), the two main stories of the play – the one of Gareth "Alfie" Thomas and the one belonging to some young people in Bridgend – productively join together as the actors on stage physically move two benches into a parallel line reflecting the creative gesture that artificially edited them together:

> DARCEY: I ran out without paying, went home, straight up the stairs without stopping, opened the bedroom drawer, took out the tablets, and swallowed the lot.
> ALFIE: 'You know what I did?' I climbed into Jemma's wardrobe, covered myself with her clothes, sprayed everything including myself with her perfume....her signature smell...and sat there rocking back and forth. I thought telling Jemma would make things better, but, with her gone, the energy was draining out of my life [...]
> ALFIE: Got to the top...took the jumper off and folded it very carefully; and my boots [...] I took my socks off slowly and folded them very careful, very neat, and put them inside my

119 These annotations correspond to the sources that Tanika reproduced in her *Gladiator Games*. In the play, we see these in the body of the text, as in what follows : "RANDHAWA : [*] Mr Mubarek you've been hurt and we've called the nurse. [V1]" (100).

> boots, and put them next to the jumper.
> DARCEY: I walked into the kitchen and lay on the floor...a cream-tiled floor...it was dark...I hadn't turned the light on.
> ALFIE: There was no noise but the wind, and the sea on the rocks three hundred feet below.
> DARCEY: It was silent except for a clock ticking and the voices in my head and they were really loud now [...]
> DARCEY: I unscrewed the cap. There was a moment of thinking, 'This time you're not going to get rescued...no more charcoal stuff.'
> ALFIE: And I just stood there [...] and I was thinking of my grandparents [...]
> DARCEY: I drank half a bottle of Domestos. At first I didn't feel anything.
> ALFIE: You know if you put your foot on the edge and a bit crumbles, I thought I would jump on the edge, hoping the whole thing would crumble [...]
> ALFIE: I jumped on the edge.
> DARCEY: I span off the edge of the world into the void. (37–39)

Third, the conventions of realism take precedence over the other conflicting aesthetic strategies normally associated with verbatim performances. Unlike *A State Affair* that rarely featured conversations, *Crouch Touch Pause Engage* does not exclusively rely on direct address to convey the verbatim narratives. The ensemble of actors as an embodiment of the rugby-enthusiast community of Bridgend helps create and reinforce the intended perception of a documentary realist structure by weaving the verbatim narratives together from beginning to end. As such, the play opens with a rugby ball on stage and the cast singing and ends with them performing a rugby training session, "[t]he curtain call includ[ing] another rugby try" (72). Importantly, there is also a sense of dramatic progression with the end of Part 1 culminating in the tragic dual suicide attempt and the end of Part 2 recreating the meeting scene between Alfie and "some young people at Bridgend College" (66) as a kind of realist denouement that draws the various verbatim narratives together. Last but not least, documentary realism is crucially reaffirmed in the realist rendering of the Welsh accent through a partnership with National Theatre Wales. Strikingly, the cast itself consolidates a sense of documentary realism. One of the actors proudly reveals in the paratextual section that she is a native of Bridgend, another that he combines his acting career with intense rugby practice and explaining that "it's in his blood" (xvii).

2.4.4 Conclusion

Coming at documentary realist aesthetics from this new perspective via the brief study of some "borderline" cases, one can suggest that the articulation between documentary realism and verbatim theatre is a complicated one. On these grounds, these plays decisively remove documentary realism from the domain of (intrinsic) aesthetic property in order to enshrine it to the domain of discourse and its particular system of possibilities. This change in the poles of the debate over what constitutes documentary realism marks a change in the status of verbatim theatre itself. In the same vein, researcher Jenn Stephenson in a recent blog post on her ongoing research into upsurges of the real in performance even conceives of the possibility of verbatim theatre without words:

> The text then is not 'verbatim' in the usual understanding of the genre where text that is taken directly from witness testimony or transcribed documentation. And yet, the process of verbatim where something 'real' is captured/recorded in the world ('in the wild') and transported to the theatre for mimetic re-performance where we recognize it as real and pay homage to its 'authenticity' does apply. ("Words" para. 2)

It is therefore fair to say that to pursue documentary realism in this period is to watch it – in each new oscillation – lose coherence as an aesthetic model, and manifest itself in ways that might seem completely unrelated.

2.5 General Conclusion of Chapter 2

The two companies under analysis in this part, the Tricycle Theatre and Out of Joint, have been at the forefront of the recent growth of verbatim theatre in Britain. Although *Justifying War* and *A State Affair* by no means form a homogeneous category (except for the fact that they were made in the early stages of the current British verbatim resurgence), they represent an important moment in the history of verbatim theatre. Not surprisingly, some of their aesthetic components – if not all – have been challenged and, to some extent, reverted to by the performances I shall discuss in the following pages of this volume. What I have attempted to describe in Chapter 2 was the necessary first stage and iterations of an ongoing investigation by contemporary theatre-makers of the potential of the verbatim methodology in performance. My strategy was thus to isolate temporarily the plays that demonstrated the strongest documentary realist impulse in order to grasp more clearly their deep aesthetic principles and carve out the terms of their theoretical positioning.[120] Then, ultimately, this volume will reinsert these works into their primordial environment on more legitimate theoretical grounds. More than anything, what emerges as vital to the verbatim theatre en-

120 Each of the "documentary realist" verbatim performances draw upon different styles that are interwoven in practice but they share a more prominent documentary realist aesthetic frame.

terprise is its endless, long-imagined search for aesthetic forms. The fragmented possibilities of other aesthetic directions are, it seems to me, very much present within the performances themselves, but also bleeding out into new works as the next chapters will demonstrate. There is a certain quality, a certain sense of tone and pace within these documentary realist productions which the following performances under scrutiny respond to, engage with which, and extend and refine. These echoes and traces are always incorporated in subsequent verbatim pieces to encompass new meanings and aesthetic experiences.[121] In this sense, documentary realism is absorbed into the British bloodstream of verbatim theatre and becomes a provisional process operating as an abstract standard with which new performances always remain calibrated and in relation to which any of their variations can be measured. More specifically, it is my contention that, in verbatim performances, the problem of a realist aesthetic is exacerbated to the point where many of these practices begin to (once again and at a more compressed tempo) question, challenge and, to some extent, revert to the dominant theatrical traditions of post-war British theatre. In this respect, the following part of this book will mark a primary shift to the main parameters hitherto discussed and ask how much the observed critical discourses on verbatim theatre have affected the constructions of these new practices.

---

121 This is very clear in a verbatim play like *Deep Cut* where its author Philip Ralph openly declares having the two models of documentary realism discussed in Chapter 2.2 and 2.3 in mind when he created the piece in 2008: “My aim, as I waded through the mass of documents relating to the deaths at Deepcut, was to combine the two forms” (qtd. in Radosavljević 142).

3 New Realism(s)

3.0.1 Paradoxes of "New Realism"

This part marks the passage from one kind of treacherous realist ground to another that arguably challenges the aesthetic of documentary realism, the terms of its discourse having hitherto dominated verbatim theatre in Britain, as I have repeatedly suggested. In much of what follows, I argue that verbatim theatre's position here is more nuanced and ambivalent as it seemingly and irrevocably bypasses the commonly held distinctions between form and content, the immediate and the mediated, fact and fiction, authentic and staged, performers and characters as well as between primary and secondary sources.[1]The term "new realism" also removes any perplexing vestigial ambiguities from what I have called "documentary realism" as these performances remarkably attempt to bring realism into alignment and tension with the avant-garde, experimental area of performance practice, a tension that cannot be resolved. To consider "new realism" as a mere aesthetic rectification of documentary realism will not do either. However, due to their relative contemporariness, there is a certain amount of natural overlap between the two strands of verbatim practice and, in many ways, the "new realist" group of verbatim plays picks up where the documentary realist breed leaves off. Very summarily and crudely, one may suggest that the complexity and problems of the documentary realist process become part of the "new realist" verbatim performances. One of them being, as elaborated by Joe Kelleher, "that the theatrical event, however engaged or engaging, however sharp its analysis, however 'authentic' its elements, and however "real" its representations, can only ever be less real and less significant than the event itself" (9).[2] Indeed, the promise of "the real" in verbatim theatre is always set against fictional productions, for "the real" within its performance frame can only ever be the appearance of authenticity.

1 Documentary theatre-maker Roy Nevitt considers the following to be primary sources: "any document (such as a letter, a diary, minutes from a meeting, a newspaper report) which was contemporary with the event in question. Criminal records, any kind of paper work such as might reside in a public transport records office, anything that pertains to an actual event, or anything that was produced within an event. A photograph is a primary source [...]. We also include as primary sources what people tell us in reminiscence [...] a song that was composed within an event" ("Theatre of Fact," 1).

2 This is often expressed by the more familiar view stating that "a representation can never be identical with that which it represents" (Morris 4).

https://doi.org/10.1515/9783110715767-005

Yet, the most striking characteristic of these "new realist" works is rather their insistence on the (evaluative) distance which separates them from a unified documentary realist picture of reality at its most totalising theoretical heights. At this level, let us grant that "new realism" is riven with paradox and departs in important ways from one of the common cores of realism that would square with a shared purpose "to show things as they really are" (Williams, *Drama* 218) through an aesthetic of nonintervention. Indeed, the use of self-reflexive devices and the foregrounding of rhetorical conventions have almost become a *passage obligé* for the theatre-makers I am considering here, in order to more or less acknowledge the specific relationship and mismatch their constructed verbatim artworks presumably maintain and arbitrate with a posited "authentic" real so that (part of) their own particular processes become visible.[3] In other words, this is the moment when these practitioners break through into the verbatim material, pushing further and further into the reaches of what has been categorically and ruthlessly dismissed by their documentary realist peers on the British stage, a situation that undeniably appears (on the face of it) as profoundly satisfactory, ingratiating and palatable as it sounds for the postmodernist within us.[4]

Now it might be thought, I myself would plead, that the notion of "realism" would certainly not come within the compass of verbatim performances presented as such.[5] While it is hoped that this move away from documentary realism is, and will be pretty clear as I turn my critical gaze to a new body of verbatim works, it may be significant that the practitioners themselves said or half-said that they "yearn for more control over the drama" (Blythe qtd. in Hammond and Steward 102) or that "a verbatim play is no different from any other [...] [i] t has to have a metaphor" (Hare "Webchat"), placing the emphasis on breaks and discontinuities and pushing this paradox to breaking point. Let me hasten to add that these breaks and discontinuities no longer happen on the fringes of the verbatim events portrayed, nor are they masked by the subterfuge of docu-

3 By this, I want to point out that these verbatim productions depend upon more processes than is acknowledged. Lib Taylor rightly notes, regarding the headphone-verbatim strand to be studied in Chapter 3.2, that "its aural complexity [remains] inaccessible to the audience" ("Voice" 371).

4 It should be noted that David Edgar saw in this a certain narcissistic effect or self-indulgence that he calls "the meta-textuality of verbatim" (Megson and Reinelt 387), "the fact that verbatim theatre became increasingly about the process of interviewing" (Megson and Reinelt 387).

5 Sara Soncini seems to make a similar observation when she discusses what she calls "the vibrant metatheatrical vein" that infiltrates "even the strict realist orthodoxy of some documentary forms" (*Conflict* 225).

mentary realism.[6] Therefore, it will be necessary to raise the question of the nature and status of the concept of realism within these performances. To do this, I might simply start by positing that these works engage somewhat more deftly with the objections that I have listed in Chapter 2.1 as they adopt some forms of conscious scepticism towards their own processes, while displaying at the same time a curious engagement with the notion of truth.

At this level, it will be held that the verbatim performances in this category present all the marks of a realist verbatim assertion, except that the conditions in which they are uttered leave (more) room for the irregularities of empirical reality. On the whole, the author(s)’mediations are highly counter-intuitive as they do not preclude a mimetic impulse, the verbatim reality being plural. Importantly, these works frame a questioning of the performance’s ontological status and increase the need for further research that was perhaps lacking from the seemingly self-sufficient documentary realist presentations of verbatim excerpts; although it is a fair inference that this does not always have this outcome in practice and the verbatim performance might still be treated as a substitute or equivalent for the reality it seeks to document.[7] Arguably, all verbatim performances rest upon a contradiction between what they presumably promise to deliver and their need to fit into a conventional dramatic fiction. In a sense, what I am calling “new realism” is simply releasing its importance, overstating the documentary realist case by sharpening its inner conflicts and bringing them to the fore. *Ergo*, such very different ways of working that are Brechtianism and Stanislaskianism can be reconciled on this “new realist” verbatim foundation.[8] This unlikely coupling might rightly be greeted with some bemusement, though it is hoped that evidence springing from the case studies shall provide the scaffold-

---

6 This was apparent in Chapter 2.3 through the discussion of David Hare’s *The Permanent Way* which was in some senses a borderline case.

7 One example that springs to mind rests within some of the reactions to David Hare’s *Stuff Happens* – one of the case studies – that were reported by the former BBC chairman Gavyn Davies as placing more “credence in the Hare version of history” (Cook and al.).

8 These two approaches to theatre-making have long been regarded as irreconcilable since Brecht advocated an acting practice of demonstration opposed to Stanislavsky’s embodiment practice. However, it should be noted that some critics such as Eric Bentley have rightly pointed that this was perhaps an oversimplification. As Margaret Eddershaw argues via actress Angelika Hurwicz (who worked under Brecht’s direction), Brecht started his rehearsals with “what Stanislavsky called the super-objective of the actor, and, in effect, required the actor to go a stage further, to move beyond the realistic portrayal that Stanislavsky sought, to add a socio-critical dimension to that ‘real’ character” (Eddershaw 286). Sometimes equally overlooked by critics is the fact that “Sanislavski expound[ed] a duality in which part of the actor’s awareness is focused on observing the character created” (Cantrell, *Acting* 38).

ing for this tentative argument. In order to construct this argument, it may be necessary for now to revisit Brecht in the light of the aforementioned remarks and seek to establish his relevance to the "new realist" verbatim performances I am attempting to map out.

3.0.2 Bertolt Brecht and Brechtianism

An influential version of realism is to be found in the work of revolutionary theatre reformer Bertolt Brecht who advocated a modified kind of realism that would not be an end in itself, freed from the delusions of inevitability and fixity that plagued "bourgeois theatre." As Fredric Jameson's 1998 book title suggests, Brecht's practice was a method[9] and not a mere array of theatrical devices. On close inspection, one detects a central paradox within Brecht's position as "his notion of 'realism' actually requires an avoidance of the illusion of 'the real'" (Halliwell 373–374). Brecht's theory of epic theatre[10] – as an aesthetic practice committed to social change – experiments with the conventions of realism and reception by directly addressing the spectator as well as fostering in the audience a position of critical detachment against identification and "the passive empathy of the spectator" (Brecht qtd. in Willett 57), while still relying on the concept of an existing and knowable reality.[11] His *Verfremdungseffekt* that addresses itself to the spectator's conscious thought and awakens his or her intellectual disposition – showcasing the underlying contradictions in the capitalist social process of an alterable world – is to be understood as "[a] defamilarized illustration [...] allowing the object to be recognized, [while] at the same time mak[ing] it appear unfamiliar" (Brecht qtd. in Rouse 300). In practice, this was meant – without suppressing emotional investment altogether – to make it more difficult for audiences to place their sympathies within one specific

---

9 Brecht's method – informed by his Marxist worldview – pervaded all aspects of theatre-making from production to reception, including how a theatre is organised and the division of labour within an institution, the work ethic, emphasis on process over product, the rehearsal and acting process, the aims for performance, etc.

10 Brechtian theory is by no means a stable theory as his theories constantly evolved over his lifetime and this part of the book only engages with an Anglocentric understanding of Brechtian terms and concepts. This book also chooses not to engage with the polemics surrounding Brecht's translations that are perceived to be largely beyond its parameters.

11 Brecht writes in his *Short Organum for the Theatre* that the "attitude is a critical one" (Willett 185).

character or another.[12] This process is summed up by Jacques Rancière in the following manner:

> the spectator must be released from the passivity of the viewer, who is fascinated by the appearance standing in front of him and identifies with the characters on the stage. He must be confronted with the spectacle of something strange, which stands as an enigma and demands that he investigates the reason for its strangeness. He must be pressed to abandon the role of passive viewer and to take on that of the scientist who observes phenomena and seeks their cause (272)

Although Brecht's influence is "still discernible in the recurring trends of twenty-first century practice" (Tomlin, *Acts* 20), it might be more suitable here to invoke the post-Brechtian paradigm[13] since his theory corresponded to a specific ideological project[14] and, as Philip Auslander demonstrates, relied on "logocentrism and certain concepts of self and presence" ("Self" 58) and thus "stop[ped] at the very threshold of postmodernism" (Pavis, "Modern" 19).[15] Furthermore, if one follows Florian Malzacher's reasoning, "nowadays [...] Brecht's notion of the theatre as a model for actions in the outside world does not work anymore" (132). Ideological reasons aside, the sophisticated audience of today has become so accustomed to the Brechtian rhetoric, that the devices appear self-conscious of themselves, drawing attention to themselves before all else and abstracting

---

12 For instance Brecht writes in his "Short Description of a New Technique of Acting which Produces an Alienation Effect" that the actor "in his efforts to reproduce particular characters and show their behaviour [...] need[s] not renounce the means of empathy entirely" (Willett 136). As far as the audience is concerned, Brecht did not want to suppress emotions but to ultimately make them feel different emotions from those enacted by the characters on stage.

13 David Barnett identifies the first post-Brechtian performance as early as 1971 "with respect to a production of Brecht's *In the Jungle of the Cities* [...], which premiered at the BE" ("Definition" 335). Similarly, Laura Bradley has observed that "a decisive shift towards post-Brechtian theatre occurred around 1970, when a new generation of directors departed from the model and began to use Brecht's methods selectively and subversively" (224).

14 Brecht's project presupposed that there was a reality that latently lay concealed behind appearances and was also against another specific ideological project : industrial capitalism. It hardly needs stressing that this type of capitalism is pointedly different from our current consumer capitalism. Importantly, Marx and Brecht's understanding of capitalism were also slightly different. Today the use of Brechtian devices are often divorced from their social reference and purpose.

15 This is not an unaninmous view. See Elizabeth Wright's *Postmodern Brecht* or Fredric Jameson in *Brecht and Method* who regards Brecht as "a deconstructionist *avant la lettre*" (1) and as someone who "was sceptical of what Jean-François Lyotard called the great narrative" (1).

any political underpinning.[16] By invoking a post-Brechtian vocabulary, one essentially attempts here to reconsider his theatrical techniques (e.g. montage, epic narration, the visual V-effects etc.) at the present time and perhaps even go beyond Brecht or complicate his project, whilst remaining in the Brechtian tradition.

For Stanton B. Garner Jr., the term "post-Brechtian" can be applied to describe a variety of practices after Bertolt Brecht's death in 1956:

> the postmodern radicalizing of Brechtian aesthetics and its subversion of dogmatism and "presence" in Brecht's theory of performance; the simultaneous appropriation and revision of Brechtian political theater on the part of feminists and non-Western writers; experiments in decentring dramatic and theatrical authorship; stagings of Brecht's plays that open dialogues with Brecht's own theater practice (1994, 160)

David Barnett uses the term in a manner close to Garner's first exemplary post-Brechtian practice to suggest "what happens to Brecht's method when it is put under pressure by the conditions that brought about postdramatic theatre [...] in the light of postmodern epistemology" ("Dialectics" 48). He then goes on to articulate the perceived difference between post-Brechtian theatre and postdramatic theatre: "Postmodern uncertainty pervades both the_post-Brechtian and the postdramatic, but in the former, it does not extend to the mechanism by which change occurs, only to the means by which it is articulated and expressed" ("Dialectics" 66). In a different article and much more explicitly, he went on to propose five theses with regard to the updated reinterpretation of Brecht's method that he refers to as a post-Brechtian performance paradigm:

> 1) epistemological uncertainty.
> 2) Dialectics are preserved, and thus Brechtian stage-craft is modified and not rejected.
> 3) The Brechtian emphasis on showing is retained.
> 4) Criticism of Brecht is limited to criticism of his interpretive system.
> 5) The stage is no longer concerned with interpretation but association ("Definition" 337)

As will be shown, whilst the verbatim performances under study in this part may not fully engage with Barnett's theses, they seem to dovetail with some key post-Brechtian elements. For instance, Barnett's fifth thesis – if transposed into the context of new realist verbatim theatre – advocates a position that does not interpret the verbatim material in any definitive way, contrary to the suggestiveness

16 David Edgar was already writing in 1988 "of the degeneration of Brecht's techniques to the conditions of a theatrical cliché" (43). However, for Sean Carney – writing in 2013 – this is mostly the case with the Brechtian *Verfremdungseffekt* and not "the less clichéd technique of Brechtian *gestus*" (7).

of rigidity to be found in documentary realism. However, following Barnett's understanding of post-Brechtian, "it is only when all [the five theses] are considered as a whole that the post-Brechtian emerges" ("Definition" 353). For this reason, and to avoid further complications, I will refrain from using this terminology. Instead, I will acknowledge that these case studies clearly derive some of their devices from Brechtian tenets, but not necessarily in the manner of the post-Brechtian performances.[17]

As I shall demonstrate in the following chapters, "new realism" in verbatim theatre engages with Brechtian concepts of stagecraft in numerous ways and some of its strategies arguably prompt this reading. Let me put this more productively here by giving an example of these strategies. In Gregory Burke's "massaged" verbatim piece, *Black Watch*, the history of the legendary Scottish regiment is told in the following manner:[18]

> Music. A red carpet rolls out, and as Cammy narrates the following history of the Black Watch the other soldiers manoeuvre him around the stage dressing him into and out of significant and distinct uniforms from the regiment's history (30)

Quite strikingly, this type of performance holds the curious view, for instance, that performers are somewhat divorced from characters. If documentary realism typically made it clear that characters were dramatic constructs divorced from real people (albeit heavily inspired by them), performers were representing these dramatic characters along realist lines. A good example of what I am calling "new realism" can be found in Chris Goode's *Monkey Bars* that "is strictly verbatim" (22) but these words – distilled from private conversations between dialogue artist Karl James and 72 children – are spoken on stage "not by children but by [...] [a]dults" (27). Similarly, *On Ageing* (2010) by the company Fevered Sleep had the exact words of eighty-year-olds spoken by young children on stage at the Young Vic in London. This Brechtian process thus adds an extra tension or rather highlights what African American performer/playwright Anna Deavere Smith calls "the gap" which is what viscerally separates the self of the actor from the other he or she is trying to reach in performance: "[c]haracters

---

17 Perhaps in this sense, one is in fact again in the realm of post-Brechtian theatre if one follows Lehmann's undertanding of the concept as "precisely not a theatre that has nothing to do with Brecht but a theatre which knows that it is affected by the demands and questions for theatre that are sedimented in Brecht's work but can no longer accept Brecht's answers" (*Postdramatic* 27).

18 The term is used here to designate a verbatim performance that is sedimented with fictional components. A more thorough definition is given in the next chapter.

live in the obvious gap between the real person and my *attempt* to seem like them. I try to close the gap between us, but I applaud the gap between us. I am willing to display my own *unlikeness*" (*Fires* xxxvii–xxxviii, her emphasis). This concept is evidently inherent to the verbatim practice but, within the new realist verbatim performances, it is no longer possible for the spectators to ignore it.

3.0.3 Outline

Finally, this part shall provide the reader with glimpses of the "new realist" ground of verbatim practice from different vantage points and I am not claiming here to have appointed David Hare and Alecky Blythe as the two most significant "new realist" verbatim theatre-makers but, rather, as the best expositors of this observed phenomenon in verbatim theatre.[19] This would evidently ring hollow in the context of David Hare's long career in the field of new writing and it shall become especially understandable when one reaches Chapter 4 of this volume only to realise, with dismay, that Alecky Blythe has also briefly flirted with the last category of verbatim practice I have identified as "post-realist". In the following chapters, I want to establish the possibility that some contemporary verbatim works are a new realism quite distinct from documentary realism. New realism, in particular, marks the point of transition between two observed paradigms – documentary realism and post-realism – that constitutes a new verbatim reality in its own terms. More specifically, these chapters investigate how the aesthetic of "new realism" produces certain effects and experiences when combined with verbatim material in performance. As specified in the last part, new realism is an aesthetic system which I am going to isolate, somewhat artificially, in order to discover how its conflicting requirements function in practice. Chapter 3.1 will address the "massaged" verbatim practice – a kind of verbatim work that wears its theatricality on its sleeve – through David Hare's *Stuff Happens* (2004) and Chapter 3.2 will then engage with another discernable pattern in contemporary British verbatim theatre, "headphone-verbatim" (for want of a better term) – a productive "encounter between digital audio technology and verbatim material in performance" (Garson, "Remixing" 50) – via Alecky Blythe's

19 I do not claim either that "new realism" encompasses their entire oeuvre, especially since their oeuvre is still changing as I write. Some might consider Hare and Blythe as two unexpected participants in this new realist discourse.

Come Out Eli (2003).[20] As I intend to show in the following pages, these two case studies are representative of a typical manifestation of "new realism" in British verbatim performances.

20 Australian scholar Caroline Wake also uses this term as it appears to be "generic and not associated with any particular company" ("Headphone" 321).

3.1 "Massaged Verbatim Theatre": David Hare's *Stuff Happens* (2004)

As already seen, verbatim theatre can be understood as the mask of seductiveness a performance text assumes as it convinces its readers/spectators that it has conformed to the laws of documentary realism within a scene of constraint.[21]Every verbatim performance enters into a relationship with documentary realism – even as it is championed or contested – and one way of conceptualising "massaged verbatim" would be to consider it as a subtext introducing a tension or contradiction that modifies the documentary realist narratives I have charted in Chapter 2. In this sense, the Tricycle Theatre's *Called to Account* was already on the path towards "new realism" by including what documentary realism forbids. Admittedly, verbatim theatre is always a site of struggle over the real and its representation and, if documentary realism seemed to brush it under the carpet, "massaged verbatim theatre" as a "new realist" practice is aware of its own artifice and becomes a site of multiple aesthetic experiences. Before attempting to articulate the logic of this position, I will now define what I mean by "massaged verbatim" theatre.

3.1.1 Definition of the Strand

As its name suggests, "massaged verbatim" supplements the verbatim material with other material and announces its own position from the start, thus destabilising the sense of documentary realism normally associated with this type of theatre and interrogating the_verbatim real it represents. Here, a distinction is clearly maintained between the original verbatim material and the performance and, to understand verbatim performance as one way of configuring the "real" is to open it to a continual remaking, an "otherwiseness." The term "massaged verbatim" will be preferred, in the following, for reasons I shall give in this introduction. As ever, the reader may have encountered different terms to describe such a verbatim phenomenon and the one that would come closest to what I intend to

21 In this case, verbatim theatre almost assumes the power of the original and gives the audience the illusion of having unmediated access to the real. However, being documentary realist and true are not necessarily synonymous (i. e. *Called to Account*) therefore a verbatim text can be documentary realist and not be true and conversely, a verbatim text can lack documentary realism and be true.

suggest is Jenny Hughes' third category of verbatim performances,[22] referred to as "the 'composed'"[23] (93) to encompass "plays that combine 'found' speech drawn from documented records and 'made' speech devised by playwrights" (93).[24] The effects of the massaged genre are not quite those of the *faux* verbatim I mentioned in the first part of this volume[25] – although they may collide in practice – as these plays tend to be based on supposedly irrefutable facts.[26] They are also not exactly what Patrick Duggan calls "proto-verbatim theatre" (149) either or those of fictional plays inspired by "true stories" such as Tim Price's *The Radicalisation of Bradley Manning*,[27] described by the author in the following manner:[28]

> The incidents, characters and timelines have been changed for dramatic purposes. In some cases, fictitious characters and incidents have been added to the plot, and the words are

---

22 Sophie Bush in *The Theatre of Timberlake Wertenbaker* uses the term "docudramas" to refer to the works of playwrights that have adapted the verbatim "trend," "combining recorded speech with their own imaginings" (21). Looking specifically at Hare's *Stuff Happens*, Jeanne Colleran uses the phrase "hybrid theatre" (135) while Jane Gaines speaks of the notion of "documentary fiction that [...] challenges an old calcified division" (45) in the context of films. However, due to the fact that these terms are already overloaded with various meanings, I have decided to look for something more specific to my object of study.

23 Arguably, the borders of her categories are never sharply drawn and what she calls "the composed" is therefore more inclusive than our understanding of "massaged verbatim." In practice, then, Jenny Hughes unproblematically considers Michael Frayn's *Democracy*, a play consisting of "imagined speech based on extensive historical research" (114) as representative.

24 The first two categories that she has envisaged for contemporary verbatim theatre are "the 'forensic'" (Hughes 93) which refers exclusively to the tribunal plays of the Tricycle Theatre (see Chapter One) and "the 'exceptional'" (Hughes 93) that "refers to the staging of testimony from spaces of exception" (93).

25 *Faux* verbatim designates theatre works that are not based on the exact words of real people but adopt a specific language and strategies that make the audience think it is verbatim. A good example of this practice – that I mentioned in the last part – is Dennis Kelly's *Taking Care of Baby*.

26 Pamela Carter explains of her mock-documentary piece, *Paul Bright's Confessions of a Justified Sinner*, that she never "considered that audiences might believe [their] show was a documentary. It hadn't been [their] intention to 'trick' anyone" (141).

27 Duggan uses the phrase "proto-verbatim theatre" to refer to those performances based on the words of real people that are not accompanied by a "claim of exact truthfulness" (149). He then adds that the companies in question do not suggest "in promotional literature or on stage, that the production is anything other than a piece of theatre" (149–150).

28 The National Theatre Wales's production also had several actors – male and female – performing the role of Bradley Manning. Some of these performances were broadcast live on the internet, thus opening up ad hoc communities within the space of the theatre.

> those imagined by the author. The play should not be understood as a biography or any other factual account. (3)

However, such palimpsest arrangements are reminiscent of those that have been termed "faction" but I would suggest that the latter is less tainted by a reliance on "the real."[29] As Janelle Reinelt and Gerald Hewitt have jointly argued, factions are by and large "completely fact-based but not necessarily tied to history 'exactly as it happened'" (47), a statement that would seem to precisely run counter to Hare's contention in *Stuff Happens* according to which "[w]hat happened happened" (vii). In other words, massaged verbatim is waywardly imaginative but not imaginary.[30] I would also want to suggest that "massaged verbatim" is perhaps inevitably more aware of itself as a newly burnished faction.[31] Perhaps, in this sense, it is close to the notion of "authentic fictions" (2010, 106) advocated by the Germano-British performance group, Gob Squad. If, for Carol Martin, verbatim theatre is rather different from "fictive theatre" (*Dramaturgy* 17), in massaged verbatim theatre, the verbatim real and fiction collide in an interdependent relational context.

As should be evident from the foregoing discussions in the last several chapters, verbatim material and theatrical performance can align in a number of different ways. I intend to show in this particular chapter that there is a growing fictionalising propensity most visible in recent years within the British spectrum of practice that goes against the standard perception of documentary theatre as "a strand of contemporary performance practice that eschews fiction in favour of something supposedly more authentic or 'real'" (Reid 137) and as one which only

---

29 Not surprisingly, the term "faction" has also been used in a variety of ways and contexts. Its coinage has wrongly been attributed for instance to playwrights Tom Stoppard (Demastes 40) and David Edgar (Palmer 7) but it seems that the term was in fact first coined by Truman Capote to describe his novel *In Cold Blood* (1966). In his Introduction to *Asking Around*, a book of edited interviews that provided the background for his National Theatre trilogy *Racing Demon*, *Murmuring Judges* and *The Absence of War* on the structures of the Church, the Law and the Labour Party in Britain, David Hare defended his purely fictional approach and vividly expressed how he "distrust[ed] faction" (4).

30 Interestingly, Aleks Sierz refers to these processes (opposed to the typical verbatim or tribunal play) as "an imaginative verbatim account" in the context of Paul Unwin and Sarah Beck's *This Much is True* (2009).

31 Rib Davis still uses the term "faction" to describe works such as David Hare's *Stuff Happens* that he defines in the following manner (beyond stating that they mix fact and fiction): "The words of the public figures that are on record are used unchanged, but the background scenes are fiction, though compatible with what is known of the characters and situations" (*Dialogue* 201).

thrives on the supposedly contemporary distrust of the subversive power of the imagination.[32] One must readily acknowledge, however, that this is a perspective on verbatim that has already garnered some recognition in numerous publications. For instance, Mark Chou and Roland Bleiker recently wrote that “Verbatim Theater use facts to create fictional representations” (240) and pioneer documentary practitioner Roy Nevitt considers that “[l]ike any other play, they are fictions. Their documentary substance has been selected, organised and arranged for theatrical effects” (“Archive” 151).

To come back to the definition of “massaged verbatim” theatre, it is helpful to refer to Australian verbatim playwright Alana Valentine, who coined the term in the first place. In August 2009, on the occasion of the Alex Buzo Memorial Lecture at the University of Sidney, Valentine introduced to a wide audience – drawing from her own practice – a variant label to designate a subdivision of verbatim theatre that she saw as the opposite of “pure verbatim” and that I have understood as comprising two main characteristics: “the use of verbatim techniques to create fiction and the shaping of the interviews around an invented narrative structure” (Garson “Kindred”). The term “massaged verbatim” carries in itself the promise of the unexpected as it stretches verbatim theatre in a way it had not been before where “[t]here is no pretense that this is anything more than a fiction” (Valentine). What is clear, from her articulation of the term as well as her observations, is that “[v]erbatim is diversifying and changing and growing to the point now where [...] it is beginning to reach again for all the skills and tools of the classically trained dramatist” (Valentine).

Valentine then admirably proceeds to sort out these seemingly insurmountable antinomic statements that one could have otherwise mistakenly interpreted as the introduction to an orientation diametrically opposed to the verbatim discourse within verbatim theatre itself, perhaps an orientation that would maim verbatim theatre in that some say that it might equally contribute to its own undoing. The most cynical among us would even say that it is nothing more than a mere trick of perception passing itself off as verbatim. Yet, when one starts thinking about “massaged verbatim”, these formulations do and do not hold good. The overtly period term carries within itself, within the aforementioned broad outlines, a kind of distancing disclaimer from what is generally understood as the province of verbatim theatre, but cannot necessarily be equated with an over-technically conscious verbatim phenomenon (if one may call it such) and

32 Literary theorist Wolfgang Iser describes what he calls “the fictive” as “an operational mode of consciousness that makes inroads into existing versions of the world” (xiv).

its entailed ontological transgressions.[33] For it is the degree of acknowledgement of such a distance that creates the very possibility of massaged verbatim theatre when a more paradoxical trajectory is followed. This means that the danger of tipping over into this alleged (outrageous) category of works is always present, as today one could push Valentine's term much further and say that all verbatim works are "massaged" to some extent, even when they adhere to the nobility of "pure" verbatim protocols. This is a position often expressed by playwright David Hare who is so passionate an advocate of such a conception of verbatim theatre and argues that, in a sense, "there is absolutely no difference between the writing of a good documentary play and the writing of a wholly imagined play" (Hammond and Steward 59).

Put more simply, if my reasoning were correct, there would even be no tipping over at all. The most rigorous and minimal type of massaged verbatim would thus consist in taking a known speech or conversation in the public domain – to some extent what Gérard Genette calls a "hypotext" in his *Palimpsests* (5) – in order to give it a new significance by putting it on stage and conducting some minor editorial work (for Genette, the creation of a "hypertext" (5)). More incisively, the term certainly clears the stage for the existence of verbatim, irrespective of disconventions but also inevitably carries a certain number of risks and I will come to some of these in the following pages. Massaged verbatim, then, productively brings together two genres that are traditionally conceived as distinct, and does so in a manner that may be perceived as problematic (even when the playwright is supposedly acting in good faith). By this, I mean that the whole concept provokes an uncertainty of genres that unsettles conventional understanding of verbatim theatre and these particular mutations need to be fully addressed. I shall take two brief examples to illustrate the "dishonest" process at hand. In Davey Anderson's *Blackout* (2008), a play about the true stories of a young offender from Glasgow who had committed a violent crime, the author tells us that his text is not really:

33 David Edgar, unhesitatingly, calls for the existence of a "post-verbatim" (Megson and Reinelt 386) phenomenon when fictionalisation is involved. I do not negate that such a flicker may be produced but I would suggest instead that a process of fictionalisation does not necessarily allow us an entry into the "post" realm of practice.

> a verbatim transcript – although most of the words are his [the interviewee] rather than [the author’s]. The play that emerged is a fictionalised account of the events surrounding the crime this young man committed, told with a fair amount of creative licence and, crucially, with certain key details left out. (vii)

With enviable ease, Peter Morgan states that his *Frost/Nixon*, “having met most of the participants and interviewed them at length, [...] is an accurate representation of what actually happened” (vii). However, he pursues the thought in his Author’s Note and tells us that he feels “most comfortable thinking of this as a fiction – a creation” (vii), adding that “it is a play, not a historical document and that [he has] on occasion, perhaps inevitably, been unable to resist using [his] imagination” (vii). A subtler example is to be found in David Hare’s *The Power of Yes* (2009) which comprises a fictional *alter ego* of David Hare in the figure of the “author” who opens the play in the following manner:

> **Author** This isn’t a play. It’s a story. Or rather it’s only partly a play. It’s more properly a story. (3)

These, one can suppose, would come under the label “massaged verbatim” as envisaged by Valentine. It follows that this type of verbatim work is a hybrid whereby fictional and verbatim material blend into one another. Furthermore, and this is a far more important point for this study, these massaged verbatim performances bear witness to a new way of conceiving verbatim theatre, a new field of practice, what I am calling here “new realism.” In recent years, there has been a steady outpouring of such transgressive works. Examples of this practice could be multiplied *ad infinitum*.[34] As I have demonstrated elsewhere, works such as Gregory Burke’s *Black Watch* (2006), Peter Morgan’s *Frost/Nixon* (2006), Duncan McLean’s *Aalst* (2008), Rani Moorthy’s *States of Verbal Undress* (2014) or Owen Sheers’s *The Two Worlds of Charlie F.* (2012), “foreground[ing] both their verbatim origins and imagination processes” (Garson

34 Importantly, the mixture of fictional/composed and verbatim material in performance is not necessarily enough for a piece to be considered “massaged” in this volume. Indeed, the video ballads of Banner Theatre in the UK such as *Wild Geese* (2006) – based on the stories of migrant workers, asylum seekers and refugees coming to the UK – that weave archive interviews on film, songs and live music do not enter the category of massaged verbatim theatre. Admittedly, the video ballads belong to a very distinct genre of performance practice and, as such, do not engage with the tropes of documentary realist verbatim theatre (which I see as a crucial feature of massaged verbatim theatre).

"Kindred") create a whole range of formal possibilities.[35] However, following strict logic, the adroit introduction of the fictional is not a mere merging of two different kinds of discourses, but in fact, a matter of a new discourse that stimulates a vigorous and fruitful debate regarding what a verbatim play can do.

3.1.2 Critical Reception

The main thrust of the attack on the massaged verbatim theatre-makers is that the connection between their shoddy verbatim constructions and their real referents is severed (for better or worse) as they become significantly detached from their source and (even) more open to doubt, thus pre-empting and short-circuiting any role for verbatim theatre in political intervention. So goes the argument, at least. However, I intend to suggest in this study that, as a vital reinvention of the realist project, massaged verbatim theatre does not really escape the problem of reference and perhaps only ingenuously attempts to display it within the performance text so as to ultimately rescue it.

At the present time, criticism divides between those who have welcomed and extolled the virtues of such experiments within the possibilities and arena of an already fatigued verbatim theatre, and those who are deeply concerned that dramatic interests can ever be allowed to prevail over a sense of duty for verbatim accuracy and facticity, that the "rules" of verbatim theatre can so mercilessly be broken. As regards the latter, Bernhard Klein writes that this new approach in *Stuff Happens* "compromises verbatim theatre's unique selling point" (214) while enabling the author "to show his characters in genuine dialogue and confrontation, interacting with each other in dramatically staged encounters" (214). One is consequently bound to admit that such perceptions of "new realism" shift the ground of attack but, ultimately – as the cynic may observe – they fare no better than "documentary realism" and its sterilised aesthetic citadel of verbatim certainties supposedly immune from all contamination by the intruding subjectivist bias of its authors and the perils of theatricality. This is arguably disingenuous as in my view, the elements of fictionality in the massaged format cannot cut short the verbatim status and make us lose sight of its "real" effects. Indeed, such a shift entails a new conceptualisation of political verbatim theatre no longer seen as a mere reflection of some external reality, but as reality itself just as material as other realities constructed by an impeccable sleight of hand within

35 The show is adequately subtitled "an unauthorised biography of the legendary Scottish Regiment."

the dominant ideology. As such, this strand more than deserves our scrutiny. This leads one to one of the heated paradoxes of "massaged" verbatim theatre that Donna Soto-Morettini articulates, in reference to *Stuff Happens*, as "both self-reflexive as well as self-effacing" (314). Moving on, it is now time to explore how this new realist paradox applies in practice, that is to say in David Hare's *Stuff Happens* (2004).

3.1.3 David Hare

David Hare, "Britain's most stridently political living playwright" (Inman 222) – sometimes referred to as "the national playwright" (Deeney 430) – has spurred more academic criticism than any other British verbatim theatre-makers in this period and has been famously praised by Richard Boon for his success in "capturing the *zeitgeist* – the spirit of the age –" (2003, 6). Today, a double Olivier Award-winning playwright he is still very active. His recent exploration of verbatim theatre would seem, one is tempted to say, to confirm exactly Boon's contention. Hare has indeed resumed his place at the National Theatre in the 21st century with no less than three verbatim performances: *The Permanent Way* (2003), *Stuff Happens* (2004) and *The Power of Yes* (2009). He has further explored the theatrical potential of verbatim material through his recent adaptation of a documentary-resourced book written by American journalist Katherine Boo, *Behind the Beautiful Forevers*, which premiered in the Olivier auditorium of the National Theatre on 10 November 2014.

Most significantly and most relevant, if one goes back to David Hare's pre-2000s writings,[36] many of the elements of a "massaged" verbatim practice were already pre-figured before he was "caught out" as it were: his realistic inventions in *Fanshen* (1975) on the early progress of the Chinese revolution during his "Joint Stock" days, to some extent also *Brassneck* (1973) that he co-wrote with Howard Brenton and which incorporated some documentary elements, and his ninety-minute monologue *Via Dolorosa* (1998) on the Israeli-Palestinian conflict, which he performed himself.[37] In fact, it is often forgotten that this odd and curious blend is by no means novel. However, arguably, its presuppositions need to

---

36 As Markus Wessendorf has argued, we can also find numerous examples of earlier practices of "massaged" verbatim theatre such as "Heiner Kipphardt (*In the Case of J. Robert Oppenheimer*), and Peter Weiss (*Vietnam Discourse*)" (340) in the 1960s but today's usage of these methodological practices has a very different intent.

37 The two-act play was an adaptation of a 600-page book by American farmer William Hinton entitled *Fanshen: A Documentary of Revolution in a Chinese Village.*

be thoroughly examined. In an interview with Duncan Wu, David Hare already expressed his ambivalence towards a certain conception of documentary drama: "I don't like documentary drama, I distrust it. Indeed, I can't see the point of reproducing on the stage something which fails or succeeds by its similarity to something which is already in the audience's mind" (Hare qtd. in Wu 170). In another interview, he describes himself as "no longer [...] a classic documentary writer" (Zeifman 5). Let us therefore examine in more detail how one of his exemplary works, *Stuff Happens*, disrupts the documentary realist conventions of verbatim theatre.

3.1.4 An Overview of *Stuff Happens*

David Hare's *Stuff Happens* (2004) is, I believe, an exemplary new realist verbatim performance that exuberantly departs from the documentary brand of realism discussed in Chapter Two, notably David Hare's *The Permanent Way* (2003) that "ha[d] much in common with the 'tribunal play'" (Dorney and Gray 196) and, as Hare himself archly stated regarding the play, "[t]he illusion is that I'm not present, but it's an illusion" (Preston).[38] Apart from the appearance of British politician John Prescott in both plays, the performances could not be more different.

Stuff Happens is also very much an exemplary massaged verbatim piece and the words used by Donna Soto-Morettini to describe it come quickly to mind: "'[t]he verbatim' employed in *Stuff Happens* is never off the cuff or unmediated – it is endlessly *massaged* by political speech writers and designed to be delivered and received as authoritative" (314, my emphasis).[39]Anderson and Wilkinson even intend to demonstrate that "the 'massaged' testimony [of *Stuff Happens*] is exactly

38 Again, despite the consensual claim that "Hare's *The Permanent Way* exemplifies the verbatim play" (Kritzer 223), some of its elements were not typically documentary realist. Hare was still present in absence through some interviewees' remarks. For instance, a British Transport Policeman says "[b]ut then you work on from there, David" (26) and, perhaps more obviously, a Senior Rail Executive addresses David Hare as he explains his role: "[b]ecause, David, what I do now is manage contracts, monitor the contracts between all the myriad different parts of the operation" (19). To some extent, as I have suggested in the introduction of this book, there are some overlaps between the categories I am attempting to map. As such, Robin Soans' *Talking To Terrorists* that I have discussed in Chapter 2.3 contains within its very architecture – in a less explicit way – some of the features of new realism, as Jeanne Colleran has rightly remarked: "[i]n *Talking to Terrorists*, the interviewer does not appear but is implied" (141).

39 Here my argument goes against the view that *Stuff Happens* is "[i]n a category of its own" (Morgan, "Epilogue" 693).

the right text to be hearing in the context of plays about political manipulations, persuasion and hypocrisy" (155). David Hare himself confessed his uneasiness with the category of the "documentary play" with regards to *Stuff Happens* since "two-thirds is made up" ("Mere" 77) mostly corresponding to private moments between the political figures. The remaining third is a recitation of the record – mostly excerpts from the speeches of politicians – by performers.[40]

Stuff Happens – "the best-known of the Iraq War plays" (Colleran 144) – was Hare's thirteenth play at the National Theatre and opened on the Olivier stage on 10 September 2004 with a cast of 22, under the direction of Nicholas Hytner.[41] The play has continued to be revived internationally, receiving its US premiere at the Mark Taper Forum in Los Angeles in June 2005 followed by a production at the Public Theater in New York in March 2006. It notably received its Australian premiere at the Belvoir in Sydney in 2005 as well as a Canadian production in Toronto six years after the National Theatre performances, a production in Japan in 2006 and another one in Poland in 2008. In France, the text was translated by William Nadylam in 2009, staged at the Théâtre des Amandiers in Nanterre and broadcast on the national channel France 2. More recently, one of its scenes between George Bush and Tony Blair was revived for a special performance on the occasion of the National Theatre's 50th anniversary in November 2013. At the time of writing, David Hare has also announced plans to do a public reading of *Stuff Happens* at the National Theatre, on stage at the Olivier, on the very day the Chilcot Inquiry will be published (6 July 2016).

The title of the play is taken verbatim from a press conference given by US Secretary of Defence Donald Rumsfeld in response to television pictures of the looting of the National Museum in newly liberated Baghdad on 11 April 2003: "Stuff happens [...] [a]nd it's untidy, and freedom's untidy, and free people are free to make mistakes and commit crimes and do bad things" (Watson 194).[42] Within two acts and 24 scenes, Hare is portraying a precise moment of international history that is to say how and why the diplomatic and political processes that triggered the American-led invasion of Iraq happened and how the United Kingdom became a collaborator in those processes. In the context of the almost palpable need for a neo-imperial power to reassert itself after it had been wounded, *Stuff Happens* is also about the subjects of idealism and disillusion, uncov-

40 Besides the public record, Hare also sometimes draws on private sources.

41 The play opened only 18 months after the invasion of Iraq by U.S. and British troops in March 2003 and on the eve of the third anniversary of 9/11.

42 This very extract is reproduced as early as Act I, Scene Two (3–4). The famous quote is also reprised at the very end of the play, in Act II, Scene 24 when commented upon by an Iraqi Exile (119).

ering the growing disconnect between evidence and action in the pursuit of the flamboyant dream of reordering the Middle East. It is also as much about the often-informal ways in which key political decisions are taken behind closed doors, the rushing in of measures during seemingly "exceptional times", about how the political class in Britain has stopped listening to the very people they are meant to represent[43] and also how Britain's status in the world has chillingly diminished.[44]

Consequently, it should be of some interest to see how Hare's verbatim enterprise compares to and differs from the spate of documentary plays produced in response to the war and, especially, the Tricycle Theatre's *Justifying War* that was the object of my analysis in Chapter 2.2. Similarly to this latter, *Stuff Happens* proceeds in a chronological, causally linked and linear manner, condensing three years of recent history (2001–2004) into a three-hour performance with a 20-minute interval. However, *Stuff Happens* focuses much more on the Bush administration ; the obvious reticence in portraying high-profile figures like Tony Blair in *Justifying War* did not seem to inhibit Hare's verbatim enterprise, which also featured the 43rd president of the United States, George W. Bush as well as Saddam Hussein, Ariel Sharon and Jacques Chirac, to name only five of them.

David Hare has availed himself of all the verbatim material at his disposal (with the help of Dr Christopher Turner, then visiting scholar at Columbia University), the truth as it is reported, to create *Stuff Happens*. He shows us what happens when the verbatim words have been exhausted and also the questions that arise when one attempts to go beyond such material, creating an aura of doubt around his verbatim credentials. In fact, as seen earlier, only a quarter or a fifth of the play is direct reportage and some conversations are pure invention, such as the private one between Blair and Bush at the latter's Prairie Chapel Ranch in Crawford, Texas (that is known to have taken place) or Bush's private meetings with his staff, but this would seem enough to make it enter the realm of the "quasidocumentary plays" (Colleran 120).[45]

In some sense, the Secretary of State at the time, Colin Powell – depicted as a man of integrity without being unmitigatedly liberal – is the central figure in the

43 The country was overwhelmingly against the invasion of Iraq but, still, Blair managed to involve Britain.

44 Theatre critic Aleks Sierz deplored this situation in his *Rewriting the Nation* and added in dismay: "poor Blair doesn't even make it onto the poster advertising the play" (74).

45 In terms of language, these invented private conversations are in fact heavily influenced by the verbatim technique as David Hare based them on patterns of speech he listened to so as to get the timing right.

play and Hare similarly offers a radical interpretation of Bush that treats him with seriousness and respect, thus significantly departing from mainstream representations.[46] Colin Powell, who has a sound military understanding of collateral damage and sees war as an admission of failure, went along with something which he did not approve of, or rather something he approved of whilst being against the manner in which it was carried out. In other words, in Hare's version of the story, Bush politically outmanoeuvred Powell.

The play is also about Shakespearian themes of power. Tony Blair, portrayed as an intellectually powerful man, thinks he can be in command when opposed to an apparently less intelligent man, George W. Bush. The latter is perhaps represented as less intelligent than his British counterpart but demonstrates a cunning strategic side. For example, in Act I, Scene 10, a crucial exchange between the two men is turned into a demonstration of Bush's power as he intentionally lets his British interlocutor do the majority of the talking until he becomes nervous:

> **Bush** It's me that'll make the decision. I'll make the decision. I'm the President.
> **Blair** Yeah. To me, it's an opportunity. The UN is an American-built institution. America built it.
> *Bush doesn't reply.*
> Internationally – well, in Europe, in Russia, I can help. I think I can chip in with a good deal of personal persuasion – with Chirac, with Putin. My relationships are excellent. One of the advantages of being a bit longer in office ...
> **Bush** Sure.
> **Blair** Knowing the people. Knowing the personalities. I have a history, remember? Sierra Leone ...
> **Bush** Sure.
> [...]
> *Again, Bush doesn't answer.*
> There's a speech of mine, in fact, I made in Chicago ...
> **Bush** I know that speech ...
> [...]
> *Bush nods slightly, non-committal.*
> [...]
> **Bush** I'm going to talk to my people. (*He smiles.*) You're always eloquent, Tony.
> *Blair is tense now.* (40 – 41)

46 As new evidence was being released, Hare's view on Powell changed by the time the play was to be produced in America and he decided to more or less update the piece on this occasion. This volume only deals with the original British production.

At the end of the play, therefore, the shrewd and inarticulate George W. Bush – the "brains" behind the operation – manages to get what he wants and the intelligent and articulate man, Tony Blair, loses. If it would be stretching the point to suggest that David Hare is using here "a method of reversal which is notably Brechtian – strong characters turn out to be weak [...]" (Hare qtd. in Wu 255) –, it may seem reasonable to attribute a Brechtian influence to *Stuff Happens*. Numerous interventions or disturbances are performed to break up the logically unfolding narrative, a technique that Christopher Innes calls "continuous commentary" ("Mainstream" 443) to describe the particular kind of distancing effect at work in the play. In the preface of *The Political Unconscious*, Fredric Jameson writes that texts are always apprehended "through sedimented layers of previous interpretations" (9). A case might well be made that the interpretative context of *Stuff Happens* hinged upon Hare's very creative act.

The performance, constructed in several episodes and spanning different locations and countries, is framed by both a prologue and an epilogue as well as a narrator figure called "An Actor." *Stuff Happens* begins on 30 January 2001 with a National Security Council briefing on Iraq's WMD ten days into Bush's first term followed by a cabinet meeting with Bush that took place four days after 9/11.[47] At the very end of the first act, the well-documented scene – discussing the pros and the cons of the Iraq crisis – that took place on the evening of 5 August 2002 between US administration figures Colin Powell, George W. Bush and Condoleezza Rice comes to life dramatically. In another notable scene, Scene 15 in Act II, at the Hotel Pierre in New York, the French – as permanent members of the United Nations Security Council and represented by Dominique de Villepin and Jean-David Levitte – seem to take immense pleasure in being suddenly, and perhaps again, at the centre of diplomatic affairs and world events. The performance ends, in Scene 24, on the point of view of the Iraqis: an Iraqi exile comes on stage, alone, and faces the audience in direct address.

3.1.5 The Contract with the Audience

> Unlike "imaginative" theatre, verbatim theatre can only draw on those stories that people are willing to tell. (108)

47 Bush's re-election in November 2004 happened during the National Theatre run of *Stuff Happens*.

In this quotation, Gail Cameron refers to a familiar and deeply imbued *sine qua non* that is always covertly operating when one approaches this type of theatre. In massaged verbatim theatre, as I intend to demonstrate, these kinds of questions are pushed aside and the guiding rule is posed differently, asking one instead to pay attention. As has been suggested in previous sections, *Stuff Happens* strikes all of these chords from the outset since – as a massaged verbatim piece – it calls more attention to the verbatim words' former real world existence, the interviews that underlie them as well as to how theatricality is effected in the first place.

Indeed, very much like all verbatim plays, *Stuff Happens* negotiates with the audience's prior knowledge and expectations. Every audience member sitting there listening has inevitably been bombarded with incessant images and discourses from disparate sources online, from the radio and the television. However, unlike the tribunal plays whose exact content was less likely to be known by the general public, the audience therefore expects more than a mere condensed replay of events in this context and the massaged format of verbatim practice seems particularly suited to the confounding role. For instance, Gregory Burke's *Black Watch*, another massaged verbatim piece on the Iraq war, chooses to engage with the audience's response to those events from the very start:

> **Cammy** A'right. Welcome to this story of the Black Watch.
> [...]
> See, I think people's minds are usually made up about you if you were in the army.
> [...]
> They poor fucking boys. They cannay day anything else. They cannay get a job. They get exploited by the army.
> [...]
> Well I want you to fucking know. I wanted to be in the army. I could have done other stuff
> [...]
> And people's minds are made up about the war that's on the now ay?
> [...]
> They are. It's no right. It's illegal. We're just big bullies. (3–4)

Stuff Happens also addresses this strenuous "horizon of expectation" in three main ways. First of all, as the massaged verbatim format is not entirely tied to the exact words of real people, it can more easily forestall the questions that normally arise. Appropriately, then, David Hare explains the title of the play at the beginning of Act One and even engages almost directly with what is on the audience's mind by rhetorically asking: "So where to begin?" (4). In other words, the new realist aesthetic framework seems to look for a different engagement with the spectators as they are confronted with a familiar verbatim quote from the public record, a re-enactment of what is considered mainstream history or widely accepted "facts." In fact, and this is the second crucial element in the contract, Hare has anticipated the audience's reaction to such an extent that he felt the need to provide a note – in both the National Theatre's programme and the published playtext – to clarify what the text/performance does in relation to the public verbatim material.[48] Due to the unusual nature of the verbatim performances I call "new realist", that depart from both the more familiar documentary realist strand studied in Chapter 2 and non-verbatim theatre, it will be argued that these performances tend to feature a more formalised contract with the audience.

A clearly signalled introduction to the type of contract Hare intends to chart for his audience is provided in the first few pages of the published text and such a contract is continuously reaffirmed throughout the performance:[49]

> *Stuff Happens* is a history play, which happens to centre on very recent history. The events within it have been authenticated from multiple sources, both private and public. What happened happened. Nothing in the narrative is knowingly untrue. Scenes of direct address quote people verbatim. When the doors close on the world's leaders and on their entourages, then I used my imagination. This is surely a play, not a documentary, and driven, I hope, by its themes as much as by its characters and story. (*vii*)

However, instead of giving a sense of clarity to the piece, as one would expect here, one finds oneself faced with a more ambivalent dilemma than what is commonly the case when treading on verbatim territory. The last sentence is particularly puzzling. Accordingly, Hare's kaleidoscopic expository prose drastically mounts a state of tension between two positions for the spectator: one horn of the dilemma being pinned down to a willing suspension of disbelief safely and joyfully keeping everything from leaping out of the overarching whole, and the other, more prudent horn, critically confronting the extent and seriousness of Hare's intervention beyond the fetters of the verbatim fallacy. It is important to stress at this juncture that the playtext remains unmarked, less manageable and stable than the scrupulous tribunal plays in Chapter 2.2 or Tanika

---

48 The massaged verbatim strand of practice is arguably more problematic than the "pure verbatim" format and it may be the case that a clarification regarding its processes is seen as more urgent. An extreme example of the need for such practices occurred in the US in 2012 when Mike Daisey was found guilty of having misled the public regarding the documentary nature of his work, *The Agony and Ecstasy of Steve Jobs*, and had to subsequently apologise on air. See Madison (2012) for a more thorough overview of the case and what is seemingly at stake as regards the documentary art form in a wider context.

49 For instance, we are told when the words are less likely to be verbatim such as when "An Actor" explains that the quote from George Tenet, Head of the CIA was uttered "in private" (64).

Gupta's *Gladiator Games* (2005) encountered in the previous part, and one may readily concede that, logically, it might be expected to virtually attempt to hold both horns at the same time if at all possible. In other words, the spectator is both forced to believe and not believe what he or she is seeing. Without much protest, this makes it possible for Charles Spencer of the *Daily Telegraph* and for a substantial number of the London audience – *dixit* Hare – to claim that "going to see the play [is empowering]" (qtd. in Smith "Feature").

There are quite specific and complex issues to be faced by the audience in relation to this unrestrained and elusive combination of verbatim and fictional materials, which is yet reluctant to completely let go of the verbatim nomenclature and its freshly assigned security guard, dramatic realism, that makes everything digestible. As I have argued, David Hare makes an appeal to our complicity, explicitly to what we already know, but more implicitly, the play depends on an appeal to realist conventions shared between audience and theatre-makers (developed outside of the particular framing discourses of the performance under analysis in this section). The well-worn case of the appearance of the playwright in the performance, (and this is the third specificity of the contract – in *Stuff Happens* disguised under the name of "An Actor" and in David Hare's *The Power of Yes* more openly as "The Author") – primarily appears as a violation and marring of the unwritten verbatim theatre contract before seamlessly merging into the unified texture of the massaged narrative. It follows, I think, that Hare intends with *Stuff Happens*, first and foremost, to play with both the official narrative of the key precursory events to the Iraq invasion and the closed terms of verbatim theatre that it disputes.

It would be foolish not to recognise the successful comic side-effect of Hare's authorial interference and one may speak of a comic *verfremdungseffekt* that reminds one that one is a theatre audience member. Clearly, the kind of humour that is generated by these ongoing comments on the verbatim material – such as the following at a press mob with Dominique de Villepin, Colin Powell and Donald Rumsfeld – cannot be overlooked:

De Villepin Gentlemen, gentlemen
An Actor The incident is known in diplomatic circles as 'the ambush'. [...]
De Villepin We believe today that nothing justifies military intervention. Military action is a dead end. Nothing justifies an American adventure. Nothing! Nothing! [...]
An Actor In response, the American people go into a frenzy of French-bashing. French tourism, French wine, French fries. [...]
An Actor Happy to see the row escalate, Donald Rumsfeld fans the flames when asked about European dissent:
Rumsfeld You're thinking of Europe as Germany and France. I don't. That's old Europe. [...]
An Actor As a deliberate provocation, a few days later he proclaims:

> **Rumsfeld** There are four countries that will never support us. Never. Cuba, Libya and Germany.
> **An Actor** Asked to name the fourth:
> **Rumsfeld** I forgot the fourth. (95–97)

Back in 1998, David Hare assumed the role of "David Hare" in *Via Dolorosa* among the thirty-four characters he portrayed and seemed to have happily reprised the part in *Berlin* (2009) and *Wall* (2009); in 2003, a certain "David" was mentioned in passing in *The Permanent Way.*[50] In Hare's radio play, *Murder in Samarkand* (2010), the narrator Craig Murray tells us the following at the very beginning of the play "My name is Craig Murray, I used to be a diplomat [...]. My story starts in 2002, five years later, I wrote a memoir. What you are listening to now is a version of that memoir. Just to be clear, the usual disclaimer: these are the facts but some of the names have been changed". This is a process that was to reach its apotheosis in David Hare's verbatim piece *The Power of Yes* with the character of "the author" – this time played by Anthony Calf in the National Theatre production – disclosing his kinship with the real-life David Hare as early as Scene Two, casually referring to "his previous verbatim plays, *Stuff Happens* and *The Permanent Way*" (4). A similar device is detectable in Gregory Burke's *Black Watch* with the often-ridiculed naive character of the "writer" who is seen "*getting a tape recorder from his bag*" and conducting interviews with the soldiers throughout the performance.[51]

Now I would like to suggest that, in the thick of the fiction, these deliberately planned contradictions only work within the predominantly realist narrative directly borrowed from an undisputed but known and selective reality. In fact, one is very much inhabiting a realist structure which imposes both coherence and control over what could have appeared at the time as an otherwise more chaotic sequence of events. With the benefit of hindsight, Hare employs the familiar dramatic devices that taint the verbatim material with a vivid sense of the arc of the story and of inevitability, as well as massaging it into a satisfying narrative progression that unfolds towards a climax and conclusion, as the mainstream playwriting conventions intended.

---

50 In an interview with Richard Boon conducted in 2002, David Hare plainly claims that "Stephen Daldry and [himself] had created a character called 'David Hare' who wasn't actually [him]".

51 At a public Q and A at the New Zealand International Arts Festival in Wellington in March 2008, director John Tiffany admitted that the character of the playwright in *Black Watch* was in fact "modelled on David Hare" (Carney 303).

More understandable in the light of Hare's statement in his Author's Note is the *de facto* connection between this kind of verbatim theatre and the established genre of historical fiction that Bernhard Klein calls "[v]erbatim history plays" (218) and Amanda Smith refers to as "instant history play". As Klein and Smith seem to suggest here, verbatim theatre cannot perhaps be readily assimilated to the genre of the history play, being devoid of a safe historical distance. Interestingly, American verbatim playwright Anna Deavere Smith writes about her understanding of this particular genre in the following manner: "[h]istorical plays, as *reorganizations* of history – interactions *with* history, reactions *to* history, metaphors *for* history" (her emphases, *House* xvii). Described as such, the genre has clearly not very much to do with a documentary realist representation of verbatim history that would leave history to unfold seemingly of its own accord and therefore appears more attuned to my conceptualisation of "new realism" that abhors the naturalisation of what is not natural. The massaged verbatim strand, as exemplified by David Hare's *Stuff Happens*, constantly interrupts the action of the play, objects to what the play is saying and sometimes even to the fact that the play is going on at all. This in turn begs the question whether these devices are robust enough to strip the dazzling aura of fetishism that attaches itself to any performance that bears the name "verbatim" or whether they merely serve as a passing distraction from the main issue. Such strategic interactions with the verbatim historical material deserve one's utmost attention and will be the specific focus of the next section.

3.1.6 Staging

What one has here, is a largely bare round plate of a stage in which Brechtian elements systematically reframe the verbatim material, even when the piece purports to be simply re-creating the "original" context.[52] Contrary to the documentary realist tribunal plays – studied in Chapter 2.2 – that gave the impression of the real places, the decors in *Stuff Happens* appear somewhat estranged. One only witnesses in this production some suggestions of location on stage by means of chairs, tables and desks to represent the White House, 10 Downing Street or the United Nations. They quote the real places such as the Congress, the UN or various press briefing rooms but do not fully attempt to replicate

52 Having missed the performance when it was on in London, I am basing this analysis on the video recording of the NT production made on 4 November 2004 and available on demand for consultation at the National Theatre Archive in London.

them. However, the Brechtian devices used in the National Theatre production such as music, lighting, and visible scene-shifting are not the exclusive preserve of the performance but come directly from the pen of David Hare, as one is all the more apt to suppose after a consultation of the published playtext.

In fact, director Sam Mendes explicitly writes that there are two strands in Hare's work, the Chekhovian strand and the Brechtian one, arguing that "Stuff Happens would be among [the latter]" (qtd. in Wood). What is more, Hare is fully cognizant of Brecht's work[53] through his university studies "with a Marxist, Raymond Williams" (Hare qtd. in Eyre 183), having adapted later on *The Life of Galileo* (1994) and *Mother Courage and her Children* (1995) and being part of that generation of dramatists growing up "in awe of Brecht" (Hare 2009, 20).[54] His *Plenty* (1978) was also "a very English-styled aestheticization of the Brechtian epic form" (Deeney 434) and both *England's Ireland* (1972) and *Brassneck* (co-written with Howard Brenton in 1973) "us[ed] actual documents in [their] staging" (Homden 1995, 29). More importantly perhaps, his *Fanshen* (1975) is described by Richard Boon as no less than "the most Brechtian of modern British plays" (29). The late William Gaskill, who worked with David Hare on this Joint Stock Company production, wrote the following regarding its Brechtian process: "It was a fulfilment of a process started in *The Caucasian Chalk Circle* [...], the process in which the actors share an understanding of the political responsibility of the play, they are not just there to serve the writer but, together with the writer, are making a statement" (136). In terms of the broader picture that interests us here, not only was Hare involved in a Brechtian visual style but he had a profound understanding of his theatrical process, socio-political purpose and world view. In an interview with Richard Eyre, he confessed that "[Brecht's] influence on [him] is simply that he's the great exemplar of politics being an adult subject, but secondly, he's an exemplar of seeing historical incidents as being incredibly complicated and not reducible to simple ideas" (Eyre 184). Perhaps more subtly, Hare's acknowledged intention to make the spectator see the ordinary "order" in which one lives in a new light, question the familiar, everything that she has internalised as such, reminds one of Brecht:

> You may think you know about something. [...] The paradox of great factual work is that it restores wonder. Thinly imagined work takes it away. 'I never knew that, I never realised that, I never felt that' is what you hear from departing audience when their evening has been well spent. Because we think we know, but we don't. (Hare, "Mere" 78)

---

53 Hare is also reported to have directed Theatre Workshop's *Oh What a Lovely War* as an undergraduate in Cambridge (Homden 47).

54 See Unwin, *Guide* 246.

To come back to *Stuff Happens*, I would suggest that such an infusion of Brechtian techniques can be viewed in the presence of actor-narrators, a casting partially conducted against type,[55] the multiple roles occupied by some actors[56] as well as an episodic structure in Brecht’s lineage.[57] Indeed, in Act I, Scene 11, one minute Bush is attending the graduating class of 2002 at the United States Military Academy at West Point, the next he is at the White House with Powell and Rice. Furthermore, the opening line of the play pronounced by “An Actor” – “The Inevitable is what will seem to happen purely by chance” (3) – seems oddly to echo Brecht’s famous words in his essay “Theatre for Pleasure or Theatre for Instruction”: “[w]hen something seems ‘the most obvious thing in the world’ it means that any attempt to understand the world has been given up” (Willett 71). Having said that, one should note that Hare’s Brechtianism is in fact very much confined to the role he has assigned to the actors, that is to say, to the kind of acting his massaged verbatim piece actively demands.

3.1.7 Acting

The first observation to be made about the acting in *Stuff Happens* is that the performers assume and do not assume their dramatic characters, forcing the audience to conceive of different relationships between the actual and the theatrical, which is typical of the “new realist” breed of verbatim performances under discussion. Thus, as I see it, the aesthetic of new realism accords more importance to the verbatim material’s relationship with the audience. However, the way in which the deployment of these particular acting processes audaciously transcend both the conventions of dramatic realism and those that have been identified as characteristic of documentary realism in Chapter 2 needs to be addressed in more detail.

---

55 Actor Alex Jennings was cast against type as George Bush and actor Nicholas Farrell in an interview with Jonathan Croall – admitted his hesitation when being offered the Blair role: “I’m not a mimic in any conventionally accepted sense and I don’t look at all like him” (Croall 219).
56 The cast of 22 actors portrays 49 characters. For instance, actor Philip Quast is portraying George Tenet (Head of CIA), Maurice Gourdault-Montagne (adviser to French President Chirac), Michael Gerson (US presidential speechwriter) and John Negroponte (Ambassador to the UN).
57 If in many respects this staging can be placed according to a Brechtian genealogy, it should be noted that not all scenic elements corroborate such a reading as the transitions between scenes are rather fast and slides are not used. Similarly, actor Alex Jennings attempted to get Bush’s “Texan accent right and […] the way he moves physically” (Croall 220).

Hare appears to have borrowed some of the defamiliarising Brechtian strategies and subjugated these to an unabashed verbatim frame of significance so as to create a new kind of practice that draws the audience into the inner workings of verbatim theatre-making, countering the opaqueness and impenetrability of documentary realism that had hitherto made verbatim theatre such a problematic category in the eyes of many critics.[58] At its core is a specific form of quotation that differs from the more traditional understanding of the term and goes beyond the simple reproduction of a pre-existing interview, extending to the perceived aesthetic gesture of the verbatim theatre-makers in their recycling of the spoken words.[59] In *Stuff Happens*, the verbatim quotations punctuate the narrative and each time an actor steps forward in a scene of direct address, another actor sets forth to caption the verbatim material and provide complementary information that is more often than not coloured by Hare's comments. When the actors address the audience directly, these are verbatim quotes (according to Hare's authorial note) and they tend to be introduced by a brief presentation of speaker and context:

> **An Actor** September 12th 2002. Gridlock on the streets of New York. A motorcade glides towards the UN headquarters on Fifth Avenue. In his hand, George Bush has the bitterly contested text of what some say will be the most important speech of his life. [...]
> **An Actor** Bush prepares to address the General Assembly, in the presence of his Secretary of State...
> *Powell puts on his headphones.*
> ... who uses headphones even though they speak a common language.
> *Bush steps up to the podium.*
> Bush makes an early claim:
> **Bush** Should Iraq acquire fissile material, it would be able to build a nuclear weapon within a year. (64–65)

In other words, the verbatim material is announced as quotation in a manner not dissimilar to Brecht's conceptualisation of the epic actor – "the actor-cum-commentator, or 'spectActor'" (Mumford 21) who "used a somewhat complex technique to detach himself from the character portrayed" (Willett 121). Indeed, the ac-

58 Here I do not claim, as others have done, that Hare's practice is completely innovative but that it reinvents the "broken" tradition to which it belongs, offering some tentative bridge-building between the origins of verbatim theatre I have briefly sketched in Chapter 2 and its contemporary manifestation.

59 Despite Hare's claim that the distinction between verbatim and non-verbatim material is clearly marked – "[s]cenes of direct address quote people verbatim" (vii) –, the oscillation of verbatim theatre towards the realm of fiction in the massaged practice indicates a departure from traditional understandings of quotation.

tors do not represent “real people” as such but present them, using a classic Brechtian exercise and structure that clearly separate the performers from their characters’ speech:

> **An Actor** When asked on February 2nd 2004 by a reporter –
> **Reporter** Do you think the country is owed an explanation about the Iraqi intelligence failure before the election so the voters have this information before they elect a new President?
> **An Actor** – the President replies:
> **Bush** First of all I want to know all the facts.
> **An Actor** Asked in 2005 whether the Americans are winning the war in Iraq, Donald Rumsfeld replies:
> **Rumsfeld** Winning or losing is not the issue for ‘we’, in my view, in the traditional, conventional context of using the words ‘winning’ and ‘losing’ in a war.
> **An Actor** In the same year, Rumsfeld boasts:
> **Rumsfeld** Before the war started, I presented the President a list of about fifteen things that could go terribly, terribly wrong. A great many of the bad things that could have happened did not happen.
> **An Actor** A journalist then asks:
> **Journalist** Was a robust insurgency on your list that you gave the President? (116)

Clearly, this would seem to fulfil most if not all three aids corresponding to his “[n]ew [t]echnique of [a]cting which [p]roduces an [a]lienation [e]ffect” (Willett 136):

> 1. Transposition into the third person
> 2. Transposition into the past
> 3. Speaking the stage directions out loud. (Willett 138)

Upon further reflection, as Brecht intended, the actor “*shows* the character, he *quotes* his lines, he *repeats* a real-life incident” (*Messingkauf* 104). This is of course a map of familiar territory, but let us see in more detail how this works in *Stuff Happens*. In alignment with its title, the play is more concerned with how stuff happens than in how it happened in the past, so as to avoid retrospective storytelling, seemingly reacting to news and events as they happen.

Having said that, these moments are somehow affected by the fictional framing of the actor-narrator, an imaginative gap that has a significant impact on the audience’s re-reception of these words and events, soliciting them in a very unusual way and yet unable to entirely erase the verbatim contract. For a few moments, one is no longer in verbatim theatre and these breaks stand out because of their stylistic differences:

> **Bush** We will work with the UN Security Council for the necessary resolutions.
> **An Actor** The unscripted use of the plural –
> **Bush** 'Necessary resolutions.'
> **An Actor** – will unleash a process of diplomacy which will last six months. (66)
> **Chirac** My position is that whatever the circumstances France will vote 'no', because she considers tonight that there are no grounds for waging war.
> **An Actor** Chirac uses the word –
> **Chirac** 'Tonight' ...
> **An Actor** – to mean he is open to argument, should the situation change. But Downing Street senses an escape route at last. (110)

The deliberate repetitions of the verbatim words by the characters of Bush and Chirac prompted by "An Actor" suspend only for a moment the affordances of the historical and verbatim contract before the narrative resumes and the verbatim material – as always already in fragments – subsists through this temporary variation. In that moment, one witnesses even more conspicuously Hare's productive framing as it unhooks the performance from the structures of realism and acknowledges the absurdity of the situation, "An Actor" clearly detaching himself from the time continuum of the supposedly ongoing event. This slight diversion from the commonalities of verbatim theatre is typical of what I have called "new realism" as it engages with conflicting aesthetic norms.

To add to the Brechtian apparatus, Hare revels in distancing mechanisms, doubling and metatheatrical games. A good example of such practices is to be found in the way Hans Blix, an expert nuclear weapons analyst, first appeared on stage, unexpectedly confessing the following: "I was an amateur actor when I was a student. Theatre teaches you the value of collaboration, of getting on with other people. As well, of course, as being damned enjoyable" (8). Another one can be found on the same page. After having been briefly introduced to Bush's inner circle – Secretary of Defence Donald Rumsfeld, Vice President Dick Cheney, National Security Advisor Condoleezza Rice, Deputy Secretary of Defence Paul Wolfowitz – as well as to Secretary of State Colin Powell, British Prime Minister Tony Blair and Hans Blix, the actor-narrator figure explains that:

> These are the actors, these are the men and women who will play parts in the opening drama of the new century. And at their head is a snappish young man, seeking his fortune in the oil-rich Permian Basin of West Texas, who will, one day, like thirty-eight per cent of his fellow Americans, say he has been born again. (8)

Yet, perhaps even more glaringly, in Act I, Scene Seven, one is told that:

> An artist had made a jigsaw of the White House with the Bushes standing in front of it. So, after dinner, the President sat with his wife and they worked quietly, putting the little bits of

their family and their house together.
Bush sits with Laura assembling the jigsaw.
Bush That looks like you, Laura. That looks like a bit of you, sweetheart.
Laura I’ll work on the columns, you work on the people. (25)

How these could be explored politically is another matter. The potential, as one shall see, for the massaged verbatim performance to be misread dramatises wider questions around the relationship between aesthetics and ideology.

3.1.8 Political Theatre

Reactions to the National Theatre’s production of *Stuff Happens* have tended to be dominated by reactions to its factual content and whether or not it matches the verification processes and interpretation of its 1,200-a-night audience members. In effect, *Stuff Happens* was such a high profile production – besides the fact that Tony Blair was officially invited to come and see it – that it caused at the time a huge storm of excitement. In the words of Aleks Sierz, “[t]here was a real buzz in the air” (*Rewriting* 74) and it became a public event itself about matters of public concern. Theatre scholar Janelle Reinelt, who had a ticket for the opening night, vividly remembers how she was then “surrounded by buzzing spectators, experiencing a true sense of hype” (“Review” 303). To research the production, therefore, is to encounter two versions of *Stuff Happens:* the performance and the event.[60] It is also interesting to speculate why such an event was deemed “too controversial to be fully professionally staged in many US cities including Washington, DC” (Pressley 77–78). If, by Aleks Sierz’s account, the performance “was more satisfying as an event than as a drama” (2011, 75), let us further explore in what capacity this might be the case and whether the massaged verbatim signifier could have prompted this reading.

To begin with, liminal to that event, or rather constituent of it, was the fact that *The Guardian* newspaper decided to send a small army of commentators and reporters to the first preview of *Stuff Happens* on 1 September 2004. Advance reviews by non-theatre critics[61] were thus published before the play officially

60 Here my understanding of “event” differs from Josette Féral’s as she considers that a performance reaches this status though spectacular onstage scenes that “bring art, particularly theatrical art, out of its theatrical framework to create the event” (Féral 52).

61 As seen in previous chapters, this is a common feature of verbatim theatre reviewing. However, as far as Hare’s work is concerned, the same phenomenon occurred a decade earlier in the context of his trilogy of plays – on the Church of England, the judicial system and the Labour Party – at the National Theatre.

opened, breaking the usual embargo.[62] Almost as the first performances were occurring, the production was thus marketed as a political spectacle. More specifically, *The Guardian* had "assembled an expert panel, including politicians, leading commentators and a man who actually fought there, to give their verdicts live from the first night" (Cook et al.). Those invited to contribute were two MPs, one UN weapons inspector, the former head of the Board of the BBC as well as a British Foreign Secretary. This special feature in *The Guardian*, along with the response of other British papers even propelled theatre critic Charles Spencer at *The Telegraph* to begin his 2004 review by stating that:

> [s]o many members of the Special Pundit Service (SPS) have infiltrated previews to David Hare's new play that the theatre critic reporting for duty on press night feels like some hapless war correspondent arriving to cover a battle only to find that it is already over.[63]

These conflations of "staged" media events and theatre reviewing seemed to echo the reporting-fictionalising of Hare's "the actor" – and by extension the one conducted by the media that has intensely permeated the popular imagination either side of the millennium – in a perpetual feedback loop. This culminated when, passing through London, the Swedish arms-control inspector Hans Blix, even reviewed himself for *The Guardian*, not finding much fault with the overall theatrical rendering of these events, albeit admitting he "didn't quite recognise [himself] in the play." The fact that the first published and aired responses were from non-theatre critics created a media framing that heavily circumscribed the reception of the performance, thereby conferring the status of a political event on *Stuff Happens*. The walls between the real and the theatrical cracked and crumbled and in that collapse it became virtually impossible to distinguish the performance from the performative event.

Now, it is all well and good to speak of how political the event called *Stuff Happens* was, but it seems to me that one needs to critically examine both the play and the performance so as to determine to what extent they engage with the category of political theatre and augment or diminish this reading. On one level – and this may come as a surprise to the reader – David Hare's *Stuff Happens* is a "conventional-political" performance in the older serious vein, in the

62 For instance, Andrew Gowers published a review for the *Financial Times* a few hours before the official opening night of *Stuff Happens*. Arts editor Sarah Crompton explains in the context of *Stuff Happens* that normally "the established etiquette [...] allows a period of previews before an official 'first night' when the press are invited and the critics give their verdicts."

63 Theatre critic Michael Billington was also reported to have become "disgruntled at having his judgement of *Stuff Happens* pre-empted by politicians, political journalists" (Callow).

sense advocated by Janelle Reinelt and Gerald Hewitt as its "central characters or events are political in the most ordinary sense of the term – government officials or rulers, engaged in decision-making, elections, international negotiations, war and large or small scale institutional matters" (4). Indeed, *Stuff Happens* corresponds to the third function of documentary theatre put forward by Carol Martin as it "reconstruct[s] an event" ("Bodies" 13), that is to say the creation of a secondary elaboration of an event that may depart from the official master narrative of the event in question. However, I would add that such a realistic or "truthful", organised reconstruction is only infused and concerned by what people have said about the past. This is also a view supported by David Bennett who argues that "[o]ften, 'political theatre' is taken to mean 'theatre with political content', and includes plays like David Hare's *Stuff Happens*" (32). By the same logic, despite Hare's self-proclaimed anti-war position[64], *Stuff Happens* even-handedly includes the case for the opposition (so as to avoid too much of an explicit stance against the war) and seems, at the same time, to have a thesis: 9/11 was exploited by an opportunist group of White House officials to launch the Iraq invasion, a war that was in fact planned years before Osama bin Laden struck.[65] To sum up, the play is seen by many as an epitomy of "the revival of political theatre" (Sierz "Service").

On another level, Hare's political play is not about a finger-wagging condemnation or the corrections of injustice, it is not an emphatic or assertive rallying play even when Hare places himself/his process at the centre of bold stage images or verbatim presentations of "real" events. His massaged verbatim piece is a practice of doubting, of not making explicit, or rather a practice of making much more explicit the disparities, contradictions, inconsistencies, ambiguities and even outright absurdities of the global mainstream narrative. Hare's massaged verbatim assemblage is at once a provocative and political act, a sort of affront to both the traditional fictional political play (if such a thing exists) and verbatim theatre. As was seen, the performance is anything but a historically

---

64 After the Author's Note and before the text proper, David Hare uses a quote from Dwight Eisenhower that openly makes the case against the very notion of a preventive war (1). More overtly in 2003, Hare published an article in *The Guardian* where he clearly expressed his views: "[e]ven the transgression of a territorial border and the murder of innocent citizens cannot compare to what is being claimed here: the right to go in and destroy a regime, at whatever cost and without any clear plan for its future, not because of what anyone has done, but because of what you cannot prove they might do."

65 Multiple views are expressed, such as the one from an angry journalist who seems to wholeheartedly condone the invasion of Iraq or the one from a New Labour politician who embraces a more midway position.

accurate theatrical document that would proceed *as if* the historical past could appear in front of an audience without being hidden. Even so, the play is questionable and this lessened the impact on some audiences. It would seem that for some, the arguments levelled by the critics of the documentary realist strand remain in force and that verbatim theatre is not prepared to acquire its dramatic licence at the cost of factuality. It seems justifiable, however, to say at this point that the specific rhetorical procedures and aesthetic structures that I have called "new realism" largely condition a certain vision of the political reality Hare seeks to (re)present. Although they continue to work their way through the performance, opting for a massaged verbatim experience keeps the audience on their toes in a perpetual state of wondering. For Tricia Hopton, the spectator in *Stuff Happens* is never totally absorbed in the unfolding narrative if Hare's piece is treated as an adaptation, encouraging "active thought and dialogue" (27). A similar proposition is offered by Jeanne Colleran who sees in *Stuff Happens* the production of "a complex spectatorial position" (140) that "asks the viewer to look at and through surfaces more critically" (141).

The final monologue pronounced by an Iraqi exile, one of the five fictional characters in the play, attempts to convey what regime change has meant for the Iraqi people, the latter having been mostly ignored by those in power, as the Americans failed to provide the much-vaunted "liberation" and ensuing better life, not to mention the Blairite belief in "humanitarian intervention."[66]Far from issuing a total indictment of the American intervention or a simplistic assessment of the situation pertaining in a region more volatile than ever, he makes a poignant point about the responsibility of the people for political reconstruction after Hussein's removal:

> Iraq has been crucified. By Saddam's sins, by ten years of sanctions, by the occupation and now by the insurgency. Basically it's a story of a nation that has failed [...] to take charge of itself. And that meant the worst person in the country took charge. A country's leader is the country's own fault. [...] Don't expect America or anybody will do it for you. If you don't do it yourself, this is what you get. (120).

For well-versed as one is now in the workings of verbatim and documentary performance practices as well as the existing radical cultural cynicism towards the events of news coverage, it appears that somehow one's craving for certainty makes it incredibly difficult for one to accept the absence of distinctions, that is to say an ontological equivalence between the verbatim and theatrical real

66 Other fictional characters include a New Labour Politician, a "Brit" in New York, an "angry" British journalist and a Palestinian academic.

within a piece of performance. British journalist Kate Kellaway betrays this inescapable desire when she confesses in her *Guardian* article how she would be “fascinated to go through Stuff Happens to learn the provenance of every line.” To come back to Brecht – who, as a touchstone, has greatly informed my thinking in this chapter – the absence of a fully-endorsed representational apparatus as well as the overlapping of verbatim quotes and fictional comments did not offer a sound political alternative to the mass-media mediation of these real events or uncover the truth behind them in a Brechtian manner. Instead, *Stuff Happens* seems once and for all to disrupt the unity and reification of the verbatim experience and its enduring manifestation of authority as it shuffles one into different and contradictory aesthetic patterns and cultural narratives, complicitly challenging one to continuously readjust one’s views. Such an intervention into the supposedly reliable referential systems that pass for the real world puts facts and fiction into such a close proximity that it seems that the hierarchies that underlie them have been entirely reversed. In other words, the form ultimately matches the content, as Hare acknowledged a year before publishing the play:

> From the moment it was first mooted, this was, for me, the impenetrable war, the war wrapped in mystery. […] All the answers that are supposed to tell you everything […] in fact seem to tell you nothing. […] Not one single speaker with an analysis that struck to the heart, that made any sense. […] I understand no more than anyone, no more than this: at some level I believe this administration does not even know why it chose Iraq. I believe it cannot even remember the reasons. The reasons have changed so many times – at least in public – and make so little palpable sense that it is, of course, tempting to believe, as conspiracy theorists will always believe, that there is some hidden reason which is being kept from us. But to me, the more frightening possibility is this: what if no such reason exists? If there is indeed, no casus belli? (Hare “Reason”).

If the attempt to both cohere and resist an overwhelming plurality that runs through *Stuff Happens* may seem to reduce the potency of verbatim theatre, it may also be the case that verbatim theatre in its massaged format equally acquires another quality through its sweeping vistas as it ceaselessly demands of us a more flexible social imagination.

3.1.9 Conclusion

The massaged verbatim practice is perhaps the most elusive one, a frustrating not-quite-real deadlock that combines a series of structural conflicts and apparently opposite determinations: realism without documentary realism, a polyphony which is nonetheless unapologetically filtered through a writer, an inherent fragmentation and yet an overbearing urge to comply with some dramatic conventions, a verbatim real of simultaneous constitution and menaced dissolution, multiple instantiations of the fictional and yet an engagement with the tropes of verbatim theatre. In other words, these performances offer two incompatible images to our gaze and one of my central tasks here has been to partially refute this interpretation, showing instead the interpenetration of these notions in practice into an increasingly dense, ambiguous and complex art form that still makes strong realist claims and performs its "realness" in distinction to fictional plays. As previously discussed, massaged verbatim performances acquire energy, fluidity and a dynamism that is characteristic of what I have termed "new realism", an aesthetic project that allows the singularity of these practices to emerge unhindered.

Massaged verbatim theatre appears as a transitional form of theatrical practice that falls between already established theatrical categories and represents a clash of sorts that raises almost as many problems as it solves, bringing to the debate its very own set of concerns.[67] At the heart of these apparent and necessary distinctions between the practices of verbatim theatre and those of the new writing genre is the debate over the traits and the purposes of contemporary British playwriting that would seem to have favoured an oppositional narrative, highlighting the way in which compositional approaches threaten to become ideological.[68]In massaged verbatim theatre, the superficial opposition between fact and fiction is overcome and it asks one instead to pay attention to the phenomenon under study. It goes even further than traditional fictional plays as its experience is of a performance in a situation, the verbatim material being only one of the terms in this new breed. As seen in this analysis of *Stuff Happens*,

---

67 The playwright apparently relieves verbatim theatre of some of its burdens. He avoids some of the shortcomings of a documentary realist reconstruction as he can imaginatively create scenes that rely on incomplete documentation (e.g. there were no witnesses to record the meeting between George Bush and Tony Blair in the woods at Crawford, Texas that Hare recreated in *Stuff Happens*) but this often annihilates what verbatim theatre supposedly stands for, as audiences can rarely tell where verbatim theatre ends and where fiction begins. In other words, the fictional elements make the boundary and definition of the verbatim material they touch problematic.

68 For Sierz, "new writing" is to be understood as "the category of text-based theatre, where an individual playwright is at the centre of the theatre-making process" (*Rewriting* 50). In his *Rewriting the Nation*, Sierz also explains that "[n]ew writing is thus distinguished from other forms of theatre, such as physical theatre (where the emphasis is on the visual), devised theatre (here a group collaborate on a text), adaptations of books or films (where the story and dialogue depend on the source) and verbatim theatre (where the words of real people are used)" (50).

the tension between verbatim and non-verbatim material is not conceived as one between two principles entirely disconnected from each other. The experience is what matters and, as such, massaged verbatim claims a new field of response in relation to the general sense of confusion following a series of cultural crises. In other words, the massaged strand draws its power from its very difficulty.

Stuff Happens was a fitting *entrée* to the “new realist” spectrum of practice that modifies the tacit and ensconced documentary realist contract and explores its limits with panache. This study strongly argued against the view that saw in the massaged strand a mere dwindling of the verbatim impulse. More importantly, the verbatim impulse is strategically made more uncomfortable but it still accepts and performs within a restricted realist framework that endures. In effect, realism has never quit the verbatim stage but it has edged towards what I have called a new realist aesthetic that also embodies answers to its possible objections. More visibly, in Scene Two of David Hare’s *The Power of Yes: A Dramatist Seeks to Understand the Financial Crisis* (2010), the figure of the author directly responds to the issues raised about making the play, such as the one made by a Chair of Mortgage Lender: “Why tell people what they already believe? What’s the point?” (4). In massaged verbatim theatre, stuff does not simply happen but it exposes and pursues lines of action that run contrary to the verbatim impulse without being given free rein. To sum up: massaged verbatim suggests a new realist revaluation of verbatim theatre, one which partially goes against its grain.

I now want to turn to the problems raised by what is, perhaps, an even more difficult verbatim terrain, as one moves to “one of the most unusual documentary performance approaches” (Cantrell, *Acting* 168).

3.2 "Headphone-Verbatim Theatre": Alecky Blythe's *Come Out Eli* (2003)

Developments in media technology were integral to the possibility of verbatim theatre from its inception in the UK (Paget "Oral History") and this reliance on technology has arguably increased in the intervening years until today, as the period under study is seen by many as a breakthrough in verbatim techniques. Such a particularity never appears more clearly than in the rise of a new verbatim specimen – "headphone-verbatim"[69] theatre – that was to create through "cutting-edge technology" (Long 309) hitherto unattainable documentary effects on the British stage and beyond.[70] The term "headphone-verbatim" was first coined by Australian performance-maker Roslyn Oades to designate "a paperless form of verbatim theatre featuring the faithful reproduction of speech patterns" (84). This type of performance requires actors to wear headphones or earphones throughout – a device that is not typically part of the conventional semiotic equipment of verbatim theatre – and repeat a beat behind, not only the verbatim words,[71] but also the way in which they were said in the original interviews, thus differing greatly from the verbatim performances exclusively based on "transcribed tape-recorded material" (Cheeseman, "Peter" 15).[72]The reproduction is extremely vivid and detailed. If the conventional definition of verbatim theatre discussed in the Introduction seemed to revolve around a word-for-word accuracy, the "headphone-verbatim" technique brings verbatim to a level of "breath-for-breath" accuracy (commonly lost on the printed page) whereby every cough, sneeze, gulp, chewing sound, snort, guffaw, giggle, chortle, snigger, slurring, hiss, sniff, sigh, personal tic, pause, dead end, gabbling, grammatical glitch, stutter, verbal inflection, emphasis, intonation, stumble, accent, cadence and even the incessant interjection of "you know" are included within the actor's de-

69 Once again, other terms have been used to characterise the same process: "verbatim headphone theatre" (Cantrell, *Acting* 9), technique of "recorded delivery" (Cox, *Noncitizenship* 33), "recorded-delivery verbatim technique" (Blythe, *London Road* vi), "[t]he Recorded Delivery Writing and Performance technique" (De Jongh 2008) and finally, "the technique of unlearnt delivery" (Hingorani 62).

70 Peter Cheeseman who pioneered a certain approach to verbatim theatre in Britain still believed in 1998–1999 – at a time when the headphone-verbatim technique was in its infancy and relatively unknown in the country – that "[i]t is impossible to handle the material from tape recordings until it is turned into some form of writing" ("Peter" 16).

71 I only use the term "headphone-verbatim" theatre when performers wear the headphones. Another verbatim strand, the "audio-verbatim" play also uses headphones but they are instead worn by the audience and therefore constitutes an altogether different practice.

72 This process is sometimes called "trailing" amongst actors (Cantrell, *Acting* 153).

livery. In other words, the focus moved from performing verbatim what was said in interviews to performing how it was said.[73] The actors thus keep the earphones throughout to avoid smothering out or exaggerating further the idiosyncrasies or even bringing in their own idiosyncrasies. The self-imposed verbatim rules – what Elisabeth Angel-Perez refers to as "constrained theatre" ("Artifice") or "constrained writing" ("Games" 79) – are still in place in this strand: the words cannot therefore be invented to ease the dramatisation process[74] (as the massaged verbatim playwrights of the last chapter have done so "unscrupulously") and under no circumstances should the playwright attribute "the words of one person to another" (Costa). Kristine Landon-Smith also adds that the audio editing[75] has to be done chronologically (2012) but this rule is not always rigorously followed.[76] Contrary to what was seen in Chapter 2.3, the interviews tend to be conducted by a single person ("the author") instead of a group of actors and the latter have not necessarily seen or met the real people behind the recordings. In the eventuality of a revival, the verbatim audio that corresponds to the play text is always provided, as the play cannot be produced without it.[77] Also, as opposed to all the other strands of verbatim practice mentioned earlier that did not necessarily originate in the UK and in the period under study, the headphone-verbatim method is a fairly recent British phenomenon and accounting for its exact origin will therefore be the object of the next section.

73 In other words, this is the exact opposite of BBC Radio 4's *Just a Minute* where participants must speak without deviation, hesitation or repetition.

74 Alecky Blythe is so scrupulous for instance that she does not allow herself "to make a point clearly because her speakers had failed to do so" (Costa). However, she allows herself to change action or context if deemed necessary from a dramatic point of view.

75 The headphone-verbatim playwright tends to compile his or her audio script via a digital process of editing based on software such as Digidesign ProTools or GarageBand. This process also enables one to ensure that the final audio script has an overall good sound quality for the actors to use.

76 Alecky Blythe's *Cruising* for instance combined for dramatic purposes conversations that happened in a different place and at different times.

77 Similarly for the audition, a CD is sent, as opposed to a script, and the actors are explictly told to listen to it on earphones and never on loudspeakers.

3.2.1 On the Origin of a British Strand of Verbatim Theatre[78]

While the origin of any theatre style is difficult to define, the literature on "headphone-verbatim" would seem to be even more replete with a significant portion of erroneous statements that uncritically toe the mainstream line. This is arguably due to the fact that we are dealing here with a very recent type of theatre-making rather "uncharted in the academy" (Cantrell, *Acting* 168) and whose origins rest within "the little-known, little-documented Non-Fiction Theatre company" (Haydon, "2000s" 46–47). Because of British verbatim playwright Alecky Blythe's mainstream breakthrough and success, it has generally been accepted that "Blythe is the only British documentary theatre-maker to employ headphones in performance" (Cantrell, *Acting* 173), that the "style [is] unique to Blythe's company" (Dorney and Gray 212), and more problematically that she directly "learned the techniques from [Anna Deavere Smith]" (Hur 216) or that she pioneered the technique herself (Brown, "Beyond" 4, Butler "Alecky" and Innes, "Mainstream" 436). In other words, one witnesses again (albeit on the level of the microcosm), one of the most contentious features of British verbatim history as it collides with the new reality of the high-stakes of globalised established culture in the twenty-first century. It is of course beyond the scope of this volume to adumbrate or to establish a definitive list of British headphone-verbatim productions. This section will thus restrict itself to outlining in a drastically simplified way how the technique originated and developed before changing gear and determining to what extent the technique can be said to disrupt the documentary realism of verbatim theatre and is therefore a new realism of sorts.

3.2.2 From the US to the UK: 'Headphone-Verbatim' Theatre as Travelling Practice

Before shedding some light on the American origins of the technique, I will now clarify in what sense one can consider the headphone-verbatim technique as a travelling practice. First, travelling means physically leaving a place of origin to go elsewhere and this is exactly what happened to the technique which

---

78 The origin of the strand is very contested, especially if one researches American performances. For instance, the Wooster Group in *LSD* in 1984 seemed to include a process very similar to "headphone verbatim theatre": "The Group also conducted its own research, tape-recording interviews [...]. During the performance, actress Nancy Reilly listened to the tape on headphones, and repeated lines from it" (Auslander, "Concept" 31).

moved from the US to the UK, and then, from the UK to Australia and New Zealand, but the focus here will not be on these last two. Second, travelling means negotiating with cultural and social difference and the technique, with its strong focus on language and listening, exactly embodies this openness to others and to the possibility of otherwiseness.[79] Third, travelling also means being in between places, indicating a sense of instability, and I would like to suggest that the headphone-verbatim technique has a strong propensity to welcome the aesthetic experience of new realism. Now, in order to fully grasp the complex intertwining of verbatim theatre, of "the technology that gave linguists to study naturally occurring speech" (Dorney 225) and the aesthetics of new realism, one must take a slight detour and turn to the specific and remarkable performance practice of American playwright, actress and virtuoso vocalist, Anna Deavere Smith who carries out interviews herself, compiles them and performs them in her own one-woman verbatim shows.

For many critics, Smith "has invented a new genre" (Kondo 315) at the crossroads between "[j]ournalism, documentary theater, shamanism, one-woman show, play ethnography" (Kondo 316), political performance, enacted oral history and verbatim theatre. Her world renowned and startlingly original performing technique that turned the exact words and sounds of real people into dramatic monologues was elaborated in the context of a series of investigative shows entitled *On the Road: A Search for the American Character* that she started in 1982 based on the principle that she could "create the illusion of being another person by re-enacting something they had said *as they had said it*" (Smith, *Fires* xxvi, her emphasis) and that "words could also be the doorway into the soul of a culture" (Smith, *Talk* 12). Anna Deavere Smith, the author of the acclaimed performances *Fires in the Mirror* and *Twilight: Los Angeles, 1992* notably explored urban racial unrest in Crown Heights, Brooklyn, and in Los Angeles through the words of individual citizens. Her kind of theatre is illustrative of my aesthetic concept of "new realism" in the sense that she constructs "verisimilar portraiture" (Favorini, "Introduction" xxxvii) whilst "her foregrounding of performance virtuosity [...] constantly diverts the referential vector of her presentation" (Favorini, "Introduction" xxxvii). Her technique mostly differs from the now familiar case of actors playing multiple roles – Smith playing as many as 52 in one production – as her extreme attention to the individuality of voices in all of its detail

79 Anna Deavere Smith who originated the technique wrote of acting as "the *travel* from the self to the other" (*Fires* xxvi).

makes the audience feel as if her own voice is not mediating others.[80] However, contrary to the headphone-verbatim practitioners I am about to discuss, she does this without the need of the headphones in performance and always performs barefoot to "signify the idea of walking in someone else's shoes" (Smith, *House* 6). The headphones are only used in rehearsal[81] and, through a laborious process of intense listening, she is then capable of performing the verbatim words in a way that impressively recreates the musicality of the original voiceprints, that is to say the exact idiosyncrasies of an individual's speech patterns or, in Steven Connor's words, "the vocal signature we all possess" (9).[82]

These voiceprints, "including non-words such as 'um's and 'uh's [...], [...] create the scansion of the text" (Smith, *Easy* 4) following the unique grain of people's voice. Indeed, on the page, her transcriptions are reminiscent of free verse, highlighting the accidental poetry of the ordinary that one tends to miss in everyday interactions:

> Race, um –
> of course
> for many years in the history
> of African Americans in this country –
> was synonymous with community.
> As a matter of fact
> we were race women and race men.
> Billie Holiday for example
> called herself a race woman
> because she supported the community
> and as a child growing up in the South
> my assumptions were
> that if anybody in the race
> came under attack
> then I had to be there

---

80 Smith explains in an interview that she is also helped by a vocal coach in order to "get the exact nuances, down to the pauses and the 'ums' and 'ahs'. It is useful to have the coach there to help me pinpoint the moments when what I am saying differs from what I am hearing" (Luckhurst and Veltman 134).

81 A similar process was also sometimes deployed in the context of the previous wave of verbatim theatre (starting in the 60s in the UK) as "the tape rather than the transcript was used in rehearsal" (Paget, "Oral History" 331–332).

82 According to scholar Caroline Wake, the term was first used in the context of verbatim theatre by Australian playwright Paul Brown to equate "fingerprints in the sense that they are utterly unique and almost impossible to reproduce" ("Definition" 3). It also differs from the concept of "linguistic fingerprint" based on idiolect as it involves both writing and speaking (see Coulthard 432).

to support that person,
to support the race. (Smith, *Fires* 27)

Importantly in 1998, award-winning English director Mark Wing-Davey directed Anna Deavere Smith's new show *House Arrest: A Search for American Character In and Around the White House* examining the American presidency and explored the methods Smith had pioneered though a four-week series of workshops on the "press and presidency" in collaboration with her at the Mark Taper Forum in Los Angeles.[83] Immediately upon his return to the UK, Mark Wing-Davey, then artistic director of the Actors Centre in London, introduced a modified version of these techniques to British actors in Covent Garden. Wing-Davey's first workshops entitled "Drama without Paper" in the late 1990s, however, did not include a demonstration and the actors in attendance were left to "explor[e] [the technique] from scratch" (Wallinger 2015). British actor and playwright James Albrecht (now director at Fane, a production company in London), for instance, drew on the technique to make his own one-man show with the Royal Shakespeare Company in 2000, *John Wayne, Mom, Apple Pie and Other Myths*, based on recorded interviews he had conducted about the American dream whilst living in Atlanta. In the piece, he brought five people to life including two black men and one white woman. However, it was not until the creation of Non-Fiction Theatre that the verbatim technique in question would become far more adaptable and therefore make a significant impact on British theatre history.

3.2.3 Non-Fiction Theatre

Non-Fiction Theatre was founded in 1999, the company consisted then of two core members – Jo Harper and founder member Louise Wallinger – and a changing third person.[84] The redevelopment of Smith's process through the use of

83 The show differed from her previous pieces as it had a multiracial cast of 14 actors instead of having Smith portraying all the roles but it kept the spirit of the series by having actors "play across lines of race, age, and gender" (Kondo 81). Albeit a purely verbatim performance text, the material was, nevertheless, not exclusively drawn from interviews conducted by Smith and included historical documents.

84 While Non-Fiction Theatre ceased all activities in the first half of the 2000s, Louise Wallinger has continued to use the headphone verbatim technique in her solo shows since 2003 such as in *Cut me and I Bleed Elvis* (2003) and *Annoying the Neighbours* (2012). She has also used the technique in the context of One-to-Ones (solo performances for an audience of one). For instance, in

headsets in performance and the inclusion of "duologues and conversations between three people" (Wallinger 2015) were at the time unique to the group.[85] Their very first show with actress Stacia Keogh – based on people they met within 200 yards of the Actors Centre – was put on at the Tristan Bates Theatre in London. At the time, the Non-Fiction members were working from recordable Walkmans in performance, with only their own characters on tape. From a practical point of view, this means that they had to press "play" each time it was their turn.

Sex 1: Death 2 (pictured above with Louise Wallinger, Rachel Spence and Jo Harper), their third full-length work and the first piece to tour more widely, premiered on the occasion of the Edinburgh Festival Fringe at the Pleasance Theatre on 1st August 2001 in a production directed by Mark Wing-Davey. The 60-minute performance was comprised of 33 vignettes involving one, two or all three performers and featured an opening voice-over, as the technique was very much unknown to most theatregoers at the time. The cast of three – Jo Harper, Rachael Spence and Louise Wallinger – channelled the voices of over two dozen ordinary people they had previously interviewed at work or on the street (on the topic of birth, sex and death) including bikers, sex workers, mothers with prams, a midwife, funeral directors and tanning girls with the assistance of minidisc players and earphones. The piece was praised at the time by Welsh actor Jonathan Pryce for being one of "the most exciting, witty and imaginative approach[es] to theatre" (2001) and, following its success, ended up being revived at the Soho The-

---

The Dream Teller (2013), Wallinger records one audience member telling her about her dreams for five minutes and then she puts her earphones in and performs it back to her, verbatim. Lexi Strauss, who worked as a third member after the departure of Stacia Keogh, started creating pencil portraits of the verbatim subjects she used to interview with Non-Fiction (another way to integrate the energy of the other beyond the use of voice and body) – has become a fine art artist who continues to explore the headphone-verbatim technique (but within a different context leaning towards therapy, education and conflict-resolution). In effect, her work today creatively mixes the two strands. For instance, some of her paintings and still works are ventriloquised in order to tell verbatim narratives, her installations use projections that enable objects to breathe, speak and/or sing with moving mouths and recorded material overlaid. Mark Wing-Davey was also part of Non-Fiction. The other members of the team met with him to perform long edits of the gathered material and he oversaw the process whilst also directing their shows.

85 As has been suggested at the beginning of this section, some critics and scholars have made erroneous statements concerning this practice, mostly due to the relative absence of publications on the subject. For instance, Caroline Wake wrongly attributed the invention of "scenes with several speakers" ("Headphone" 326) to Alecky Blythe and Roslyn Oades, and Alice Jones wrote that "Wallinger started out working on real-life dramas with London Road creator Alecky Blythe" (10).

Figure 2: Non-Fiction Theatre's *Sex 1: Death 2* (2001)

atre and at the BAC in London in April 2002.[86] The same year, theatre critic Maxie Szalwinska would perceptively write about their new show on the topic of deception, *We Haven't Said a Porky Pie Yet*, that "it could open up all sorts of possibilities for investigative and documentary theatre" (2002).[87]

Typically, the headphone-verbatim actors within Non-Fiction Theatre, armed with a Dictaphone (instead of a portable tape-recorder), in a manner not dissimilar to the one come across in Chapter 2.3 (in the context of Robin Soans' *A State Affair*) conduct interviews (sometimes in groups of two or three) that are to be edited onto minidisc players. Minidisc players were used at the time as the technology was less sophisticated than today. In performance, the trick for the actors is not to look as if they are listening to the disc. To further distract the audience

86 I am reproducing the quote as it appeared on the original promotional flyer for the show. Non-Fiction was then using recordable minidiscs and binaural microphones, as the technology developped.

87 "Pork pie" or "porky pie" – often shortened to "porkies" – is Cockney rhyming slang for "lie." Therefore "telling porkies" means "telling lies." The title of the show is a quote from one of the interviewees in the piece that fits this particular performance very well, but also what Non-Fiction's work in general was about in the sense that we all think we speak the truth but we are – to some extent – all unreliable witnesses. The subtitle of the show, "Deception: Self: Deception", was chosen by director Mark Wing-Davey to convey the idea of the lies we have to tell ourselves in order to deceive others.

from the technology, the audience, and at times one audience member only, is sometimes cast as the interviewer. This is noticeable for instance when the interviewee suddenly asks a question or challenges the interviewer on something. In their 2003 piece, *See-Through Soho* for instance, Louise Wallinger plays an ex-criminal who says "You're not a policewoman, are you?" and then pauses before saying, as she looks at the audience member, "I don't care. I don't care" before singing "I should care, I should care".[88] When it comes to duologues or three-way conversations, the headphone-verbatim performers have to make sure they are listening and responding to the other actor as well as listening to the disc to avoid responding to the other actor's character on the disc before the actor has actually got to that point. For some actors that have tried the technique, this may constitute an argument for leaving the earphones behind in performance. However, for Louise Wallinger, it is instead an opportunity for paying closer attention and being as exact as possible.

Another important characteristic of Non-Fiction's verbatim material was that they always tried to interview people while they were doing something to avoid having a mere tale told in retrospect. Sometimes these things happen by chance when the interviewee is taking a phone call – as was pointed out in Chapter 2.3 – or when a noise or some music is heard in the background and the interviewees start reacting to it or even when another person comes along while the interviewer is there. A last point to be made about the technique is that it changes both the nature of dramatic dialogue – as one shall see more clearly in this chapter – and, to some extent, that of "normal" everyday dialogue. For instance, Louise Wallinger told me in an email sent on 30 July 2015 that, sometimes, the interviewees were competing for the interviewer's attention and not listening to each other, a feature that could be seen as highly theatrical when transposed into a performance setting since the headphone-verbatim performers are seemingly competing for the audience's attention. To reproduce such "scenes" verbatim with as much exactitude and timing would be virtually impossible without the earphones, as actors might inadvertently listen to each other too much.

Now, it might be time to come back to a name that has already been mentioned on several occasions without a proper introduction.

88 All quotes are from a private email I received on 30 July 2015 from Louise Wallinger.

3.2.4 Alecky Blythe

Alecky Blythe, whose name appears on numerous occasions in the following pages of this volume, started her career as an actress and was initially found struggling to attract suitable roles for herself.[89] She thought at the time that her only option would be to write these roles herself, but she lacked experience in the field and was unsure about how exactly to go about it. One day, however, she signed up for British director Mark Wing-Davey’s influential course “Drama Without Paper” at the Actors Centre in the first half of 2001, an experience that would genuinely kick-start her artistic career, have an impact on the conventions of verbatim theatre-making and even on those of mainstream British theatre.[90] From that experience, she discovered both playwriting and acting in a new way and it was not long before she would start experimenting with the exciting possibilities of verbatim theatre.

2003 would reveal itself to be a crucial year for both British verbatim theatre and Alecky Blythe. Indeed, that year she was asked by Mark Wing-David to join Non-Fiction Theatre as part of the creative team/performer for *See-Through Soho* after the daparture of Rachel Spence. The performance, directed by Mark Wing-Davey and whose focus was on the memories of Soho’s less visible residents (gangsters, bingo players, rickshaw riders, etc.), premiered on 5 November 2003 at the Soho Theatre on the occasion of their fourth Writer’s Festival. *See-Through Soho* looked into the diverse club scene Soho is known for, such as Ronnie Scott’s place, Private Members clubs and strip clubs but also the less obvious clubs connected to the community that live there (Weight Watchers, the over-sixties club, etc.). Blythe then went on to create her own theatre company, specialising in headphone-verbatim productions, Recorded Delivery.

The same year, *Come Out Eli*, “her breakthrough piece” (Rebellato, “Writing” 159), which will be the main focus for the remainder of the chapter, was the first play she created experimenting with the technique while being mentored by Mark Wing-Davey and looking for material on the topic of fear.[91] At the time,

---

89 In fact, Blythe describes herself as a “struggling actor” (162) in her first play *Come Out Eli* (2003).

90 Roslyn Oades also signed up for the class in the 2000s and it sparked an entire career in headphone-verbatim theatre in her native Australia.

91 Alecky Blythe has to date produced a total of thirteen headphone verbatim plays : *Strawberry Fields* (2005), *The Days of All Days* (Flight 5065 Festival, London Eye 2005), *All the Right People Come Here* (2005), *I Only Came Here for Six Months* (2005), *Cruising* (2006), *The Girlfriend Experience* (2008), *Do We Look Like Refugees ?!* (2010), *London Road* (with Adam Cork, 2011),

she followed very closely the guidelines she had learnt from Wing-Davey's *Drama Without Paper* workshop which involved two main ideas: "choosing a good talking point and going to interview a wide variety of people on the subject" (Blythe qtd. in Forsyth, *Anthology* 118). However, she departed slightly from the kind of theatre the workshop seemed to produce and that was "more of a collage of anecdotes and fascinating stories" (Blythe qtd. in Forsyth, *Anthology* 118), in order to find an "overarching narrative" (Blythe qtd in Forsyth, *Anthology* 118) and structure. Today, the actor-turned-playwright Alecky Blythe is regarded as one of Britain's "chief exponents of verbatim theatre" (Spencer "Girlfriend"). This analysis of *Come Out Eli* should be understood in the context of a whole series of headphone-verbatim performances that continues to this day.[92] In other words, *Come Out Eli* is taken here as a paradigmatic example to examine the way in which the headphone-verbatim methodology can fundamentally affect the core documentary realist principles of contemporary British verbatim theatre.

3.2.5 An Overview of *Come Out Eli* (2003)

A "rough cut" (Blythe qtd. in Hammond and Steward 88) with 91 "scenes" of *Come Out Eli* was first performed in April 2003 at the Tristan Bates Theatre (Actors Centre) and Blythe was helped by playwright Judith Johnson with the editing as well as Sara Powell who directed it. *Come Out Eli* was then performed at the Arcola Theatre, London in September 2003 and won the *Time Out* Award for Best Performance on the Fringe before being transferred to the Battersea Arts Centre as part of the *Time Out* Critics' Choice Season 2004 as well as a showing in Cork, Ireland. The final 72-minute play – with its 41 characters (played by five actors including Blythe) and 72 "scenes"[93] – focuses on the surrounding community's

---

Voices from the Mosque (as part of Headlong's *Decade*, 2011), *Where Have I Been All My Life ?* (2012), *Friday Night Sex* (with Michael Wynne, 2013) and *Little Revolution* (2014).

92 Besides the UK practitioners and performances that are mentioned in this part, many others have used the headphone-verbatim approach, such as Dan Murphy in his *Jaywick Sands* (2015), Curious Monkey in their *Leaving* (2016), Lizzie Roper in her *Peccadillo Circus* (2006) and even Philip Ralph (the author of the verbatim play *Deep Cut*, mentioned in Chapter 2) performed in a headphone-verbatim performance, *Doin' Dirt Time* in 2013.

93 In the piece another actress, Miranda Hart, plays Alecky Blythe and the real Alecky Blythe plays 11 characters in total including a research criminologist (a tall man), a character called "toffee woman" (a West Indian grandma who continually chews toffees), a Londoner called Mary, Mrs Field (a lady in her early sixties), Anthony (a middle-aged theatre designer), a KFC Manager, a French girl in her mid-twenties, an SM19 army officer in his late forties, Hetty (a char-

reaction to the sixteen-day Hackney Siege (the longest armed siege in British history) and its aftermath in the East End of London as it unfolded just after Christmas 2002, on Boxing Day.[94] In other words, the work was less about the tragedy and more about the ripples that affected the exasperated residents as roads were closed, people were evacuated from their homes or trapped inside their own homes, severe restrictions on movement were imposed, the press and other interested parties were increasingly present, specialist armed officers were deployed in a round-the-clock operation, nightly air-raid sirens and shots were heard, and shops could not function as normal during the otherwise busy Christmas season. The conversations with policemen, pensioners, a politician, a publican etc. were mostly recorded at the cordon of the siege in the Graham Road area of Hackney via a clip-on microphone and minidisc recorder by Alecky Blythe who happened to be a Hackney resident herself. Besides the residents' ongoing quest for compensation, the unfolding of the story seemingly takes a sinister turn when "the character" of Alecky Blythe is being chatted up by the traumatised hostage who then tries to blackmail her into unwanted sexual intercourse in exchange for an interview for her verbatim play.

For Caroline Wake, *Come Out Eli* is a "social crisis play" ("Headphone" 325) – one of the three dominant genres of headphone-verbatim theatre she has identified – in the sense that the collected materials "emerge from moments of social crisis" ("Headphone" 325) "such as during or after a riot, siege, or murder" ("Headphone" 333).[95] As the title suggests, the play dramatically revolves around the collective desire for Eli to come out and stop the siege as peacefully as possible. By that point it had almost become a part of the local scenery, until the eventual denouement on Boxing Day when his body was removed from the blackened flat by the police.

The performance opens with Alecky (played by performer Miranda Hart), on the phone to the former hostage of gunman Eli Hall – a student who happened to live in the same building – as she repeatedly refuses his advances and ends on a

---

acter with a gentrified Jamaican accent), a male cyclist in his thirties with a strong Newcastle accent and a builder in his mid-fifties, Brickie (a male character with Romanian accent).

94 The siege tragically ended in 2003 with the death of 32-year-old local resident Eli Hall as he shot himself and set fire to the bedsit after the surprising escape of the hostage.

95 The other two are the social justice play that "emerges out of a more general concern about social justice and the rights of marginalized subjects – migrants, the elderly, and sex workers among others" (Wake, "Headphone" 333) and the social portrait play that "emerges from a playwright's general interest in a person, place, or community" (Wake, "Headphone" 333).

rather comic note as he tries again, rather awkwardly, to receive her favours after the interview has been conducted.[96]

3.2.6 The Contract with the Audience

There is something about the making of a headphone-verbatim performance that unavoidably affects the production and reception of the finished product. Like the new realist contract of Hare's *Stuff Happens* discussed in the last chapter, the headphone breed of verbatim theatre presents a more ambiguous relationship with its audience. First of all, one expects absolute accuracy as "[n]othing [...] is written or made up" (Blythe, *Revolution* 5) and that the performance will be exactly the same every night. Indeed, even the "corrupting" impact of the spectators is reduced to a minimum, as within this approach it is virtually impossible for the actors to accommodate an audience's response or bend to the liveness of theatre whilst wearing the headphones. Second, in contrast to my first remark, the author in these pieces does not appear as a mere remote observer or witness but as a fully-fledged participant, very often making it into the character list in the performance. In the case of Alecky Blythe's productions, this reaches another level of complexity, as she tends to also act in her performances. Third, each performance might be delivered to an audience, some of whom could happen to know the "real" people portrayed by the headphone-verbatim performers even though these are rarely people of repute.[97] Fourth, the means by which the verbatim conventions work are never hidden from the audience, so that they are constantly reminded that they are seeing a different kind of performance. Fifth, as a result, contrary to the documentary realist breed of verbatim theatre that was only interested in what the performance was referring to, the headphone-verbatim strand seems equally interested in how such a reference is being made in the first place, thus powerfully restoring the double nature of theatre that tended to be silenced within the aesthetic framings discussed in Chapter 2.

---

96 Alecky Blythe never plays herself on stage in *Come Out Eli* and impersonates instead different characters in the piece. At one point in the piece she even plays the role of a builder who talks to the character of Alecky Blythe.

97 Contrary to the tribunal plays, the real people behind the verbatim words are ordinary people who simply happened to have crossed the path of a verbatim theatre-maker. To my knowledge, only Alecky Blythe's *All the Right People Come Here* (2005) featured a celebrity in the person of Roger Federer.

In effect, the observed "new realist" framing of audio verbatim material consists in a series of aesthetic strategies that are mutually interdependent, albeit with their own specific effectiveness, fleshing out each headphone-verbatim performance. The differences between this particular variant of realism and documentary realism can be partially traced to a different performance contract with the audience that marks out the boundaries within which it exists. First of all, headphone-verbatim performances tend to be very clear from the start regarding their process. In her latest headphone-verbatim play *Little Revolution* (2014), Alecky Blythe (who acts as herself) starts by telling us the following: "My name is Alecky and I made the play. The play is all created from real-life conversations that I had with people at the riots."[98] In *Come Out Eli*, the exact contract is less formally displayed although a few minutes after the start of the performance, the character of Alecky – on the phone to the former hostage – explains to him that: "All [she] require[s] is to get [his] story, record[s] it, and that's what the actors work from in the show, they hear it playing through earphones and they copy [him] exactly" (134).[99] Blythe used a different, but equally effective stratagem in *The Girlfriend Experience* (2008), the first lines of the play conjuring up her verbatim methodology:

> ALECKY (*voice-over*). I feel like I should explain – what I'm doing with m-microphones an' stuff like that [...] I kindof make (*Beat.*) – um (*Beat.*) – they're sortof documentary plays. (*Pause.*) But – I don't – *film* anything (*Beat.*) – I just re*cord* – hours and hours of-of – audio. (*Pause.*) Um (*Beat.*) – and I – edit it (*Beat.*) – and then, um (*Beat.*) – those (*Beat.*) – so (*Beat.*) – people's *real* words your real words – then become the words that the actors speak in the play – and they, they – *hear* – your voice – speaking – through earphones – and then they *copy* – exactly your intonation, accent [...] And it's – it's really weird, kindof very true – obviously *so* so true to *life*, kindof thing. (5)

In *The Girlfriend Experience*, this was combined with having a visible technician, both a sound and lighting desk on stage, "conspicuous headphones" (Cantrell, *Acting* 150) and thus "avoid[ing] a naturalistic set" (Blythe qtd. in Cantrell, *Acting* 149). The audience is therefore constantly aware of the process.

In *Come Out Eli*, the minidisc players are strapped to the actors' waists, they wear visible earphones and they have to plug into each other's machines when a

---

98 Notes taken on 26 August 2014 on the occasion of the first preview of *Little Revolution* at the Almeida Theatre.

99 One may have assumed that by now the audience would be so accustomed to the headphone-verbatim practice that they would not need to be so formally reminded of the process. In fact, quite the reverse is true, as some members of the audience – despite the popularity of Alecky Blythe's work – always end up asking why the performers are wearing headphones, assuming that it is in fact a fake strategem and they are not really repeating the words.

dialogue takes place.[100] The process is also made more explicit through "metatheatrical" references triggered by the presence of Alecky as a character in the play. As Cantrell has argued, Blythe's presence is "constantly foregrounded, in contrast to the way in which Norton-Taylor and Soans striv[e] for absence" (*Acting* 148) in the performances I analysed in Chapter 2. Or to be more precise, one is to regard this strategy as one of the key features of the new realist verbatim performances that have cleared a space beyond documentary realism. Likewise, in Tamasha Theatre's *The Trouble with Asian Men*[101] (2005) – another headphone-verbatim piece, this time dealing with the issues of immigrant communities and identity – the whole process is explained to the audience by actor Amit Sharma before the performance proper starts:

> So we went round the country and we interviewed and we recorded these interviews of people from all types of backgrounds, whether they were Asian, non-Asian, men or women. And once we had finished recording those interviews, we got back together and realised we had 106 hours worth of interviewed material. So we realised that was a bit too much. So we kinda edited down [...] all these interviews [...] into an hour's worth, which is what you gonna see here tonight. Some of you might be wondering why I'm holding this black box and some of the other actors are holding theirs. These black boxes are receivers [that] [...] are linked to a transmitter. Now, when I give the signal, our stage manager [...] at the back there will press play and those recorded interviews, the edited ones will transmit themselves into our receivers up our earphones and into our ears. Now, when the actors hear the interviews, they gonna be saying exactly what they hear, verbatim. Word for word. Now, the other thing to remember is once I give the signal to [our stage manager], there's no going back. The interviews are back to back and last [for] [...] an hour.[102]

100 As the technique has been honed over the years, the actors no longer need to plug into each other's players using headphone-splitters and only press play simultaneously at the beginning of the performance. In Blythe's later shows, the audio was even wirelessly received through each performer's headset.

101 Louise Wallinger of Non-Fiction Theatre was one of the co-creators of this show. In fact, she introduced the technique to Tamasha Theatre. This indicates, therefore, that parallel to the version of the technique adopted by Alecky Blythe and those she has influenced (especially in recent years), stands another breed of work that equally deserves scholarly scrutiny. Besides showing the technique to Tamasha via workshops, Wallinger also taught the headphone-verbatim approach at Soho Theatre, Lost Theatre and to the Youth Theatre at Theatre Royal Stratford East who then went on to create their own piece in 2009, *Mad Blud*, on the effects of knife crimes in London. One of the creators of *Mad Blud*, Philip Osment, would later be inspired to create another headphone-verbatim piece with young people, *Tales from the MP3* (2013) on the story of an ordinary group of young people.

102 Notes taken from a 2006 performance of *The Trouble with Asian Men* thanks to a DVD recording, courtesy of Tamasha Theatre.

In effect, very visibly the performance starts when the performers put their earphones on and ends when they take them off. Typically in a headphone-verbatim performance, there is no intermission and the exact duration of the piece is communicated to the audience down to the very second. For instance, Alecky Blythe's *Cruising*'s total running time is 1'11'26" (2006, 61) and Non-Fiction Theatre Company's *Sex 1: Death 2* "lasts 56 minutes 33 seconds (approximately)." Attending *Come Out Eli* is therefore a very different experience from *Stuff Happens* despite the fact that they are both working within the aesthetic frame I have named "new realism." By interpellating the audience within the very structure of the listening situation, the performance asks for deep listening, listening beyond and beneath words.

It is also my contention that, contrary to what one might expect, the foregrounding of the technique serves to some extent the same purpose as documentary realism, operating in a different manner but no less effectively insofar as – in Tom Cantrell's words – "the headphones did not function as alienating devices for the spectators, but rather became a convention in performance and did not necessarily interrupt the narrative flow or the audience's identification with the characters" (*Acting* 168). One of the New Vic Theatre's staff in the post-show discussion of Alecky Blythe's *Where Have I Been All My Life* (2012), for instance, explicitly told the audience in attendance the following:

> [e]very single word that you have heard here today was spoken by a *real* person, not made up, nothing made up, nothing added. A real person spoke it. And as you can see, transmitted directly into the lugholes of the actors and repeated by them a beat behind the tape. (qtd. in Long 310)

If documentary realism drew upon what Bill Nichols calls "the discourses of sobriety" (3), the "new realist" framing in conjunction to verbatim material in performance caters for, perpetuates and legitimises a belief and hope that so called reality can at last be transposed verbatim into the auditorium. Indeed, still all too prevalent in these productions is the typical reassuring framing announcement such as the one that featured in the programme of Alecky Blythe's *Where Have I Been All My Life* (2012):

> Every word you will hear in this performance was spoken by an actual person recorded during interviews in 2010 and subsequently edited to create this theatre performance. The acting company work from the original recordings to ensure they're absolutely faithful to what was said, listening to these recordings throughout rehearsals and even throughout this performance to replicate exactly how the words were spoken, with every cough, splutter and

> mumble intact. However, this performance departs from reality in the case of one song, for which performing rights were withheld. (3)[103]

More reassuring, perhaps, is the fact that these productions tend to play an excerpt from the edited audio recording without it being remediated by the performers or that sometimes the voices on the headsets are "accidentally" audible to the audience. In this sense, the headphone-verbatim practice overlaps with the category of the "Theatre of Real People" that Ulrike Garde and Meg Mumford define as "a mode of performance that presents contemporary people, who tend not to be trained theatre performers. These people often appear live in person, or via an audio and/or visual recording" (Garde and Mumford 3).

What is more, the anticipation of the failure to be as close as possible to the original is not present in these works. In other words, it may be hazarded that the mechanical process to which the actor is subjected seems to allow reality to imprint itself verbatim upon the actor's mind – beyond his or her conscious control and into "a symbiotic entity" (Taylor, "Voice" 369) – so that he or she merely serves as a transparent conduit or empty vessel for others' streams of consciousness.[104] Interestingly, this is what Richard Schechner seems to imply when he writes that Anna Deavere Smith with whom I started, "is to some degree possessed by those she represents" (*Studies* 206). A rather crude sketch would imply that it is only through this "new realist" aesthetic framing that (some of) the means of verbatim production can be unproblematically displayed in performance, which ironically returns one to the charge. Put another way, one must understand "new realism" in verbatim theatre as a renewal and completion of the documentary realist contract seen in Chapter 2. Having said that, such a contract might raise further complications for analysis as actors, contrary to what one might imagine, do not necessarily feel entirely subject to a documentary realist aesthetic that would guide their every move and creative decision.

This stated contract with the audience alone is not, of course, enough to identify "new realism" in all its complexity. Therefore to this first criterion must be added some others that are also used to produce particular effects in performance.

---

103 Long notices how – in the context of *Where Have I Been All my Life* – the production team "tap[s] into the 'documentary' realism of the play in order to elaborate their own concerns about human development and skills quality within North Staffordshire" (311).

104 One of the actresses in *Where Have I Been All my Life* describes this process in the following manner: "It just comes straight from them, we're like, kind of mediums, if you like … 'And through and out,'" (Long 319).

3.2.7 Staging

The kind of verbatim theatre I am describing here is arguably complex and less predictable than the documentary realist performances come across in Chapter 2. It will be my task therefore to explicate in this section some of its contradictory contours as regards to the staging of *Come Out Eli.* Of importance was the fact that the piece premiered at the Arcola Theatre, which was in fact located only minutes away from where the real siege had taken place. However, unlike the aesthetics of documentary realism which would have capitalised on this to recreate the exact location of the siege, the headphone-verbatim technique dictates another kind of aesthetic experience.[105] Indeed, on the one hand within the headphone-verbatim approach, there is no fourth wall, no realistic set, only a few props and most of the verbatim words are directed out to the audience. On the other hand, sound effects, which are typically recorded on the spot, also feature in the productions to give a sense of the original location, atmosphere, time and place. For Alecky Blythe's *Where Have I Been All My Life?* (2012), for instance, sound designer James Earls-David even visited the very pub in which Blythe had conducted a number of her interviews so as to record some background noise. Now to come back to *Come Out Eli*, at the very start of the piece, the audience hears the sound of a helicopter, police sirens and someone making a phone call before hearing the first verbatim words. However, actress Miranda Hart (playing Alecky on the phone) re-enacts this by only placing a hand over one of her ears instead of using a physical phone or prop.

It would therefore seem at the outset that this particular area of verbatim practice is accomplished with a minimum of theatrical means in a space devoid of all the usual markers of place, time and dramatic characterisation. On stage, a snowman, some dirty snow and Christmas lights adequately suggest that we are in December. However, the siege cordon itself is only represented by a traffic cone, some red warning lamps and a sign that reads "siege street" while the sitting room "*of a pretty Victorian cottage near Graham Road*" (137) is reduced to one armchair, a blanket on top and a lamp adjacent to said armchair. More surprisingly, several locations in the piece seamlessly blend into one another, sometimes at the speed of a finger's touch on a computer key. By this process, Mr Henry's Barber Shop on Graham Road becomes completely indistinguishable from the newsagents and the Vietnamese takeaway restaurant on the same road or the vegetarian café nearby.

---

105 Although the audience enters the theatre through a mock pedestrian diversion, the stage area itself is relatively bare and abstract.

Rather than following the documentary realism prescription to the letter, one's awareness of this duality is constantly being triggered by seemingly contradictory signs on stage as well as constant interruptions that disrupt temporal sequence.[106] For instance, Alecky's phone call to the hostage is constantly intercut by a series of monologues in various locations. Even the stage directions are out of kilter with what the audience sees on stage whilst being closely tied to the real interviewing situation, thus placing realism in the imagination of the reader rather than the eye of the beholder. One of these towards the end of the play indicates that the character is a "*Cyclist – On his bike*" (160), yet Alecky Blythe in her portrayal of him never uses a bike or mimics one on stage.[107] Another one a few minutes later explains that the builder Alecky Blythe is impersonating is "*wearing a pink crochet hat*" (164). The hat will be mentioned several times during the subsequent conversation but will never appear visually on stage, soliciting the audience in a particular way. In effect, it would appear that the stage directions do not really maintain or fulfil their theatrical function and serve here as descriptions of settings,[108] physical locations and interviewees to evoke a sense of the real people, place, mood and atmosphere of the interview.[109] Visually, the audience has little access to contextual information and this is designed to make

106 There are also several concepts of time running conjointly in *Come Out Eli:* the real time of the performance (which is bound by technological parameters), dramatic time (in this case, how the characters progress over time through the siege until its resolution) and the real time of the interview (for instance one stage direction on page 134 indicates that Alecky is "twenty minutes into the call" which is not necessarily perceptible for the audience in attendance).

107 At the time of the first experiments with verbatim theatre in the 1970s, Christopher Honer never used any props because the changes were happening so fast between scenes, times and places that it would have been virtually impossible to do so. However, everything had to be "*mimed*" (Honer qtd in Paget, "Oral History" 321).

108 Interestingly, in another headphone-verbatim piece by British company Curious Monkey entitled *Leaving* (2016), the conventional use of stage directions is also repurposed and each performer recites them before repeating the words of a real person.

109 This is in fact rather similar to Anna Deavere Smith's own use of stage directions. However, in the case of *Come Out Eli*, the play was not initially destined for publication following the paperless form of verbatim performance the actor is subjected to. For ethical and technical reasons, Blythe never transcribes her audio verbatim pieces. However, due to her mainstream success, she secured a contract with Nick Hern Books in 2006, coinciding with the premiere of *Cruising* at the Bush Theatre in London. Importantly, *Come Out Eli* was published within Alison Forsyth's anthology of testimonial plays (Methuen Drama) more than ten years after its first production at the Arcola Theatre. For all these reasons, it is particularly difficult for the researcher to reconstruct with exactitude what the first audience's reaction might have been, especially considering they had no access to the playtext and were completely unfamiliar with this type of performance.

them focus even more on the aural aspect of the performance that sporadically divulges these contextual clues: "when a character first started speaking, the audience would have no idea where they were until, for example, he shouted 'Bill, please!' across the stage, and it suddenly became clear that he was in a restaurant" (Blythe qtd. in Hammond and Steward 98).

Within this theatre of rapid transformations, some of the scene changes draw attention to theatricality on purpose. In scene 68 entitled "Enormous Fun" (169) between Mr and Mrs Field in their sitting room, actor Don Gilet plays the part of Mr Field, a well-spoken and distinguished retired solicitor and ex-naval officer. Without any transition to the following scene entitled "That Was It" (170), the actor in question suddenly leaves the armchair to rest his head on another actress' knees so as to visibly become a child at a Christmas party. Similarly, earlier in the play and without much pause for breath, actress Alecky Blythe goes from giving a monologue as a KFC manager in a scene titled "One Sun Newspaper" (152) to reprising one of her former roles as Mrs Field. One may also argue that these are not exactly scenes in the conventional sense of the term that would involve a realist change of location and/or characters but units of theatrical time that one could aptly name "moments", as they are fragments never claiming to build a whole. Partly, this "new realist" staging is a direct result of the headphone-verbatim methodology that resists, by its very nature, documentary realism. Partly, and this is also my concern here, it is the product of numerous artistic decisions that have embraced – and even emphasised – the inherently paradoxical enterprise of creating a verbatim performance (with all the aesthetic baggage that accompanies such an endeavour within the contemporary British stage) with the recorded delivery technique.[110] Arguably, the desired "new realist" effects on stage were even more powerful at the time due to the extreme novelty of the headphone-verbatim genre.

As I have suggested, typical of the "new realist" breed of verbatim theatre, Blythe's play weaves no realist fabric on stage but this does not make it non-representational and these Brechtian devices do not allow the dramatic to disappear completely. To some extent, one might choose to call this staging (mostly informed by the absence of visual supports, a realist set and characterisation) "Brechtian", albeit perhaps not as Brechtian as Anna Deavere Smith's performances which use slides "to announce each character and to inform the audience that the words in the play are verbatim from interviews" (Smith, *House* 5); or, as

110 In this section, the main focus is on the decisions that were taken as regards the staging of the piece. At the level of the audio script, new realism can be seen in Blythe's choice to have, inserted within the narrative of the siege, her own narrative revealing the way in which she collected some of the material.

some other headphone-verbatim productions by Alecky Blythe such as *Where Have I Been All My Life?* (2012) that featured "[m]onitors around the stage [that] juxtaposed the on-stage action with historical footage of Stoke's industrial heyday" (Long 308). In *Come Out Eli*, long extension cables were spread across the stage and the earphones as well as the minidisc players-recorders were made visible. In other words, these technological devices do not represent and only *present*, short-circuiting the possibility of documentary realism. New realism, as I see it, is manifest when the real technology, body and voice of the actor in the now compete with the voice and framing of the dramatic character on stage to become the dominant locus of an audience's attention. Why, one must ask? Before attempting to answer this question, one need to investigate another crucial element of the headphone-verbatim practice that collaborates in performance to construct the aesthetic of new realism.

3.2.8 Language

Common to all verbatim productions is a strong interest in language and discourse, perhaps even a deep and splendid fascination for the aesthetic effects these produce on a stage devoid of its usual elaborate apparatuses. However, one of the main differences between the documentary realist language found in the tribunal plays in Chapter 2.2 or the documentary realist verbatim plays of Chapter 2.3 and these plays is very much the speed of language delivery that is closer to "real life", the enacting voice almost equating with the recorded voice. Indeed, for the listening audience, every vocal utterance feels lived in the moment it is uttered. Sometimes, the characters are talking so fast, they do not even finish words: "We'll take 'im out one hit. You wanee mess wi' us- we'll fuckin mess wi' you boy ... kill ya" (159). This evidently goes against the long standing tradition of theatre acting whereby actors tend to speak lines more slowly than people speak in real life and generally place themselves on a performance level that is different to that at which the audience receives the play. The point of separation between most verbatim plays and the headphone approach then comes over the complete substitution of realist aesthetic concerns for a practice that makes stage realism a unique function of language. In this sense, headphone-verbatim's "new realism" is inconsistent with conventional appraisals of realism in the theatre. Every tiny detail of human speech is left unedited, as opposed to being airbrushed by the playwright following dramatic conventions, a process that Anna Deavere Smith refers to as the "unmannering of theatre". Sometimes, this is even to the detriment of meaning, as in the following:

Alecky About what?
Hostage About the – the – the – the. You know what I mean. You don't, no need to be still, asking me back 'about what about what?' *(Quietly but seriously after a short pause)* Think about it.
Alecky What about what I wasn't?
(Beat)
Hostage You know what I mean. I don't know. (171)[111]

Admittedly, this is conversant with the understanding of "new realism" advocated by Carol Gelderman who pointedly suggests that "[t]he dialogue in new realist plays is not a convention; it is real in the sense that it is based on actual speech with all of its repetitions, silences and pauses, the slips in language which reveal ignorance and lack of thought" (360). More than any other type of verbatim theatre and especially in comparison to the massaged verbatim format, the headphone technique is characterised by a complete absence of "idiolect" which is what Keir Elam defines as "an overall style characterized by recurrent syntactic and rhetorical patterns, lexical iteration (repetition and variation of the same words and phrases), dominant illocutionary modes" (163).[112]This raises some trenchant questions as to what kind of language is actually reproduced and performed. As Alessandra De Martino claims regarding Blythe's work, "[t]his experimental theatre has shifted the attention to 'everyday conversations', underscoring the fact that in England ordinary people often use non-standard language and have different accents, and opening a window on what may otherwise be considered a homogeneous linguistic environment with no dialects" (De Martino 179). In *Come Out Eli*, the audience hears a representative sample of the diverse linguistic richness and heterogeneity among people using the English language in Hackney: a North London accent "*with a slight lisp and stammer*" (131), a West Indian accent spoken by a "grandma who continually chews toffees" (132), a woman with "*Northern roots* [that] *can slightly be heard behind London influence on accent*" (134), a boy in his late teens who speaks with a "*London 'street' accent*" (135), the Vietnamese accent of a woman who "*stutters as a result of her struggle with the English*" and her emotions, a man with traces "*of Welsh roots* [that] *drift*[s] *into his more measured middle-class educated voice*" (137), a woman with a "*Ghanaian accent which is colourful, lively and loud*" (140), a

111 In fact, the transcript/playtext of *Come Out Eli* even admits its own failure to recreate textually the audio recording. For instance, the hostage is said to stutter through "*more than can be put onto paper*" (162).

112 Having said that, the author function does not disappear completely in the case of Alecky Blythe as she is the one conducting the interviews and editing them. In a sense, a certain coherent perspective transpires even when the language has not been written by the playwright.

man with "*a Yorkshire accent*" (140), another one with an "*RP accent*" (140) and a woman with a "*Sheffield accent*" (143) to name only ten of them. This is of course what has always been at the core of verbatim theatre, its bedrock strength, at least within its British incarnations. Indeed, at the time of Peter Cheeseman and his pioneering verbatim work in Stoke-on-Trent (from the 1960s) that I mentioned in Chapter 2, phonetic annotations were given to actors to ensure that they could painstakingly reproduce the rich textures of vernacular speech: "*at that stage they were using the methodology of linguistics to actually* annotate *the stuff, so that you can see how you speak it*" (Cheeseman qtd. in Paget, "Oral History" 319).

But more than that, at least within Anna Deavere Smith's original conception of the verbatim strand, the focus tends to be towards those colourful structures that do not fall into regular metres, structures that cannot easily be accommodated in formal pre-articulated everyday language practices, a kind of language within our normative and fortressed official language.[113] For Smith, the current hegemonic signifying apparatuses create an inherited obliterating language of some sort, colonising the unconscious, determining our conscious habits of mind and thus informing one's every action. In short, it is as if a certain language were a cover-up, keeping one from true meaning as one slips into ways of saying things, into certain forms of articulacy that buy into certain power dynamics that one is not even aware of in that moment. In the words of British playwright Howard Barker, this is identified as "the decay of language into banality, both technical and naturalistic, a banality we sense to be politically contrived" ("Body" 16). Remarkably, Joe Kelleher suggests in his *Theatre & Politics* that theatre can work politically precisely "at those points where its political messages appear to break down" (42). In other words, the political dimension, or at least one of the political dimensions of these works, offers a compelling resistance to the prefabricated. Here, any silence speaks volumes and betrays the fissures and gaps that are always lurking beneath the surface. The headphone-verbatim approach thus conceived precisely looks for these points of overload and inscribes value to what is routinely discarded, giving considerable space to the turbulent idiosyncrasies and inner poetry of everyday language over "received-pronunciation" characters, remaining as close as is theatrically possible

113 Postmodern scholars have automatically dismissed such claims. Notably, Judith Butler was reported to have "looked at [Smith] with pity when [she] said [she] believed in an 'authentic voice'" (Smith, *Talk* 260). On another occasion, when Smith told Hayden White that "[she] believed in authentic voice, it was as if [she] told him [she] still believed in Santa Claus" (*Talk* 41). In any case, whether or not Smith's argument holds, it can be argued that the partial suspension of disbelief in the course of the performance becomes, as such, a political act.

to the original speakers and their quality of typicality. At the same time, the name "recorded delivery" has rendered the process self-evident and therefore "socially and politically neutral – a merely enabling private delivery system" (Shand and Knowles 3), a point to which I shall return in the later section titled "radicalism" that specifically addresses these issues.

These differences aside, one could imagine a documentary realist construction even with a headphone-verbatim technique. The author could be silenced, leaving the audience with "almost a fetishisation of 'these are real words of non-present, or maybe present, others'" (Williams, "Putting" 112). In order to assess whether this is even theoretically possible, I must now turn to another paramount element of "new realism": the acting practice. It should finally be noted here that language also prompted "a quite fascinating new mode of character creation" (Cantrell, *Acting* 168) that I shall investigate in the next two sections.

3.2.9 Acting

First and foremost, as I hope is now clear, headphone-verbatim theatre was mostly created by actors as an acting tool. In a 2003 correspondence sent to its audience, the Battersea Arts Centre in London described the technique employed by both Non-Fiction Theatre and Recorded Delivery's *Come Out Eli* as a style "based around the actors' verbatim translations of recorded mini-disc interviews." However, my remark in Chapter 2 suggesting that "acting is often considered as the worm in the apple of documentary realism" still has some weight in the headphone strand of verbatim practice. This reticence, springing from a desire to control or neutralise the semiotic surplus of the actor, is sometimes effected here, as Tom Cantrell has pointed out through the close study of Alecky Blythe's *The Girlfriend Experience* (2008), by alienating the performers from both "a complete immersion in the[ir] role" (*Acting* 168) and from themselves as actors.[114] It can also be seen in a perception of the actor as a mere conduit or vehicle for the verbatim recording, with no contribution of his or her own or any creative freedom, that the technique seems to encourage. Indeed, the fact that the technique is so exacting, requiring the actor to concentrate intensely with only a space of two seconds between listening and speaking, fosters a greater emotional and physical restraint. Actress Debbie Chazen confessed in an interview that:

114 Importantly, Cantrell's research uncovers that this was happening only at the beginning of the process and run.

> Your brain literally can't do it all at once. You've got to listen to it, repeat it, you've got to interact with the other characters by looking at their eyes and lips, you've got your props, your actions, so all these things mean you have no time or space to do anything else. (Cantrell, *Processes* 243)

Moreover – perhaps at the height of distrust in the acting process – in the most technologically advanced versions of headphone-verbatim performances featuring a sound desk, the actors are no longer in control of cueing the audio clips as it is all conducted by stage management.

Typically, within a headphone-verbatim performance, it is important to keep in mind that performers do not at any point learn their lines, as the audio recordings are not transcribed.[115]Actors therefore only receive "a running order; the names of the characters, titles of the tracks and who's playing what part" (Blythe "How") and are told very little about the people behind the recordings. Therefore the traditional "read-through" in rehearsals becomes what Lib Taylor terms a "listen-through" ("Voice" 370) and actors are encouraged to reproduce as much as they can hear. In this way, they avoid falling into their own speech patterns and thus respond in the moment to the audio verbatim recording.[116]The technique also prevents actors from hearing themselves, the audience and each other as they are "instructed to turn the volume on their headphones to maximum" (Cantrell, *Acting* 152). The physicality of the actor on stage is also to some extent heavily contrived by the audio recording that reverberates in his or her body, as clues about the physical activity of the interviewee can also be heard by the performers (and confirmed by Alecky Blythe who conducted the interview):

> What happens is when they get better and better at the technique and as they copy the voice more precisely they have to put their faces into the shape that is actually how that person's face is, because in order to make that sound... they become them physically by becoming them vocally. (Blythe "Interview" 10)[117]

---

115 When they are transcribed for publication, it is not done by Alecky Blythe herself but by her publisher, Nick Hern Books. Alecky Blythe only does the audio editing and "of that edit [she]'ll log the timecode and who said what" ("How").

116 This indicates a clear departure from other strands of verbatim theatre that typically have "the actor's own speech rhythms [...] often located in subtle balance with the speech rhythms of the original source" (Paget 1987, 332). Gary Yershon even considered that the actor had "*a contribution to make as well, and the actor's speech rhythm* [was] *actually quite important*" (Paget, "Oral History" 332).

117 This is something Alecky Blythe calls "the pointing" in rehearsal, a process whereby she indicates to the actor who is speaking, what the interviewee was doing at that time, filling in some missing information etc.

The time constraints of the technique, as previously seen, serve as an additional guarantor of authenticity for the audience who can only assume that what they are seeing performed is an exact copy of what happened.

Clearly, this style of theatre is so unique that Tom Cantrell wrote in his *Acting in Documentary Theatre* that "the process provokes performance questions of a different nature from any of the previous case-studies" (139) he had examined in the entirety of the volume. Needless to say that the headphone-verbatim technique is pulling in opposite directions and to pursue this line of argument for a moment, passing attention must be paid again to Anna Deavere Smith. It is therefore useful to keep in mind that her entire career was allegedly created against Stanislavski or rather as a significant bifurcation from his teachings. In a 1993 interview with Carol Martin, she sums it up neatly:

> It's not psychological realism. I don't want to own the character and endow the character with my own experience. It's the opposite of that. What has to exist in order to try to allow the other to be in separation between the actor's self and the other. What I'm ultimately interested in is the struggle. (Martin and Smith 52)[118]

In a letter to Dorine Kondo she even added that "[she] do[es]n't believe when [she] play[s] someone in [her] work, that [she 'is'] the character" (Kondo 96). In other words, Smith is more interested in the other than the self, in becoming rather than being and the actor's stage persona inhabiting this liminal space is neither himself or herself nor the character he or she is trying to portray, what Richard Schechner has termed the "not-not-me" (*Anthopology* 112).[119] He or she never quite transforms into this "other" and arguably sends to an audience of spectators conflicting signals:

> As spectators we are not fooled into thinking we are really seeing Al Sharpton, Angela Davis, Norman Rosenbaum, or any of the others. Smith's shamanic invocation is her ability to bring into existence the wondrous 'doubling' that marks great performances. This doubling is the simultaneous presence of performer and performed. (Schechner, "Anna" 266)

The "not-not" is seemingly the position of the new realist actor in verbatim theatre, an actor who is no longer himself or herself and, at the same time, not yet the character he or she is trying to portray. Put differently, verbatim theatre al-

118 That is to say against the US appropriation of the Stanislavski method which proceeds from the inside out: a focus on subjectivities, self-absorption and the interiority of the psyche. By contrast, Smith's technique is one that constructs characters by looking at the outside.

119 The actor is not the "real" person either, whose (edited) words serve as a basis for the stage character.

ways betrays these gaps but the headphone approach, as a new realist practice, excavates them more fully.

Unsurprisingly, few actors have worked and trained in this manner and it is abundantly clear that this process goes against the bedrock of tradition within the British conventions of acting practice and presents numerous creative challenges, notably regarding the concept of characters.

3.2.10 Characters

The intersection of language and character is at the very heart of verbatim theatre, even more so when it comes to the headphone practice. The rhythmic phenomenon I referred to as "voiceprint" earlier is as important (if not more) as what is communicated to construct the notion of characters in these works, voice being everything but a neutral medium. However, as is common within "new realism", there is more seepage into the auditorium and the relationship between actors and characters is acknowledged. Arguably, this sometimes makes it particularly difficult to distinguish where the character begins and the "real" person behind it ends. Evidently, realism is still present within the episodic performance style of headphone-verbatim and as such, these pieces rarely dispense with characters completely, albeit making it clear at every stage that these are actors putting on these characters, constantly checking in with the materiality of the performance and therefore letting the audience in (contrary to the documentary realist works studied in Chapter 2).

More specifically in headphone-verbatim theatre, all characters are built solely on the basis of the information contained in the audio recordings (the detail of the speech patterns of the interviewees and some indications of quotidian physical activity).[120]For instance, actress Lu Corfield who was in Alecky Blythe's *The Girlfriend Experience* explains the following:

> You're given everything. You can hear by how somebody speaks; the pace of them physically you can feel how they might sit; you can feel how they talk; you can feel how they lift a cup of tea. The stage directions and everything that you might read from a normal script that help you get into character – you don't have those (Shuster).

120 To use Brechtian terminology, this could be equated to *Gestus* as both a physical and socially encoded expression. Alecky Blythe tends to capture the words of real people when they are in the midst of a particular situation and/or an activity.

As verbatim playwright Amanda Stuart Fisher writes, with this kind of theatre "there is a danger that the characters become 'trapped' forever in the time location of the script" (Hubble and Fisher 12). Indeed, despite their real-life accuracy and integrity, the characters are theatrical constructions, character-effects, or a "figuration of subjectivity in theatre" (Delgado-García 2015, *Character*), not to be conflated with the real off-stage people interviewed in the creative process.[121]In the words of one interviewee that went through the process in the context of Alecky Blythe's *I Only Came Here for Six Months* (2005), the real people are "Aleckied" (Westlake). Put differently, the way the material is edited corresponds to an invented, rounded off and coherent view of a given real person serving a larger picture that is constitutive of the staged event. In *Come Out Eli*, it is quite clear that the interviewees became "types" as they went through the dramatising process that turned them into theatrical characters with coherent psychological traits within a fixed, fictional frame. One of the police officers that Blythe interviewed in the context of the research phase of her practice therefore became "Good Cop" (131) in the piece, a "[k]ind faced" (131) character "[h]appy to oblige" (131). Even Alecky Blythe's character is testimony to the very same process so as to be reduced to three main characteristics: "Blonde, middle class, thirty" (130). "Characters" in Blythe's plays often strictly follow a realist dramatic arch, ordinary people are represented as they are caught at the centre of social turmoil (the riots in *Little Revolution* [2014], a siege in *Come Out Eli* and so on) around which all their (often contradictory) verbatim narratives can converge and be organised at will towards a climatic resolution of some sort.

Nonetheless, the headphone-verbatim technique makes it possible for the verbatim playwright to deploy a great number of characters and the changes the actors undergo at one stroke in front of the audience reminds one of Brecht's A-effect, challenging the imagination of the spectators. There is, in effect, a particular way of being on stage that is almost spectral, betwixt and between, as actors do not have the sacrosanct unity or stability their single body is suggesting and mostly play outside what the industry refers to as a "casting bracket."[122] In other words, they are not exactly what Anne Ubersfeld calls a "surrogate sign",

---

121 The reproduction of the "voiceprints" is so detailed that, for instance, actress Esther Coles who played Amber in Blythe's *The Girlfriend Experience* was told by a member of the audience who knew the person behind the recorded voice very well: "You, God, there was something sooo, you were so like Amber [...] You're so like her, but you know you don't look anything like her" (Coles 15).

122 Australian headphone-verbatim practitioner Roslyn Oades even suggests that "[i]t is almost as if the audience begins to see *through* the people on stage and to instead see something real" (Trezise and Wake 7).

that is "a presence which stands for an absence" (129). For instance in *Come Out Eli*, Alecky Blythe plays a 70-year-old West Indian grandmother and Don Gilet, a young black actor, portrays a 60-year-old white man. Instead of using the verbatim recordings as a springboard for conventional characterisation, the headsets enable an exploration of the dramatic tensions between content and expression. Indeed, all actors double up – enacting sometimes multiple roles sequentially – disrupting the expectations of a conventional frame of reference for the voices securing correspondence between voice and body, and this means that they do not need the traditional costume change that traditional realism and illusionistic characterisation or optical figuration might dictate in that instance.[123]Just as different from conventional realism is the fact that actors in the piece openly break and refute the boundaries of age, class, physical appearance, race and gender so that by the end of a given sentence (perhaps even mid-sentence), they are elsewhere again and taking us there with them.[124]Indeed, in a headphone-verbatim piece such as Blythe's *Where Have I Been All my Life*, nine actors play multiple roles to enact 47 characters. In scene 63 of *Come Out Eli*, Alecky Blythe is simultaneously audience (she was the first audience of these conversations), a performer – in the role of a builder who "*walks into the conversation*" (164) – and a character, "Alecky" (to whom she talks as a performer). In this sense, this marks a continuation with performance artist, academic and playwright Anna Deavere Smith's solo shows that constantly crossed the presumed boundaries of identity through her portrayal of a whirl of American voices and worldviews. As explained earlier in the presentation of Anna Deavere Smith, for Smith it remains crucial to perform in "[b]are feet' as it 'signifi[es] the idea of walking in someone else's shoes" (*House* 6). This idea was arguably taken on board in the UK and conveyed by way of the headphones in performance, as the actors on stage appear to literally tune into someone else's voice.

Contrary to one's impression stated many times that the acting technique in headphone-verbatim theatre was an instant replay that was in some sense hermetically sealed, the presence of the headphones as controlling, meaning-making signs are in fact not as controlling in practice and can sometimes lead to unintentional humour. In attaining competence, the headphone-verbatim performer could – without changing their naturalistic delivery – make some "creative interventions" (Cantrell, *Acting* 170). This counters the idea that actors

123 In *Come Out Eli*, the actors are always on stage and never use the back stage area. In fact, they are wearing their own rehearsal clothes throughout.

124 According to David Barnett's extensive research into the Berliner Ensemble, "Brecht did not encourage rehearsing role reversals of gender with the BE, but he did entertain other ideas designed to bring out differences between actors and their characters" (14).

in this type of theatre are overly reverent to the verbatim "real" and by the same token that there is such a thing as a fixed verbatim meaning. Revealingly, actress Debbie Chazen – who has performed in both Alecky Blythe's *The Girlfriend Experience* (2008) and another headphone-verbatim piece, Kerry Shale's *Listen, We're Family* (2014) on Jewish families in London – explains in her interview with Tom Cantrell that:

> [She] realised that although [she] had to say the lines the same, the meaning behind it could be completely different. [...] When anyone says anything, there could be a hundred different readings behind it. If I say to someone 'You look nice today' it could mean you do look nice, or you look awful, and it can mean both of those things without changing the way you say it. So [she] could play many meanings, and all could be true. By doing that [she] was able to keep it fresh each night. (*Acting* 165)

Now it remains to be seen whether the ambiguity and tension characteristic of the aesthetic of new realism have any bearing on the intentional contents of the headphone-verbatim performances.

3.2.11 Radicalism

Which radical mechanisms and transactions of spectatorship does headphone-verbatim theatre as a new realist practice activate? Can a form of radicalism arise from a prominent focus on a different part of experience and truth that is sound, timbre, tone and resonance? Far from arguing that the technology used in headphone-verbatim performance constitutes in itself some sort of radicalism, my purpose here is to further illuminate the verbatim theatre of new realism.

At the present time, it is almost orthodox to think of verbatim theatre here and there as radical or, rather, for a while it was unexpectedly claimed without reserve, in a celebrative mood, *circa* 2003 that political theatre had been summoned up at last by a fix of much needed and hallowed anti-hegemonic documentary plays, that the dissident mantras of the counterculture in nuance or extremity were stirringly rising again as a sort of atonement for the previous period. This is a shaky and acrobatic version of an argument, however formidable and beguiling it may be, that inevitably rings the bell of suspicion (or insurgence) for some and that British theatre-maker Chris Goode, in particular, vocally rejected with panache on the occasion of the British Theatre Consortium's conference, "Cutting Edge: British Theatre in Hard Times" at Central Saint Martins, London, on 25 April 2015:

> There seems to be more verbatim everywhere and I wonder whether it's been one of the lubricants of the perceived movement in theatre's centre ground […]. I think I'm supposed to like the ways in which the centre ground has shifted – as it undoubtedly has – to create a mainstream theatre language that's absolutely full of the performance apparatus of the past twenty-five years of upstream practice, which has absorbed (or is somehow mimicking) those technologies and uses them to reinvigorate more-or-less conventionally scripted narrative work […]. To most observers and commentators, this seems to register as a sort of victory for, or vindication of, those leftfield practices and instincts. For me, marvellous though those works and works like them surely are, and useful though they also are in marshalling arguments about the economic virtues of art and the many happy returns on arts funding, they also represent a kind of defeat: in exactly the way that mainstream neoliberal democratic politics learns to refresh itself by harnessing the energies and assimilating the gestures of protest and resistance to which it gives rise, and returning them to us repackaged as tolerance and freedom of choice.[125]

Indeed, when a case is made for political theatre's "renaissance," one is usually told that more political performances are being made than ever before. One may have more mixed feelings – beyond an unappeased and adamant longing – about the (foundering) state of contemporary political theatre but, certainly, it was not very long before critics became somewhat disenchanted by its updated symptoms, vicissitudes and the impasse in which it found itself. Yet, it means, nevertheless, that there is much more to be said about the obdurate potential of new forms of documentary theatre in this context and how they may indicate the prospects of new social models.

Coming from a very different vantage point in comparison to David Hare, Robin Soans, Nicolas Kent, Richard Norton-Taylor, Max Stafford-Clark or even Anna Deavere Smith[126] who pushed for "the idea of a more equitable American society" (Luckhurst and Veltman 136) and was intensely distrustful of political theatre with a capital "P," Alecky Blythe appears to be less concerned with social or political themes – even when she intends to show an antidote to the media

---

125 Unpublished paper of Chris Goode's remarks regarding the state of British political theatre made temporarily available to the followers of his Twitter account @beescope a day after the conference.

126 Even though there is a sense of shifting orientation and no dominant point of view in her performances, Smith's work is often described by herself as political, notably in terms of its search for how speech causes action, its exposure of pressing social issues and our own complicity in cultural stereotypes, its resistance to racism and its sustained questioning of the media in shaping facts. One of her collaborators, Dorinne Kondo, explains in an article that Smith's intention is to "frustrate (white) audience desire for closure, for a central unifying voice, and for a humanist reading of "we're all human beings" that always ignores historically specific power relations" (316).

representations of specific groups.[127] Quite explicitly, Blythe has expressed not having "any kind of political agenda in trying to represent the unrepresented" (Blythe, "Interview" 7). This may be the case.[128] Inevitably, critics have cheerfully jumped on the bandwagon to describe the work as "always informative but only the totally naive will find it a life-changing event" (Billington "Girlfriend") or having "no politics" (Sierz "Little"). Here, however, I will beg leave to propose a doubt as such declarations in no way impede one's reception of these works nor make them less political, though they do demand a more complex and challenging articulation of the political in verbatim theatre in ways that had not been foreseen and to which one must now turn one's attention.[129] In fact, it is part of a much wider shift that gives the present study both its impetus and difficulty. As I hope to demonstrate here one can detect a more latent radicalism, even within the unremitting claustrophobic (and lucrative) refashioning of verbatim theatre into a cultural dominant, compelling an urgent and necessary rethinking of the subject with an open mind.[130] I wish to reiterate here a question already raised elsewhere: "In what ways does the headphone-verbatim form constitute a novel and radical intervention?" ("Remixing" 51). Another one followed – before I had enough grounding to pursue a theorisation of the aesthetic of "new realism": "Is it possible to reconcile reflexive and realist modes in performance and still serve an oppositional politics?" ("Remixing" 51).

Having gained a more thorough understanding of the aesthetic nature of British headphone-verbatim theatre, I would now like to attempt to explore these issues again, taking into account the various tendencies of "new realism" that seem defiantly to pull against each other. First of all, the headphone-verbatim practice foregrounds the act of listening, throwing light on this process.

---

127 In an interview with Alice Troy-Donovan, Blythe was pleased to note that her plays are "political with a small 'p'".

128 Exception must be made regarding her *The Day of All Days* (2005) commissioned by Cafedirect, the Fairtrade pioneers, as part of the African Awareness Festival, Flight 5065 at The London Eye as well as *Strawberry Fields* directed by Teresa Heskins (2005) which focused on the work of migrants in a strawberry farm in Herefordshire and that was described as a "political verbatim drama". See http://www.ashdendirectory.org.uk/directoryView.asp?productionIdentifier=20091010_62228030.

129 As I have argued in an article published in 2014, the headphone strand of verbatim theatre "presents a political intervention that is very different in kind from other strands of verbatim works" ("Remixing" 51). This is also in accordance with the remarks of Cristina Delgado-García claiming that "the absence of a specific political agenda in a theatre text or performance does not necessarily diminish its political charge" (Delgado-García, "Time" 94).

130 This radicalism also greatly differs from the one Jenny Hugues articulates as being "most clearly identified in acts of performed violence that have obliterated life" (3).

Throughout, one watches and listens to actors listening, making one aware of one's own process of listening as audience members. Moreover, as Jean-Luc Nancy reminds us in his *Listening*, "the sound that penetrates through the ear propagates throughout the entire body something of its effects, which could not be said to occur in the same way with the visual signal" (14). And it is appropriate at this point to indicate those moments when the audience hangs on every breath, starts to truly listen and when the aural stimulus gains the upper hand at this expense of the visual signal. For instance, one such moment of audioception occurs when Hong, a meek "*softly spoken Vietnamese woman in her early forties*" (136), played by a white male actor, Philip Marshall, tells the audience about her experience of the siege: "So it –the last day suddenly we heard shooting, three shots and one shot different and we heard err the police" (161). Arguably, there might be more to it than a desire to hear and understand, within a performance setting, the voice of someone whose frequency rarely encounters the mainstream wavelength, a voice that requires more attention than is commonly and culturally paid to it within our fast-paced society.

Writing in 1999 about Anna Deavere Smith's practice and her "doubling" in performance – a "simultaneous presence of performer and performed" (Schechner, "Anna" 266) – Richard Schechner perceptively argues that "[b]ecause of this doubling [...] audiences – consciously perhaps, unconsciously certainly – learn to 'let the other in', to accomplish in their own way what Smith so masterfully achieves" ("Anna" 266). It follows from his assertion that headphone-verbatim theatre as a new realist practice, at least to some extent, favours openness to the other, if not a bridge-building attitude, as well as its replication "in ordinary life" (Reinelt, "Race" 615). For it seems, in the words of scholar Jill Dolan, that such performance practices "can offer glimpses of utopia" ("Finding" 495) for spectators and performers alike. What she means by that – beyond the "gratuitous" idea that "an identification with the other [...] is [possible and in itself] truly radical" (Hall 151) – is that Smith's performance stages a possibility rarely encountered in the world outside the theatre whereby the same space is inhabited by different bodies in a state of mutual attentiveness so complete that it may sporadically foster genuine intersubjectivity across radically opposed societal groups. As a direct offspring of Anna Deavere Smith's theatre of "civic dialogue" (Cohen-Cruz 104), headphone-verbatim might be said, then, to engender a different quality of attention, a sort of outbreak of humanity of really sharing and listening. Several such moments appear, it seems to me, when language is no longer a key mechanism of debilitating normalisation.

Verbatim theatre as a potential site of subversive interventions, we recall, is characterised by a desire to redress injustices, to broadcast marginalised, less articulate and anti-establishment voices that do not repeat the status quo in a public arena. The elements that make the basic headphone-verbatim formula – with the notable exception of Alecky Blythe's *London Road* that deserves in itself a whole separate chapter – also address the political processes of subject and narrative formation as well as the problematics of identification. Furthermore, as Caroline Wake puts it very cogently, headphone-verbatim theatre is "a form that does not so much 'give voice' as 'grant an audience'" ("Headphone" 321) as the audience lets itself be transported into different languages as well as ways of being and living, discovering another way to know. In French, the word *entendre* "to hear" can also mean *comprendre* "to understand." In this sense, the act of listening reflects how one is straining towards a possible new meaning. Perhaps even, one may hypothesise, this can lead to the kind of radical hospitality advocated by Jacques Derrida in the following:

> Let us say yes *to who or what turns up,* before any determination, before any anticipation, before any *identification,* whether or not it has to do with a foreigner, an immigrant, an invited guest, or an unex- pected visitor, whether or not the new arrival is the citizen of another country, a human, animal, or di- vine creature, a living or dead thing, male or female. (*Hospitality* 77)

This would be a kind of hospitality that radically enables connections to be made and that embodies the speculative connections between the apparently disconnected, thus momentarily transforming the social sphere as it activates a place for the rehearsal of imagination, a place for new references of what could be.

3.2.12 Conclusion: Back into the Realm of Documentary Realism?[131]

Having had occasion to mention that *Come Out Eli* belonged to a kind of realism that I saw as different to the one I have called thus far "documentary realism," it is now time to draw some conclusions; for that kind of realism is riven with paradox whilst at the same time, as one critic expressed it, speaking on the subject of headphone-verbatim theatre, it "certainly raised the bar for realism in the future" (Marlowe). Clearly, this chapter demonstrated that this type of realism pushed towards fragmentary kinds of narratives, welcoming those voices in, throwing a lot of "characters" at the audience very quickly in order to explicitly

131 Caroline Wake defines the headphone-verbatim works in terms of two main modes: "the *epic* mode" (327) when it "aims for verbal authenticity and visual alienation" (327) which more or less corresponds to what we have termed "new realism" and "the *naturalistic* mode" (327) when it "aims for both verbal and visual authenticity" (327).

prevent them from thinking in terms of conventional realist pictures of characters and narratives. In *Leaving* (2016), another headphone-verbatim piece, by the young British company Curious Money, the actors even say "1, 2, 3" before tuning in to the words of their verbatim subjects.[132] In other words, it completely changed the documentary realist engagement with verbatim theatre and its lesson in suppression described in Chapter 2 and purposefully created what Caroline Wake has described as "verbal authenticity and visual alienation" ("Headphone" 327), that is to say an imbalance that is propelled from the absence of a visual correlate. Echoing how big events (such as a siege) impact on people, the audience of *Come Out Eli* was faced with too many characters to follow. In a more radical way, Tamasha Theatre's *The Trouble with Asian Men* – another headphone-verbatim piece – even let go of characters' names completely by calling the characters Woman 1, Woman 2, Man 1, Man 2, Boy A, Boy B etc. and even "Other voice" (Bhuchar et al. 19).

Much has obviously been learnt between the first experiments when the headphone-verbatim technique was totally permeated by technology and those we encounter now more than fifteen years later. Resisting the ease with which such innovations have become so familiar that the audience may be in need of yet another innovation, the headphone-verbatim strand still puzzles in its embrace of the aesthetic of "new realism." However, as the years have gone by, it has become apparent that Blythe's work became more overtly realist when she did *The Girlfriend Experience* at the Royal Court on 18 September 2008, when she was "more into the idea of having to cast to type, almost forgetting about how [she]'d done things with *Come Out Eli*" (Blythe, "Interview" 8). Such a visual realism was perhaps to be expected.[133] Crucially, this noticeable change coincided with her move "into the mainstream" (Dorney and Gray 212) which would seem to support and confirm the hypotheses in Chapter 2.1 according to which contemporary British verbatim theatre fell prey to a market-driven tendency that reproduces ideological representations and conforms to certain patterns of identity. *The Girlfriend Experience* (2008) was less interview-driven and closer to a kind of conventional "fly-on-the-wall" realism and its familiar parapherna-

132 These notes taken on the occasion of the TaPRA Interim Event on verbatim, documentary, biographical and autobiographical performance practices held on Wednesday 11 May 2016 at Northumbria University, Newcastle after seeing an extract from their performance at the event and talking to them afterwards.

133 With some variations, Anna Deavere Smith herself underwent "an alteration in her performance style" (Martin, "Witness" 85) when she reached the mainstream and Blythe admitted in her interview that (mainstream) audiences "come to see the shows because the stories are good, not because of this quirky, gimmicky technique" (Blythe "How").

lia. Through-characters made their apparition, "costumes, scenery and props" (98) became much more prominent, the number of locations and characters was drastically reduced, direct address was less frequent, "present-tense cause-and-effect dramatic action" (Lane 67) was incorporated to create "a predominantly naturalistic dramatic text" (Tomlin, *Acts* 125). Disparate verbatim narratives were thus fitted neatly together into a recognisable fixed, linear and legitimate realist dramatic structure to the delight of the "cultural tourist" (Tomlin, *Acts* 129) in search of "real" entertainment. In this sense, Blythe was reacting against her former self, thus rendering the headphone-verbatim technique safe for global exploration and consumption.

It is of course interesting to speculate, though I suspect one can do no more than that, for British verbatim theatre is a history in the making. As Blythe admitted herself in a telephone interview with Tom Cantrell conducted in 2009, one can envisage that the headphone-verbatim technique might serve again the kind of realism she pioneered with *Come Out Eli:* "[t]he way I am working at the moment is very different from the way I started, but that doesn't mean I won't go back to it" (*Acting* 142); a point I will return to later in the discussion of Blythe's *London Road* in Chapter 4.2. As I am writing these lines, another British artist, Clare Baybutt, is also experimenting with the technique towards a form that seems to run counter to the new realism I have perceived in headphone-verbatim theatre and towards a kind of documentary realism. In her verbatim performance *Seven* (2016), Baybutt focused on seven couples who were asked to interview each other via a questionnaire and record their conversations.[134]As a result, the performance creates a seamless realism devoid of the usual disrupting features of the verbatim mode such as the direct address to the audience, the presence of the interviewer or the foregrounding of the technique within the piece.

However, the type of development of performance methodology known as headphone-verbatim may have been one of the most remarkable landmarks in the history of contemporary British verbatim theatre. This approach to theatre-making made possible entirely new ways of engaging with verbatim material in rehearsal and performance, moving this type of theatre closer to the liveness of the actual event without the artificial intensity of theatrical communication as well as the typical enhancement of performance. Headphone-verbatim is arguably more than an (already) old verbatim magic given a new technological

---

134 In the context of *The Girlfriend Experience*, Blythe also placed a recording device in the parlour so that she could record the conversations of her interviewees (with their permission) while she was away. However, even though that material certainly informed Blythe's creative process, she decided not to use it in the show.

twist and has perhaps more radical implications in stage terms than one would expect. A battery of new devices also procured an experience whereby multiple interviewees cohabitated within the body of a given performer, radically departing from the assumed documentary realist contract that verbatim theatre tends to foster in the minds of audiences at a subconscious level.

Having thus far taken the term "headphone-verbatim" in its most commonly understood sense and manifestation, I shall now – in the following chapter – propose a wider perspective so as to test the theoretical insights that have been gleaned from the study in this part.

3.3 New Transgressions

Given my insistence that all new realist verbatim theatre is transgressive, one may fairly ask: what is the point of this chapter? Indeed, as seen in the course of this part of the volume, verbatim theatre has been exploded by the aesthetic strategies of new realism and is no longer what one knew at first thought. Before drawing conclusions about the contribution of "new realist" verbatim performances and how they disrupt the documentary realist signature one associate with verbatim stagecraft, the purpose of this chapter then is to incorporate some ill-fitting puzzle pieces into the mix, those performances which do not quite fit into this analysis of the new realist strand in verbatim theatre, frustrating any simple mapping of the paradigm. In what follows, I do not wish to deny that new realism is to be found in a significant number of verbatim performances, nor is it my intention to iron out the difficulties inherent in categorising such a broad range of performance activities. As seen in Chapters 3.1 and 3.2, a new realist verbatim work is everything but a tidy performance. This chapter therefore carries this discussion of the contemporary British verbatim theatrescape into even more complex realms of investigation that the previous chapters in this part had to leave dangling.

3.3.1 Alecky Blythe's *Little Revolution* (2014)

As the audience members take their seats in the Almeida Theatre, London, on a warm August evening in 2014, the central stage area is already crowded with apparently ordinary people generally chatting and socialising, one of them being the playwright. In fact, the space has been changed for this production (from the familiar proscenium) and the audience finds themselves surrounded by walls left as bare bricks with some scaffolding, whilst the in-the-round stage is only demarcated from the auditorium by some painted and plain sheets of chipboard. As the performance begins, these apparently ordinary people – who are in fact the cast – move out of the acting area instead of into it, following the usual conventions of theatre. One of them, wearing a bright green t-shirt, however, has stayed and the audience is about to understand that this is the playwright. The lights also become brighter instead of dimming, as one would normally expect in this context and there is a lot of full light on the audience so that they can see each other.

What I have described here is the introductory moment of Alecky Blythe's *Little Revolution* until Blythe says to the audience that her name is Alecky, that she made the play and it was "all created from real-life conversations that [she] had with people at the riots".[135] The transgressions that I alluded to in the previous paragraph may not seem particularly transgressive for the reader who has already experienced some verbatim theatre and/or my own discussion of the verbatim performances in Chapters 2 and 3. Perhaps so. However, I intend to demonstrate that another transgression is at play in this specific production and that it deserves one's full attention as it may contradict – or rather complicate – some of the claims I have made in the preceding chapters. Before I consider the complexity of what could be at stake for this theorisation of "new realism" in recent British verbatim theatre, it may be necessary to conduct a brief overview of Alecky Blythe's *Little Revolution.*

3.3.1.1 Overview

Little Revolution, the latest headphone-verbatim performance from Recorded Delivery, premiered at the Almeida Theatre, London in August 2014, and focuses on the aftermath of the London riots of summer 2011, particularly on the post-riot efforts. As such and similar to *Come Out Eli*, this 85-minute piece without interval (but with numerous jump-cuts and 51 "scenes") belongs to the genre of the "social crisis" play (Wake, "Headphone" 325) and also offers a new verbatim snapshot of Hackney through Blythe's particular comic "ear."[136] The title of the play is derived from the words of a Hackney barber, Colin, who is "kind of like the voice of like lot of people" (96) in the area and talks about the riots in terms of "a little revolution" (47). This idea is reinforced by the poster for the show that features a red "Keep Calm and Carry On" mug with tea and milk being smashed by a brick. *Little Revolution*, following what I described in Chapter 3.2, also has its actors wearing visible in-ear monitors (IEMs) as well as ordinary everyday clothes and a bare, rough, minimal set and design that do not call attention to themselves, comprising mismatched sheets of chipboard on the floor, a table and a few stacks of plastic chairs round the edges of the central acting area.

135 My notes from the performance of Alecky Blythe's *Little Revolution* which I saw on 26 August 2014 at the Almeida Theatre, London.

136 Caroline Wake's article in *New Theatre Quarterly* has provided the first theorisation for the headphone-verbatim genre. Within this attempt she identified three performance modes and three genres, one of them being "the social crisis play" to designate a headphone-verbatim piece that "emerges at or from a moment of acute social crisis, such as during or after a riot, siege, or murder" ("Headphone" 333).

3.3.1.2 Realism(s)

First of all, *Little Revolution* engages with a certain kind of realism insofar as the "real" people readily fall into the category of customary archetypal characters: middle-class do-gooders, opinionated bystanders, an almost comedy vicar, a visiting foreign correspondent from Germany, hoodies, the verbatim theatre-maker energetically waving her Dictaphone around. The piece is in some sense guilty of what Julie Salverson calls "an aesthetic of injury" in the sense that it crudely "reproduces configurations of power necessary to the identity of the 'injured'" (38).[137] As Blythe says herself in the published interview preceding the playtext: "On one side of the road you've got a lovely square and on the other side there's quite a big estate, so there're quite contrasting people living on the two sides of the street" (7). Second, realism also pertains to the way the story is told and structured, from "Riots Part 1 [...] On Monday the 8th" (August 2011) to the "EPILOGUE" (95) in January 2014 when the following verdict was pronounced: "And a majority say lawful killing with two of the jurors saying there is an open conclusion and there are cries of 'No!' from the Duggan family" (97). Third, one also finds a through narrative – despite the episodic nature of the headphone technique – via the story of a newsagent's shop in Hackney (at the crossroads between the Pembury Estate on one side and the wealthy residents of Clapton Square on the other) belonging to a Tamil refugee called Siva Kandiah that had been ransacked during the riots. The main narrative therefore tells the story of how a group of concerned and well-meaning residents and shopkeepers – including the Trades Council and UK retail chain Marks and Spencer's – stood together with big smiles to help Siva re-open his convenience shop, raising £50,000 to repair the damage. Fourth, the piece reduces the usual divide between actors and characters (inherent to this type of theatre) to a minimum, as it does not involve the purposeful casting against type that *Come Out Eli* exemplified. Similar to *All the Right People Come Here* (2005) on the Wimbledon Tennis Championships, Blythe is this time playing herself throughout, or rather a version of herself, instead of portraying other characters like she did in *Come Out Eli*. Fifth, Almeida Projects (Almeida Theatre's creative learning team) recruited 31 participants aged 16–74 – mostly made up of volunteers and non-professional actors – from the area (Islington, Hackney and other London boroughs) to become a community chorus that plays the rioters and appears in some big

137 If verbatim theatre is sometimes viewed as an ideal remedy to the notion of "epistemic injustice" – a concept used by Miranda Fricker (2007) to refer to a special kind of injustice informed by feminist theory whereby the relatively powerless are unfairly denied testimonial credibility – *Little Revolution* seems to have a very different agenda which may appear as ethically questionable.

group and action scenes as local bystanders or tea partygoers.[138] These "real people" almost perform their own reality and lead the audience to assume that, if they are in the show, what one is seeing on stage must be really true.

However, on another level, *Little Revolution* also engages with the oddity of new realism I depicted in the previous chapters through the display of its own contrivance. In fact, Blythe makes herself even more central to the verbatim narrative than in *Come Out Eli* and everything is more or less refracted through her. She is a constant presence, even when she does not speak, and most of her interventions at the time of the recordings have not been edited out of the final audio script, as in the following:

> SAM. There's a chopper right up there just filming everybody ...
> It's kind of. There's a couple of by, bystanders, one with an Amnesty T-shirt on, which is a bit, bit of an irony ... Ha.
> *Laughter from* ALECKY *followed by a sigh from her, then a pause.*
> Yeah I dunno whether this is – I'm not sure we should be laughing / at this. (29)

In other words, *Little Revolution* is as interested in dramatising the constructed nature of the piece itself, that is to say, how a verbatim show about the riots is made. As such, the performance text is drenched in metatheatrical references:

> SARAH./ I-I-I'm really slightly concerned that we / might –
> ALECKY./ What?
> SARAH. That we might be the worst possible / sounding middle class.
> TONY./ Hmmm. Hmmm.
> ALECKY. No no no no. / No. I mean I'm sort of interested at looking at that that tension but I'm really not tryin' to ss-
> TONY./ No cos I think it's what what we were jus' saying there [...] I'm proud to be middle class. I'm white and middle class – that's what I am. Ya know so d- I have no concerns about being seen as that. But ya know it's what we do that matters.
> ALECKY./ Yeah you're.
> Exactly, don't worry, I'm not going to stitch you up./
> SARAH *laughs nervously.*
> Okay? Okay? Okay? I mean she will obviously have to have flowers in her hair and plaits. // But ha ha ha. [...]
> TONY. But I – cos ya know, I jus sort of always been ya know proud to be middle class. (*Beat.*) Food's much better!
> ALECKY *and* SARAH *laugh raucously.* TONY *chuckles.*
> ALECKY. That's going in. (*Referring to her Dictaphone.*) (81–82).

---

138 Each half of the group (divided up into a blue and a red team) joins the show on alternate nights, as opposed to the professional cast of twelve.

In considering the aesthetic strategies of *Little Revolution*, one quickly notices that they are in effect going against a documentary realism that would "directly" show the riots as they happened (minus the presence of Alecky Blythe who conducted all the interviews) such as in Gillian Slovo's verbatim piece *The Riots* (2011), but also – and it is my main argument in this sub-chapter – beyond new realism as I framed it in the previous chapters. In other words, self-consciousness is allowed to take precedence over the narratives in a way that the "new realist" verbatim pieces I discussed never did. *Little Revolution* is seemingly trying to be a piece about the riots or the rioters and failing, thus becoming its own metaphor. This in itself is remarkable and many theatre reviewers at the time overlooked or dismissed this crucial aspect of the production, blaming *Little Revolution* for not fulfilling its promise as a verbatim piece of theatre:

> *Little Revolution* tells us nothing we didn't know about the 2011 riots; it comes late after the events; it says more about the difficulties of recording people in stressful situations than about the events themselves; it seems to have no politics; worst of all, it has no imagination and lacks any metaphoric resonance. (Sierz "Little")[139]

To come back to my main argument, *Little Revolution* invites one to feel the shortcomings of its makers through the pulse of Blythe's patterns, often going beyond the meaning of the verbatim words. Blythe's misjudgements, clumsiness, class privilege and prejudices, self-awareness of her own inadequacy and limitations are here embraced, even celebrated to some extent. In some regard, the opening sequence is startling, not only for its comic undertone of distanciation, but also for its implicit commentary on how Blythe is in fact trapped within the phenomenon she attempts to expose (the way in which the politically progressive sector of society failed to critically grasp the riots and articulate its stance):

> ALECKY (*PA*). I don't s'pose I can speak to any of you? I make documentary plays, and I record conversations on here.
> GIRL WHO SAYS NO (*PA*). I don' wanna be on TV it's / alright. [...]
> ALECKY (*PA*). [...]
> *Half-lines heard in the soundscape:*
> I'm/ – make documentary plays, can I talk to you for two // minutes.
> GASMAN./ I'm. // I'm the gas board. I'm gas board, I'm not police.

---

139 I must concede that *Little Revolution* was promoted as a piece about the riots. Its trailer, for instance, claims that the performance brings "the voices of one North London community from the streets to the stage", triggering a certain kind of expectation and this is perhaps a more significant reason accounting for the reactions of numerous critics. The fact that *Little Revolution* was not really about the rioters was considered at best misleading and, in some quarters, as one of the most repellent extremities of verbatim theatre.

> ALECKY (*PA*). Ah ha.
> GASMAN. I'm just 'ere, I'm just 'ere doing a gas job. [...]
> TURKISH SHOPKEEPER. Dis is-dis is not Turkey, dis is not U-Afrika, dis is United Kingdom, democrat's country, okay [...]
> ALECKY. Do you think – where do you think the fighting is now? Up there? / Where do you think?
> TURKISH SHOPKEEPER. I think somebody tell me, somebody tell me Clarence Road. Looka dat.
> ALECKY. Ah, oh my God, oh my God that's pretty exciting. (14–16)

Little Revolution is therefore as much about language and voice as it is about questioning how a verbatim show is made in the first place, doing away with many conventions of mainstream verbatim theatre.

Having said that – and this is a decidedly thorny argument – *Little Revolution* does this at a level that is not overt, contrary to the "post-realist" verbatim performances I shall explore in the following part. Put another way, it is possible for an audience to read the performance in terms of the new realism I described in the previous chapters.

3.3.1.3 Conclusion

Verbatim theatre is traditionally not seen as a medium which concerns itself with a robust aesthetic and stylistic questioning, these being mainly considered as a characteristic of more fictional approaches to theatre-making. Due to this division that pervades both academic and professional discourses, many assume that it is impossible for verbatim theatre to achieve these characteristics, its scope for aesthetic exploration and ambiguity being somehow restrained by its strict adherence to the exact words of real people. With this in mind, this analysis of headphone-verbatim theatre brought this type of theatre into conversation with the rich field of experimental aesthetics and the resulting tension and dysfunction, as well as the constant process of negotiation that I perceived, were framed within what I called "new realism." More specifically, *Little Revolution* departed from the alleged direction in which it had been consistently channelled and Blythe's self-conscious subversions challenged our reading and the general assumptions that underpin the method. This case study clearly exceeded the initial understanding of the aesthetic properties of headphone-verbatim theatre, revealing that the aesthetic framing of a given verbatim performance could attenuate and perhaps even outdo its inherent style of presentation as well as its spectatorial practices. Alecky Blythe's latest headphone-verbatim performance is therefore a fascinating example of how verbatim theatre can go beyond documentary realism and, to some extent, beyond new realism, towards a new kind of elusive theatrical experience that I will call "post-realism".

Now, it is time to revisit these provisional conclusions about the massaged verbatim format via a new and problematic case study – Catherine Grosvenor's *Cherry Blossom* – so as to help one identify more clearly the bounds and limits of new realism.

3.3.2 Catherine Grosvenor's *Cherry Blossom* (2008)

I started Chapter 3 of this book with David Hare's massaged verbatim piece, *Stuff Happens*, to suggest that there were more affinities than had tended to be noted between verbatim theatre and fictional plays. Edinburgh-born writer Catherine Grosvenor's *Cherry Blossom*, written in collaboration with Lorne Campbell, Mark Grimmer and Leo Warner, as a massaged verbatim piece is no exception, but it has the singularity of taking the matter to another extreme. This production, which premiered in the autumn of 2008 at the Traverse Theatre in Edinburgh and was seen as "a landmark theatre event in Scotland" (McMillan "Language") for many critics, takes transgression as its very subject.[140] Not only does the performance address issues of language, cultural identity and Polish migration, it unceremoniously intersperses verbatim material with fiction, devised elements in a bilingual production. In other words, this 90-minute multi-media piece of theatre sits at the extreme end of the massaged verbatim spectrum, which may have some incidence on the understanding of new realism within this type of performance work. Now, before going into the question as to whether a change in the makeup of a massaged verbatim piece is a speculative concern or not, let us pause for a moment and provide an overview of this case study.

3.3.2.1 Overview

The title of the play, "cherry blossom" refers to a shoe polish and, through a play on words (polish/Polish), it is in fact a pejorative connotation alluding to Polish migrants in the 1960s.[141] *Cherry Blossom* belongs to the massaged verbatim genre insofar as it mixes actual verbatim quotes from the people that were interviewed during the long research process in Scotland and Poland with fact-based written

---

140 The show was co-produced with company Teatr Polski of Bydgoszez, Poland and later toured to Bydgoszez and Warsaw.

141 In a report of *Cherry Blossom*'s post-show discussion forum, Rachel Clements explains that the title also refers to "playground taunts aimed at Poles in Glasgow during the 1960s".

material. However, despite its roots in actuality, it differs from a piece such as David Hare's *Stuff Happens* since the vast majority of the words in the play are heavily fictional and refer to an entirely made-up story with made-up characters. One question that performances such as *Cherry Blossom* raise is whether or not the proportion of verbatim material in the final production drastically affects the reception of a given audience. Or – as seen with *Called to Account* in the section on transgression and documentary realism – whether or not the aesthetic of the performance work is a more decisive and affective element at the point of reception, enabling borderline pieces of theatre to be perceived in other terms. Fascinatingly, one could attempt a thought experiment as regards to when a massaged verbatim piece is deemed too massaged to be read as verbatim theatre, and by extension – which is going to be the basis of my investigation in Chapter 4 – at what point (if any) a theatre work based on extensive interviews and using some of these words in the final performance script stops passing as verbatim theatre.

To come back to *Cherry Blossom*, the text is made up of three core elements: a fictional story, a real story based on factual information available in the public domain and a selection of verbatim quotes directly drawn from interviews conducted by the creative team on the social phenomenon that saw 40,000 Poles arrive in Edinburgh alone since Poland joined the EU in 2004, making up 10 % of its total population. Within the space of twenty-four scenes, the real journey to Canada of Robert Dziekanksi, a Polish citizen and construction worker with no knowledge of English – tragically ending with his death at Vancouver Airport in October 2007 tasered by the police after a long ordeal of miscommunication – is paralleled with that of Grażyna, the fictitious young Polish protagonist, who moved to Scotland as an economic migrant. Arguably, the verbatim component of the piece seems to problematically validate the fictional elements, as the show tagline "Grażyna Antkiewicz lived, Robert Dziekanski died" suggests, and the author herself claims in the Foreword of the published playtext: "There are three distinct strands: the story of Grażyna Antkiewicz, which presents fiction as truth" (7).

Looking at the published text, it is virtually impossible to differentiate the verbatim quotes from the written material, except in the last scene, Scene 24, which is presented in a more poetic manner:

1. If love is money
4. If love is always
3. If the distance is too far
1. If my arms cannot stretch
2. If my heart has broken
3. If my faith has gone

4. If you take her hope
1. If he gives me hope. (91)[142]

Or, when characters in the story quote from published material such as in what follows:

> EWA (*reading from dictionary*). 'A permanently fixed item in a house, such as a bath, which would not be taken by someone when they move to a new home: 'all fixtures and fittings are including in the price.' *Figurative:* They've been at it so long he's become a permanent fixture in her life. (*CB*)

In effect, the verbatim quotations (from the interviews) have been incorporated wholesale into the "fictional" dialogues in a way that cannot be identified on the page whilst the transcript of sound from a bystander's video at Toronto Airport – describing the final minutes of Robert Dziekanski – is provided in full and in the appendices, instead of within the playtext. Furthermore, for the non-Polish speaking reader, the text – which is partly written in Polish and intended to be biligual – may give the wrong impression that the Polish elements are verbatim. One may also assume that this is perhaps the reverse for the Polish readers and spectators. Clearly, the very fact that the text is ambiguous makes it automatically engage with the conventions of verbatim theatre in the minds of audiences. The appendices after the playtext also remind one of the origins of some of the words and a whole section is dedicated to "Things People Said to Us", featuring a verbatim sample of the conversations the creative team had with "a wide range of people in Scotland and Poland, from factory workers to academics" (101). The story is also punctuated by references to the exact exchange rates between the British pound and the Polish Zloty – matching the supposed time of events in the play – which are updated in each scene:

> Scene Nineteen
> *7th February 2008*
> *Exchange rate £1 = 4.8343 PLN* (*CB* 65)
>
> Scene Twenty
> *11th March 2008*
> *Exchange rate £1 = 4.6403 PLN.* (69)

142 The numbers below refer to the performers on stage, who never play a character throughout. At the textual level, this is the only time in the piece when characters completely disappear. One can also assume that it is a poetic montage drawn from verbatim testimony.

Another, more interesting question for the purposes of this section is, what did the staging do to the elements hitherto mentioned? Did it attempt to clarify the distinction between the verbatim quotations and the fictional material in a manner similar to *Stuff Happens* or, did it make the overall experience even more confusing? Finally, did the performance as staged engage with new realism (or even documentary realism) as I have framed it or did it offer something radically different?

3.3.2.2 Staging

If one looks again at the underlying themes of *Cherry Blossom* – displacement, border-crossing, miscommunication, incomprehension, disorientation and confusion experienced by migrants and displaced people alike – the contradictions of the massaged verbatim format exactly reflect this subtext at the level of form. Now, it is time to look at the original staging and assess whether it reinforces or suffocates the original impulse of Grosvenor's text.

First of all, one needs to acknowledge that the staging is itself divided into a variety of strands that maps the audience's reception of the words.[143] The creative team behind *Cherry Blossom* was composed of Leo Warner who designed the set and coworked with Mark Grimmer on the video design, which was a crucial component of the show. Together with director Lorne Campbell, they invented a playful structure to accommodate the strangeness of the massaged verbatim text, rather than look for the kind of stage realism that would simply tell the story with clarity. In other words, their way of working was experimental and almost conceptual, was about troubling and rendering uncertain, which is arguably rather unusual for verbatim theatre. Several points need to be borne in mind. First, the acting process was informed and changed by the provocation and ambiguity of the massaged text, that is to say by its formal boldness.[144]The four actors – two Scottish and two Polish – "multi-role" all parts, regardless of age, gender and nationality throughout the performance and projections indicate their names at the start of each scene. As remarked earlier in this chapter, these techniques – which some have perceived as taking "the experimentation a needless

143 Other parameters beyond the performance such as what was on the news at that time also shape and mediate the audience's experience to a degree, but these are largely beyond the scope of this research.

144 Admittedly, *Cherry Blossom* is more provocative at the level of form than content. Indeed, the fictional story of Grażyna leaving her native Poland for Scotland to financially support her family (especially to send her elder daughter to university), and falling in love with a local more or less fits an archetypal narrative.

bridge too far" (Scott) – are reminiscent of Brecht who frequently used similar contrasting methods such as "alternating two different actors in the same role" (Barnett, *Berliner* 14). Second, the stage area was almost completely bare and devoid of props. However, multimedia devices were deployed along with moving white oblong and square panels that served as both floors and walls for the acting area and, also, as screens onto which various visual elements were projected to suggest time and space. They also helped the audience keep track of which actor is playing which character and intermittently provided subtitles in both Polish and English. Third, actors read the verbatim words related to the last minutes of Robert Dziekanski from visible clipboards, interrupting the main fictitious story. Here, there was no effort made to create a semblance of realism and, contrary to the documentary realist tribunal plays that had their actors read some of the evidence, the actors in *Cherry Blossom* in a "ready-made" fashion did not attempt to pretend there were doing anything else but the task at hand. In other words, they were performing this as actors and not characters. Clearly, the verbatim effects of the performance through such a staging were not reduced, even though they completely departed from theatrical realism.

Watching *Cherry Blossom* is therefore a very different experience from reading the playtext. As director Lorne Campbell explained in an interview, this allows "the audience to step for a moment into emigrants' shoes and feel what it's like to be excluded, to find yourself outside the natural milieu" (Grossman 245). If the performance fits at some level our conceptualisation of new realism – a marked split between actors and characters whilst having the characters in a situation of progress, constant interruptions of the narrative as well as visual alienation with obvious references to an external verbatim reality, on another level it goes well beyond its core elements. Indeed, in *Cherry Blossom* one sees emerging a more profound disintegration of the notion of character at times, a deeper sense of theatrical duality, a blatant erosion of the usual verbatim markers as well as an audacious acknowledgement of its out-of-jointness that are allowed to dominate. For this study, this means that even though the massaged verbatim strand has a stronger affinity with the aesthetic of new realism, its encounter with a stage can potentially either smooth over the existing cracks or widen them.

3.3.2.3 Conclusion

Importantly, *Cherry Blossom* has disclosed further possibilities for reflection, as it seemed to compete with and even subvert the aesthetic structures of new realism. As Elwira M. Grossman has argued, *Cherry Blossom* opens up "a very interesting dimension for the emergence of hybrid theatrical forms" (247). Indeed, the performance of the verbatim material on a stage inescapably re-signifies its semiotic and aesthetic qualities, which can appear even more contentious in the case of the jumbled massaged verbatim strand. The multimedia format chosen by the creators made the initial effects of being a spectator at this type of theatre all the more intense. The state of conflict between the various components of the massaged play (verbatim, fictional and fact-based writing as well as Polish and English) shaped the form of the performance without ever suggesting an overall illusionary dimension.

Looking at the development of verbatim theatre into increasingly hybrid and intriguing forms, it is undeniable that the widespread use of the words of real people in mainstream theatre has turned into a powerful driving force for many contemporary practitioners in Britain. In this ever-shifting context, the last case study will add another layer and complicate the discussion, as one revisits the now familiar strand of headphone-verbatim theatre.

3.3.3 BeFrank Theatre's *Do We Do the Right Thing?* (2014)

This third transgressive example is borrowed again from the headphone-verbatim area of practice. Alecky Blythe, being "in the vanguard of verbatim theatre in Britain" (Forsyth and Megson ix), has been a seminal influence in the development of headphone-verbatim theatre in the UK whether or not the new practitioners had any direct contact with her. Kerry Shale and Matthew Lloyd created in this manner *Listen, We're Family*[145] in 2013 on the successive waves of Jewish immigrants in London, "Theatre in Education" (TIE) company 20 Stories High created *Tales for the MP3* in 2014 about the performers themselves while Snippet Theatre Company uses the headphone approach to verbatim theatre-making in rehearsals only.[146] BeFrank Theatre was also heavily inspired by Blythe's work that they saw at the Actors Centre as well as at a showcase evening. The specificity of their *Do We Do the Right Thing?* (2014) – the case study in this section – is that it is a hybrid piece of verbatim theatre that does not use the headphone technique throughout the entirety of the performance and also combines verbatim material with autobiographical elements as well as sound and projections. To

145 The piece included in the cast Debbie Chazen, a British actress who was already familiar with the headphone-verbatim method through her role in Alecky Blythe's *The Girlfriend Experience.*

146 Ruth Tebby, co-founder of Snippet, said this to me in a personal email on 30 September 2015, explaining that they conducted workshops using the headphone technique as part of their creative process before composing the "final version" of their verbatim performance text.

my knowledge, the only other pieces that have partially summoned the headphone-verbatim technique in performance are Jo Harper's *Can You Hear Me Running?* that was shown as a "Research and Development" (R&D) performance at the Pleasance Theatre, London in September 2015 and *Still No Idea* (2018) made by Lisa Hammond, Rachel Spence (a former member of Non-Fiction Theatre) and Lee Simpson with Improbable and the Royal Court Theatre. In this spirit, *Do We Do the Right Thing?* is an interesting development of the headphone-verbatim technique that demands further study.[147]

3.3.3.1 Overview

BeFrank Theatre is a London-based international theatre company formed in 2010 by artistic director Tommy Lexen and specialising in real-life stories and how they can bring to the surface what might be otherwise left unheard. For some productions, the company has used verbatim material and, for some others, this material only served as a foundation to create fictional pieces that go beyond my definition of "massaged verbatim theatre." Their second production *Do We Do the Right Thing?* is also a carefully documented verbatim piece focusing on the Iraq conflict – like the other verbatim memorials *Justifying War* and *Stuff Happens* – but it does so from a very different perspective, that of Royal Wootton Bassett, a Wiltshire market town, and its local community.[148] More specifically, the show is one man's response to these events – the writer Neil Walker's – as well as the stories he heard surrounding them from serving military personnel and civilians alike. It explores acts of remembrance and society's relationship to the military through both autobiographical material and interviews with those affected by and involved with the repatriation of military personnel killed in the Afghan and Iraq conflicts. The production premiered on 3 November 2014 at the New Diorama Theatre, London before touring the UK during November 2014 and spring 2015. It was also taken to the British forces in Germany in March 2015 following an invitation by Major General John McNiven Ross Henderson CB. This analysis is based on the performance I saw at the Old Fire Station, Oxford, on 18 February 2015. The piece, comprising 24 scenes, mostly takes place in the attic of an elderly widow's house and opens with improvised dialogue as the actors in character are sorting through boxes, family possessions

147 *Do We Do* is their only performance to date that has used the headphone-verbatim technique.

148 The show is affectionately called *Do We Do* amongst BeFrank members.

and photos, spoiling any notion of documentary realism before it has even begun.

3.3.3.2 Realism(s)

Do We Do is composed of three main strands: the headphone-verbatim material that is pure (edited) verbatim, verbatim material in the form of monologues that are directly drawn from interviews or development workshops and some written elements that are autobiographical. In some sense, the performance is both headphone-verbatim and massaged. Following the study in Chapters 3.1 and 3.2, one can therefore assume that new realism is inevitable. Both these ideas of having a) visible headphones or earphones referring to the origin of the words and showing the means by which the verbatim delivery is produced and b) making visible the author(s) through the introduction of fiction might seem to tend towards an aesthetic of new realism. But, I would like to reiterate the earlier remarks throughout this volume according to which one should always proceed with caution as far as verbatim is concerned.

First, one finds the usual markers of new realism within the casting itself. The entire performance comprises a cast of four – three males and one female – playing 13 characters regardless of age or physical appearance, which is made possible by the headphone-verbatim technique. However, it is also important to note that one of the characters, the protagonist who stands for the author, is played throughout by the same actor. Andy, we are told, is a middle-aged man who grew up on military bases as the son of a career soldier. He is played by the author, Neil Walker, and largely draws on his own life, while the other three actors play the remainder of the characters (four roles each including the members of Andy's family and multiple verbatim characters). Second, as is common with new realism, time and space is produced with the simplest of stagecraft means (cardboard boxes, a few chairs, military memorabilia, old toys, rolls of tape, an old film projector). Third, one is provided with a realist narrative: the author's own journey as he revisits his childhood, and his relationship with the military as he interviews people about the ripples this has made on their life. Fourth, the audience experiences constant interruptions as the actors become different characters, the protagonist Andy adds comments in direct address to the audience and one moves through time and space very rapidly. Fifth, the experience of the audience is an ambiguous one as it is not always possible to identify what is verbatim and what is not.

Nonetheless, many other elements in the performance do not corroborate this now familiar picture of new realism I have been sketching in this part. Indeed, contrary to the other headphone-verbatim works studied in Chapter 3.2,

Do We Do the Right Thing? incorporates the technique as only one of the means of presenting the verbatim material in performance.[149] In effect, when this happens this is denoted in the script with the abbreviation 'RecDev', standing for 'recorded delivery technique' in reference to Alecky Blythe:

> **Andy:** I stayed in a pub across the road. Drank a lot of tea. Ate far too many biscuits. I really needed to know what had happened on that high street.
>
> **Anna (RecDev):** I remember going up on the High Street, doing your normal, like, the kids were at, erm, nursery and you wanted to stop people an' let them know because people were just carrying on with their lives, with the momentous thing that had happened to you, and they're, they're just carrying on and you wanted to say, no, stop, don't you know that, you know ...
>
> **Andy:** Anna was a pad brat too. She was related to one of the ten repatriated through the town in 2005. (8)[150]

On stage, one clearly sees the actors controlling their audio devices and putting on the headsets every time they use the technique. Admittedly, this would seem to further complicate the A-effect described by Brecht in his discussion of actor Charles Laughton and the character of Galileo (in a Beverly Hills production of *Galileo*) in his *Short Organum for the Theatre:*

> This principle – that the actor appears on stage in a double role, as Laughton and as Galileo; that the showman Laughton does not disappear in the Galileo whom he is showing; from which this way of acting gets its name 'epic' – comes to mean simply that the tangible, matter-of-fact process is no longer hidden behind a veil; that Laughton is actually there, standing on the stage and showing us what he imagines Galileo to have been. (Willett 194)

As I argued via Tom Cantrell in Chapter 3.2, the headphone technique was always on the verge of becoming swallowed up by the realist frame, the performers subsuming themselves into other people's voices and ceasing to experience the words in the moment. For the audience in attendance, the initial visual alienation of the technique seems to lose its effect over time as they simply accept the convention of the headphones and forget about them, captivated by the stories they are told. In other words, the headphone technique therefore ceases to call

---

149 The author Neil Walker also transcribed all the 30 interviews before proceeding to edit them, contrary to most headphone-verbatim theatre-makers who tend to directly edit the audio files.

150 All extracts are taken from an unpublished script of *Do We Do the Right Thing*, written by Neil Waker (edited by Tommy Lexen, dramaturgy and additional editing by Amanda Fromell), and made available to me courtesy of Neil Walker.

attention to itself. Here, however, insofar as the actors are required to put their headphones down at times and then back on again, the alienation effect persists. The actors can never evade the visual alienation fostered by the technique even when the realism of the narrative attempts to contain its disruptive force. For instance, in the very first scene using the headphones, Scene Three, Andy "responds as if he is listening to each of [the headphone-verbatim characters]" (8). Notwithstanding this, at times, Andy proves more disruptive as he adds information and comments in direct address to the audience about the characters in a manner similar to "An Actor" in *Stuff Happens:*

> **Andy:** I met Maggie & Steve over a pint. They moved into the town years ago. Still didn't quite see themselves as locals. (11)
>
> **Andy (to audience):** His brothers saw action. Fred didn't. '*Just National Service*'. As if that wasn't enough. (14)

These constant interruptions are seemingly characteristic of the aesthetic negotiations of the verbatim material I have called new realism. Yet, they appear to go further than the strategies analysed throughout Chapter 3 of this volume as there are visibly two characters of Andy on stage: one who conducted the interviews and one who is reflecting on them as well as on his own relationship to the military world. Scene 11 in which he meets the major is exemplary of this:

> *Andy steps into the office, the Major is still sitting. Andy straightens his back, puts hand out, makes a really firm handshake and says:*
>
> 'Hello, it's really good to meet you Major, sir.'
> *(to audience):* Did I just say *Sir?* (26)

Consequently, there is also an outright foregrounding of the simultaneity of temporal registers inherent to the theatrical situation.[151] In later scenes one encounters yet another split between his re-enacted younger and older self. In the following extract, *Do We Do* becomes a performance about imagining what might have possibly taken place, about its own undecidedness:

> **Younger Andy:** But I am still your son and I love you very much. That will never change. I just want us to talk, get to know each other. Properly. [...]
>
> *Actors leave the stage leaving older Andy alone, sitting in his armchair.*

151 This is an important component of the theatre that was denied by the mechanism of documentary realism, asking us to emotionally engage with the performance not as an edited record of a past event but as the event itself.

> **Andy:** That never happened. I wasn't that brave. I said:
> Well I'm not going. No way.
> Why would I run the risk of getting shot or killed for some place hundreds of miles away that I hadn't heard of until this week? (29)

At this point, everything one sees is forever contaminated by an air of fakery one cannot undo, despite the reconstructed instantaneity enabled by the headphone technique of performance. In other words, the precious equilibrium of new realism that underplayed the unruly nature of fiction is no more. As important is Be-Frank's decision to intermittently use the headphones, a division which brings the technological means even more clearly into view, framing the spectatorial experience away from the kind of realism encountered in Blythe's *Come Out Eli* and towards one that deconstructs its own aesthetic preconditions. Put another way, together these strategies uncover the aesthetic edges of new realism, how the experience of watching a verbatim piece of theatre can draw us in another direction and make different demands.

3.3.3.3 Conclusion

As a paradigmatic transgressive example, *Do We Do the Right Thing?* demonstrates how verbatim theatre can thrive on the collision of different materials as well as on its own instability and vulnerability in performance. Instead of constructing a verbatim illusion so complete that the audience is drawn into the fantasy of documentary realism, one witnesses a kind of verbatim theatre that starts critiquing its own language and means even as it is using them. The verbatim promise, it seems, has failed. Or rather, it is saved by another seductive aesthetic framework – new realism – that strives towards realism, enabling the audience to enjoy a self-reflexive narrative nonetheless. If new realism was looking to breach its own limits, it never allowed the verbatim performance to completely strip bare its own authoritative ground as the mask of documentary realism receded. What *Do We Do* seems to point towards is how practitioners have challenged again this boundary by showcasing within a verbatim performance the impossibility of a faithful re-enactment.

3.3.4 Conclusion of Chapter 3.3

Among the many verbatim performances that have arisen on the new realism landscape, I focused on those that set us aesthetic puzzles to figure out. These works yielded a range of new insights into the ambiguous aesthetic paradigm

I have been attempting to define in this part and, to some extent, they also foreshadowed the ways in which verbatim theatre can go beyond these parameters and become more than itself. The discussion began with what seemed to be a well-established division between the headphone-verbatim practice and massaged verbatim theatre. Except of course, that the examination of the three "transgressive" case studies quickly revealed that the two categories overlap and intermingle in a more complex interrelation of forms and media. *Do We Do?* was in effect a useful limit case which illustrated most vividly how these boundaries are constantly being redrawn by theatre-makers. Further analysis suggested that the concept of headphone-verbatim theatre was itself prone to new aesthetic experimentations beyond new realism as it overtly combines imaginative summonings and a replay of verbatim materials. To summarise these arguments at their simplest form, it may be that we are now at a point where contemporary performances increasingly question and challenge verbatim theatre from within.

Inevitably, some interesting "transgressive" case studies had to fall through the net of my categories. Another massaged verbatim performance, Neil McPherson's *I Wish to Die Singing: Voices from the Armenian Genocide* (2015), a hybrid of "history lesson and documentary drama and verbatim theatre" (8), which has projected images accompanying all the action as well as a fictional narrator is a case in point. Some massaged verbatim theatre-makers even turned verbatim theatre against itself. In Stuart Slade's *BU21* (2016), the words of real people are used to imagine the disquieting aftermath of a terrorist attack in July 2016 at the heart of London while further confusing the audience by having the "actors' real names [replacing] [the] character names" (4) and by updating the year in subsequent performances to make sure that the July mentioned in the performance corresponds "to the July immediately after the date of performance" (4). Similarly, Alice Bartlett's *Not in my Name* (2009) "varies from most verbatim scripts in that it is a story: a hypothesis of what *might* occur in the event and aftermath of a localised terror attack rather than a reconstruction purposefully based upon any previous disaster" (25). Fascinatingly, on the headphone-verbatim spectrum of performance, Dan Canham's *Ours Was the Fen Country* (2013) – on rural ways of life that are dying out – productively and playfully mixes dance, sounds and the headphone verbatim technique, having real people's voices embodied and then disembodied, exploring the relationship between words and movements.

It is also startling that even the verbatim tribunal play that served as the archetype for my theorisation of the concept of documentary realism in Chapter 2 can engage with new realism, especially when it moves from the page to the stage. In Richard Norton-Taylor and Matt Woodhead's *Chilcot* (2016), based on the Chilcot Inquiry investigating Britain's role in the Iraq War, the auditorium is set as a courtroom and it seems at first glance to perfectly recreate the atmosphere of the Inquiry. However, very different strategies are also at work as the performance also features the staging of verbatim interviews with key political players, Iraqi refugees, veterans and military families as well as a sound and video design that constantly interrupt the progression of the reconstructed Chilcot Inquiry. Furthermore, contrary to documentary realism that aims to minimise the gap between the original context and the performance, in this production actors are sometimes voluntarily cast against type whilst the same characters are even played by different actors at different times. In effect, despite their attempt to give a flavour of the original speaker in their delivery, visually, we see for instance actress Sanchia McCormack in the role of chief UN weapons inspector Hans Blix. Last but not least, *Chilcot* uses a voice-over narrator and includes mentions of the redacted/omitted passages in the publication of the transcripts of the Inquiry:

> NARRATOR: This reply has been redacted.
> ALLEN: ***
> * ***
> ******
>
> CHILCOT: Thank you. One could go on, but I think I'll turn to Roderic Lyne at this point. (1)

Having discussed how new realism materialises in contemporary verbatim performances (notably through its offering of a glimpse of its own making process that cleverly avoids smothering the kind of realism associated with verbatim theatre) and analysed a series of problematic case studies riven with contradictions, I must now sum up and conclude.

3.4 General Conclusion of Chapter 3

When a theatre-maker intrepidly decides to turn the verbatim material he or she has collected into a performance, he or she is faced with many dilemmas. One of them is whether or not to abide by the rules laid down by documentary realism or allow the work to distance itself from the original contexts in which the words of "real" people were uttered by melding crafted elements – be they of a performative or textual nature – with what was then raw, spontaneous, immediate speech, actions and events generated in the spur of the moment. In the verbatim productions studied in Chapter 2, everything, including the set meticulously recreated the original context.[152] Even *Called to Account* (studied in Chapter 2.4) which featured some transgressive elements was identical to the interview room as actress Diane Fletcher – who played Clare Short – explained in a telephone interview with Tom Cantrell: "We even had the camcorder onstage that was used to record the interviews – to act as a reminder to the audience that these were real interviews" (Cantrell and Luckhurst 71).

In this second Part, I have therefore attempted to look at what happens when practitioners start to depart from the first aforementioned position and forecast the audience's expectations, reflecting on their own attempts to make sense of events.[153] More specifically, new realism marked the simultaneous collision between two opposite aspirations and the task here has been to look at certain ebullient contemporary developments of verbatim theatre from the point of view of their aesthetic contribution, taking the strand beyond the circumscribed and narrow documentary realist position by which it has largely been bound in common parlance. I then attempted to study the effects of this new realist aesthetic on the verbatim material to assess whether it could temper/impact on reception or whether its initial sense of shock would ultimately be absorbed by the comforting and familiar knowledge that we are hearing and seeing what really happened in the words of real people.

Although this category occupies a fairly narrow part of the general British verbatim spectrum in comparison to documentary realism,[154]these variations are of considerable significance, not least because of their greater elasticity, their engagements with both realist and reflexive modes in performance – for instance in the way these works show the process by which the original testimony is redirected into the voice of a "character" – and their indebtedness to Brecht.[155] In a sense, these practices also recall Peter Cheeseman's own kind of verbatim

---

152 For instance, in Philip Ralph's documentary realist verbatim piece *Deep Cut*, the front room of Private Cheryl James' parents was accurately recreated on stage.

153 According to theatre critic Lyn Gardner, this position is problematic for the spectator as "we quickly lose sight of that mediation. [...] We accept what is presented to us as true without questioning how statements have been selected and organised, or even how interviews that elicited the 'evidence' were conducted" (Gardner "Talk"). She then adds that this position "functions in a way that cajoles you into accepting the piece's particular bias as the truth and nothing but the truth" (Gardner "Talk").

154 According to David Hare – in a lecture delivered to the Royal Society of Literature in April 2010 – only "a handful of [playwrights] are attracted to mixing fiction and reality in unfamiliar ways" ("Mere" 77). However, this book does not consider "new realism" as a brief moment of instability between two sets of aesthetic conditions.

155 As I have argued, new realism is influenced by Brecht (disjunctures and disharmonies in the performance apparatus) and yet not simply determined by his theories.

theatre whereby "the actor does not pretend to become another person, he or she becomes the advocate for that person. The actors address the audiences themselves, saying: 'I am not this person, I am speaking for this person.'" ("Peter" 17). Also, if the documentary realist verbatim plays tended to be staged within a physical setting that echoed the context in which the original verbatim words had been uttered (for example, courtrooms), this is no longer a requirement for the new realist breed. But most significantly, new realism discloses the verbatim material in two ways at once as it mediates between two levels of reality, two opposite but complementary theatrical languages. This brings clear pressures to bear on the conventions and emotional register of verbatim theatre through its staging of critical interruptions of the mimetic that temporarily interfere with the verbatim spectacle. In so doing, nothing is "innocent" anymore, these performances challenge the spectator to look behind the scenes, to be more self-conscious and to read between the verbatim lines. The theatre here is not an obstacle to overcome but a truly organic part of the process shown that pervades all levels of the performance, both marking and transgressing the verbatim referent.

Within an accelerated blurring of strict genres, new realist verbatim works clearly point in another more ambiguously placed direction between wholeness and fragmentation – an aesthetic that foregrounds its operations and makes itself recognisable – but their (passive) self-reflecting devices are never bold enough (or willing) to operate as a postmodern engine of strategies deconstructive of ontology. For all their aesthetic unusualness, they are nevertheless functioning within the conventional measures of verbatim theatre. Put differently, these devices are never allowed to dominate so as not to undermine the basic realist qualities of verbatim theatre which would inevitably turn a given performance into another aesthetic system, "post-realism", which will be our particular concern in the following part. The study of *Stuff Happens* and *Come Out Eli* makes it abundantly clear that these performances draw on a more visually complex system of signs that furnishes the viewer with a different theatre language in the palpable negotiability of its verbatim substance with an overlapping fictional realism. More specifically, in *Stuff Happens* this rift was playfully and coyly negotiated by a fictional actor-narrator, standing for David Hare's rendering of the events leading up to the war in Iraq. While in Blythe's *Come Out Eli*, the deliberate visual disjunction between the "original" contexts of the interview and what appears on stage opens the verbatim component of the piece to suspicion, destabilising the certainties of an original verbatim source.

In other words, the "new realist" verbatim theatre-makers are determinedly more experimental than their documentary realist peers as the figure of the actor becomes crucially sought after and cleverly used to achieve this new type of realism. Viewed another way, there is an important shift of focus from the "self-evident" documentary realist meaning to the way that meaning is constructed which highlights the dangers and limitations of verbatim theatre and its supposedly unproblematic referentiality, keeping a certain dose of uncertainty in oscillation. Clearly, the final performance is deliberately suffused with the presence of its own process of fabrication (for example, recording devices in *Come Out Eli*, the presence of the deceiver, the deluding author/interviewer in David Hare's *The Power of Yes*). These verbatim theatre-makers seek not only to engage with the contemporary world but, by the same token, to investigate their own role and means of production and perhaps even to present the process for what it is: a mutually sustained verbatim edifice. The impossibility of reaching a pure state of documentary realism allows for a breadth of interpretive freedom that is here celebrated, the question of representation being inserted into its very fabric.[156] Now might be a good moment to recall the numerous acts of interference or downright verbatim sabotage that make one disengage from an initial "documentary realist" spectating position. The character of "An Actor" in *Stuff Happens*, the character of Alecky Blythe in *Come Out Eli*, the acting process in *Stuff Happens* that leans more towards "presentation" by the performers instead of "representation" and "characterisation", the self-reflexive headphone-verbatim technique used by the actors, the lack of visual supports in both *Stuff Happens* and *Come Out Eli*, all concur to disrupt the lionised aesthetic edifice of verbatim theatre (as seen and discussed in Chapter 2).

Yet, one should not exaggerate the break, as realism, far from ceding its rights, still retains its pull – though at first it might have seemed to lean in another direction – and it is to be hoped that this chapter has shown that "new realism" both impedes and invigorates the documentary realist project and that verbatim theatre demands it. Under the guise of contesting documentary realist verbatim theatre, new realism – as a new aesthetic meta-narrative to which all else defers – also successfully expands its parameters. In fact, curiously enough, the realist tendencies in "new realism" seem to coexist quite comfortably with what may have seemed to exclude or violate them, their terms constantly collapsing into one another. One needs only to have a look at David Hare's *The Power of Yes* (another massaged verbatim piece), for instance, which follows a realist epistemological progression, as the reader/spectator is assumed to gain knowledge of the financial crisis throughout the narrative of David Hare's own metatheatrical journey into this world. Clearly, realism, even as it escapes us, is still the aim

156 Even so-called "documentary realist" verbatim works tend to see their immediacy/rawness wane after repeated performances.

and is simply arrived at by a different route. This strongly suggests that the noted variations from documentary realism see their bearings on the verbatim performance being relatively tempered and mitigated, never entirely breaking the realist semblance of unity. Laudable as these variations may be, they only move realism from a documentary impulse to truthfully recreate an original and real event to a new kind of realism that realistically stages part of the process of documentation and how it interacts,[157] supplements and gives access to real events. One could argue that "new realism" works to maintain and harbour two seemingly contradictory positions in verbatim theatre: a retreat from the anti-theatrical prejudice of documentary realism on the one hand, and an insistence on realism on the other. In this picture, realism is thus not weakened but extended in a way that confers total power to the verbatim theatre-maker or playwright as his or her figure, for instance Alecky Blythe in *Come Out Eli*, serves as a means of getting all the dissonant elements into the aesthetic frame, mediating between extremes. In other words, "new realism" continues the tradition of documentary realism rather than marking its decline.

In the course of Chapter 3, we saw that new initiatives led to a wider scope and a broader use of verbatim material in performance beyond the currency of documentary practice considered in Chapter 2. These also constituted a certain turning away from Brecht's legacy and the use he made of distanciating devices in performance. More theatrically enriched and sophisticated forms (re)appeared and this provides the backdrop for the last instalment of this investigation that completely dissolves the documentary realist framework, taking verbatim theatre's willingness to experiment to the extreme, to the very limits of its manifestation. However, the ideas and aesthetic concerns voiced in this part – such as allowing the doubts and conflictedness related to the making of the work to be part of the stage presentation – also live on in the works that make the case studies of the following part. In other words, documentary realism has sometimes suffered irreparable damage and, as such, "post-realism" grows out of the hybridity and tension inherited from "new realism" that threaten to exceed and cancel its verbatim "identity."

The postdramatic current has been liberating in its effects of breaking decisively the invisible links which restrict verbatim theatre to a documentary realist practice so as to provide a radical testing of its parameters, which have yet to be explored. It follows that verbatim theatre has necessarily found itself engaged with a variety of performance practices and stage techniques that are apparently

157 The editing process is not staged, nor is the process of the reappraisal of existing verbatim evidence.

very distant from its original concerns and premises. These inexorably push verbatim theatre to extremes, successfully taking it further along the amorphous road beyond documentary realism, reinvigorating a critical discussion around aesthetics which had been rendered dormant in some quarters. That was of course inevitable and it might be considered a mark of the exhilarating vitality of verbatim theatre that this investigation does not end here and firmer conclusions must await the end of the next part.

4 Post-Realism

What are the means by which verbatim theatre has become post-realist? Can realism ever be overcome in verbatim performances? Might it be useful to invoke Lehmann's theorisation of the postdramatic to account for the new relations of aesthetics that some verbatim performances have recently adopted in Britain and further afield? If so, what are the fundamental points of convergence and divergence between the two? And if not, what other theoretical approaches may one test against the idea of the "post-realist"? Finally, the questions that preoccupy this chapter are the following: what happens when the confirming power of the verbatim meets the evocative force of the postdramatic and what hints of realism (if any) reside in post-realism? How does verbatim theatre perform in such a scenario? Can a verbatim performance be categorised as verbatim theatre without its author(s) overtly dressing it up as a piece of verbatim theatre?

4.0.1 From Playful Antagonism to Outright Aggression within the Frame

Realism and its many aesthetic variations that seemed to systematically adorn and guide verbatim materials in performance are disappearing from view on the British stage, having a profound and troubling impact on its core definitions and principles.[1] In the 21stcentury, British verbatim theatre has indeed pushed itself to become more indeterminate and unwieldy, more absorbent of experimental and devised representational strategies as well as live art practice. In his description of the contemporary theatre landscape, Dan Rebellato observes that:

> What characterises the new century in theatre, if anything, is its diverse ecology: there are verbatim theatre makers, live artists, political playwrights, devising companies and physical theatre makers – and, increasingly, no clear boundaries between any of them. ("Introduction" 5)

For this reason, it is now possible to encounter pieces such as Hyphenated's *Home is Where …* (2016) that productively combine verbatim material with "music, dynamic ensemble movement, and multimedia design […], pushing the boundaries of what verbatim theatre can be'.[2] This new context has therefore

1 From this perspective, this investigation of the phenomenon of "new realist" innovations in contemporary British verbatim theatre is still considered as part of that tradition.
2 This quote is taken directly from the director (Amy Clare Tasker)'s website that promotes the show here: http://www.amyclaretasker.com/hyphenated/.

https://doi.org/10.1515/9783110715767-006

made possible what one might think of as the first verbatim performances of sustained post-realist intent and absolute distrust of linearity, meaning that critics and audiences alike no longer know what to expect at the unruly edges of the disciplines. In these performances, there is a clear formalistic break with the documentary realist as well as with the new realist model. In other words, verbatim theatre as one has come to know it, that is to say with "very little in the way of clutter, or of creative mediation" (Kinghorn 16), becomes a withered organism in this new format and the people behind these shows are no longer to be called verbatim dramatists but plainly verbatim theatre-makers. For instance, in Lucy Ellison's *Torycore* (2014), billed as nothing less than "a verbatim record of a government ripping society's guts out", the spectator encounters stage elements that are typically foreign to the conventional verbatim theatre apparatus, such as large speakers, guitar amps, mixers and a music stand. Quite stunningly, in *Torycore* the verbatim words taken from Tory speeches – such as George Osborne's entire 2014 budget speech – are set to death metal music (with Chris Thorpe and Steve Lawson on guitar). The level of noise generated by the improvised musical score unsettles the verbatim material into near incomprehensibility, throwing the words back on the audience stripped of their usual official lustre and elegance and giving the audience a sense of the (perceived) lived impact of those words. Similarly, towards the punchy end of the verbatim spectrum, British artist Hannah Silva in her solo show *Opposition* (2011) turns the verbatim words in the form of cut-up speeches of politicians such as Ed Miliband, Tony Blair and David Cameron, that is to say political language and rhetoric, with the help of a loop pedal, repetitions and vocal techniques into a verbatim soundscape, a kind of sound poetry that experiments with these language constructions. In the following excerpt, Silva turns David Cameron's speech on the "Big Society," dating from 19 July 2010,[3] into what I would call here a "post-realist" verbatim collage:

> And yes cutting the
> And yes cutting the nat
> And yes cutting the national
>
> And yes cutting the cutting the cutting cutting
> The cutting cutting cutting cutting cutting cutting
> cutting cutting cutting cutting cutting cutting cutting
> cuttingcuttingcuttingcuttingcuttingcuttingcuttingcctctctctct
> tcctctctctctctctctctctctctctctctctctctctctctctctctctctctctctctctYes

3 Silva's version can be compared with the original transcript on the UK Government website here: https://www.gov.uk/government/speeches/big-society-speech.

Cutting the national deficit
Falls into that camp
We're happy about that (61).[4]

In the course of the performance, Silva also directly works from live Twitter feeds (coming from the people she follows via the account she has specifically created for the show here: https://twitter.com/oppositionsilva), demonstrating how her particular process works in real time with verbatim sources. In Point Blank's *An Evening with Psychosis* (2009), this time the verbatim narratives from mental health workers, patients suffering from psychotic episodes and their families are relocated inside a science-fictional spaceship. Clearly, these three examples bring out a new kind of tension, pointing not only to the act of making the verbatim show and/or the exploration of the practitioners' own subjective confrontation with the verbatim material (in the manner of the new realist verbatim dramatists), but also to the absence of any ounce of stage realism. Here, the spotlight is no longer cast on the fantasy of accuracy and exactitude and the editing techniques openly manipulate the recorded material through all kinds of energising distortions, greatly adding to the verbatim repertoire of devices in performance.

It is my contention, therefore, that these devices entail a change in the dominant paradigm of verbatim theatre – that is to say, the mimetic re-performance of recorded verbatim elements from the real world on a stage – since the further back the strand goes in this direction, the more it becomes audaciously anomalous as it foregrounds its own instability and ceases to exist as verbatim theatre, becoming a witness to its own oblivion. Or to be more accurate, the more its inherent verbocentric premises do not cohere with its inflated phenomenological aesthetic dimension in performance, apparently thriving and luxuriating in their dissonance. In this changed atmosphere, whereby new organising principles entirely remodel the verbatim theatre architecture, dispensing with mimesis and its minutiae of "accurate" details, the documentary component tends to be disavowed and evaporates into a complex and constant dispossessing aesthetic reformulation, as if driven by an unabating and roaring compulsion to exceed verbatim reality, to deconstruct once and for all the very last emanations of

---

4 The distortion of words as well as the "excessive" use of non-words in a verbatim performance is not necessarily post-realist. For example, Chris Goode's *Hippo World Guest Book* (2007) features language conglomerates such as "fddljdafaweuihεjdfhajhllkjautnufgiuuigaiguliaerulehu-blapajghjifhh498uhuuhjhljkhugiahguoiudhyaudyfuifyuifyiudsfyuidfyuifyiudfyuhfjojdh-fuuuuehkjhdfuieuliujdll;ufhfhulf" (qtd. in Rebellato, "Writing" 161) that are a documentary realist reproduction of actual internet messages.

the referent. In other words, in this chapter, practitioners are working against the conventions of verbatim theatre and perhaps even those of mainstream British theatre, in the sense that they no longer sustain the tension deemed so important by their "new realist" peers.

Verbatim theatre has clearly attained here its least coherent and consistent formulation as things forcefully and relentlessly fall apart in performance works that seem to document both an unsustainable withdrawal from verbatim theatre as verbatim theatre and its post-realist appraisal. Perhaps, one might even conceive that, here, verbatim theatre, unlike "regular" verbatim performances, might be no more than – in the words of Savas Patsalidis and Elizabeth Sakellaridou – "a collage of written fragments devised to fit a performance piece" (7), or to be more exact, a collage of collected spoken fragments within our present predicament. One can certainly afford to reduce verbatim theatre even further. This is what happens, it seems, in DV8's *To Be Straight with You* (2008) – the first case study in Chapter 4.1 – whose aggregation of verbatim words disappears into a dismantling and disorienting physical performance so rich that it alienates the audience from any sense of realism.[5] This "new wave" and profusion of verbatim works correspond to an imagistic and performative take on the verbatim component. Put differently, the verbatim document is perceived itself as a performance or, rather, inseparable from performance and its aesthetic corollary: that is to say that the verbatim material is in a constant state of becoming.

4.0.2 Case Studies

What is peculiar to this post-realist batch is that these productions offer a very different aesthetic transcription of the real that seems to annihilate the visceral frisson of proximity always at work in the exacting performance vocabulary of verbatim theatre. In other words, these three contributions – DV8 Physical Theatre's *To Be Straight with You* (2008), Alecky Blythe and Adam Cork's *London Road* (2011) and Jonathan Holmes's *Fallujah* (2007) – are of a rather different kind from the other case studies in this monograph. As to be seen here, post-realist verbatim theatre takes many forms along different and grandiloquent lines. Its disparate, rich, provisional and polyvalent nature with no principle of coherence or any systematic aesthetic logic, for that matter, apart from a perceived de-

5 The show opened on 6 December 2007 in Berlin, Germany, but started touring the UK in April 2008. As the case study is based on one performance at the National Theatre, I have decided to retain the year 2008 in all references.

parture from the authority of the verbatim material, defies all attempts at order and makes any critical argument particularly tentative.

Arguably, there is something troubling in the idea of "post-realist" verbatim theatre that may seem immediately like a contradiction in terms, as the opening quote by Shane Kinghorn suggests, verbatim theatre abolishing verbatim theatre or at least undermining itself.[6] This being said, and drawing from the corpus of performances, a post-realist paradigm in verbatim theatre is to be envisaged as a series of possibilities or opportunities rather than as a definite category of theatre-practice carried to its logical conclusions in current performances. In other words, post-realism is visible when theatre-makers are working at some unassuaged and untenable limit of the strand, adding contradiction upon contradiction. Alecky Blythe and Adam Cork seem to have managed it with sheer brilliance in *London Road* (Chapter 4.2) – the second case study – and so has DV8's artistic director Lloyd Newson with his series of physical verbatim pieces (Chapter 4.1). This is also clear in a theatrical moment that increasingly sees the evaporation of the boundaries between verbatim theatre and immersive practices as exemplified by the last case study: Jonathan Holmes' *Fallujah* (Chapter 4.3).

Having said that, I now want to assert the appropriateness of this particular category and its heavy load of signs for the verbatim performances studied here. In order to do so, it will be necessary to concede that the search for, and aspiration towards, a post-realist verbatim theatre have to start from what is at the centre of the postdramatic proposition – which I shall go on to expand – as these productions rely heavily on what has been considered thus far as negligible non-essentials. However, if it is clear, as to be seen, that post-realist verbatim works negotiate a strong engagement with postdramatic discourses, it remains doubtful that postdramatic theory alone will be helpful in this enterprise. In this analysis, Lehmann and his theorisation of postdramatic theatre will thus be enlisted as a means of describing the aesthetic of "post-realism" which occurs, it seems, when the physical rules of verbatim theatre are suspended, the verbatim material is no longer the rudimentary force in performance; this is the final nail in the coffin of documentary realism. Once verbatim theatre severs itself from its original signifier status and radically changes its physiognomy, it follows that its main characteristic becomes the eschewing of any realist framing. One question that automatically springs to mind would be: is this still verbatim theatre? According to Timothy Youker's understanding of documentary

---

6 On this note, Patrice Pavis rightly questions whether "one can still find it and identify it distinguish it from other genres" (*Routledge* 260) and then contemplates the possibility that verbatim theatre has become "a basic ingredient for any kind of stage cuisine" (*Routledge* 260).

theatre, which could very aptly be applied to verbatim theatre, "a work remains documentary theatre in that it centres on the self-conscious knotting-together of a performance, a body of documents, and a non-fictional subject" (219–220). The question then becomes: can it become post-realist at all within the ontology of verbatim theatre? Michal Lachman seems to have cleared the ground for us by positing, albeit with a dose of irony, that:

> [v]erbatim drama is post-dramatic as much as it is post-literary, post-fiction and post-journalism. Is it post-postmodern? Well, perhaps yes because it innocently believes that the objective truth of facts can be reconstructed from the chaotic pool of rioting, subjective voices. ("Colours" 321)

For all that, there must be, so the deduction might continue, a post-realist verbatim theatre that is clearly part of a larger conversation and it might be worth further investigating how it performs and interacts productively with the highly disjointed, visual and sensory domain of the postdramatic.

4.0.3 Verbatim Theatre and the Postdramatic

First of all it is clear that, to a certain degree, verbatim theatre – particularly in its "new realist" configuration – was already engaging with some strategies that one may choose to call "postdramatic", notably when it strove to establish a more immediate interaction with audiences (for example the predominance of direct address and monologues, the erosion of the notion of character and a constant foregrounding of the real). However, I intend to argue here that the specificity of the post-realist subset of verbatim theatre is exactly its broader and more self-conscious use of these techniques. Before being in a position to pursue this analogy any further, it is necessary to define what I mean by "postdramatic theatre."

4.0.4 Definition and Context

Postdramatic theatre began to be theorised in the 1990s, coinciding with the resurgence of verbatim theatre. In 1999, theatre scholar Hans-Thies Lehmann published in Germany the seminal work *Postdramatic Theatre*, which was then translated, in a slightly abridged version into English by Karen Jürs-Munby in 2006 with the powerful international impact it has since come to represent. Today, the term "postdramatic theatre" covers a very wide spectrum of contemporary practices in the Western world that are perceived as opposed to "dramatic theatre", a model of theatre "subordinated to the primacy of the text" (21). For Lehmann at the time, it was an answer to a theoretical problem as the development of new types of theatre and performance since the 1970s urgently required a new vocabulary with which to assess them. The term can be understood as a way of designating a new type of avant-garde theatre that eschews or undermines traditional and dramatic components such as characters, to the benefit of abstract notions such as image and sound. Importantly, and despite the popularity of the term, some scholars have strongly argued against Lehmann's definition of the dramatic which appears to be somewhat reductive – perhaps even historical – and therefore problematic in view of today's progressive dramatic output. In essence, even in its most finely elaborated format, Lehmann's theorisation of the dramatic has understandable difficulty in articulating its concepts with its implicit but no less extraordinary proposition that aligns the dramatic to a discredited logocentric model. This means that the key factor in order to distinguish the postdramatic from the dramatic model is not necessarily the medium but the broader philosophical framework. As Liz Tomlin reminds us "[t]he exclusive alignment of text-driven performance with the dramatic model has undergone significant developments since it was first proposed" (*Acts* 62). These strong ideological implications in turn have entailed an attempt to "rescue", especially within a British context, more and more text-driven theatre works from Lehmann's bold attack on the dramatic.[7]In other words, more and more text-driven pieces were called postdramatic in Britain such as those by David Greig, Martin Crimp, Simon Stephens, Dennis Kelly, Caryl Churchill and Tim Crouch. It is significant that Lehmann himself considered the work of British playwright Sarah Kane (*Postdramatic* 18; "Word") within the realm of the postdramatic even though her work did not technically fulfil one of his key criteria for eligibility. Arguably, even if Will Hammond and Dan Steward only conceived verbatim theatre through its engagement with the dramatic in their *Verbatim Verbatim* (2008), it has been relatively exempt from such an ideological debate and to understand why this has been the case, one must look into its relationship with the postdramatic project in more detail.[8]

---

7 See for instance David Barnett's 2008 article "When is a Play not a Drama? Two Examples of Postdramatic Theatre Texts".

8 For instance, they define verbatim theatre by stating that the words of real people are "edited, arranged or recontextualised to form a dramatic presentation" (9). Hammond adds later on in relation to the tribunal plays that they "can be described in dramatic terms, involving protagonists and conflict and classic dramatic structures" (153). Obviously, much has changed since this was written in 2008.

4.0.5 A Fluid and Complex Relationship

First of all, in a most interesting way, it should be remarked that such a *rapprochement* is in fact not novel at all since documentary theatre is itself dispersed into the escalating activities of the postdramatic and has entered into resonance with the concept ever since its inception, notably by destabilising the author-function and by being somehow detached from the dramatic logic of conventional plays.[9] Put in another way, the idea of "postdramatic" theatre implies a decisive break with the conventions of drama, something which is – as seen earlier – in the marrow of verbatim theatre. To cite Lehmann again, one of the characteristics of the postdramatic is precisely its blurring of "the distinction between the spheres of the aesthetic and the real" ("Tragedy" 99), a feature built into verbatim's very mode of being. As a result, many scholars have underlined the relationship between the postdramatic project and the forefathers of contemporary British verbatim theatre. For example, Marvin Carlson has recently claimed that "Piscator's rejection of psychological and character concerns make his work an early example of what Hans-Thies Lehmann has called 'postdramatic theatre'" (23). One can also recall that Lehmann himself writes in his *Postdramatic Theatre* that "[t]he genre of *documentary theatre* developing in the 1960s also points some way beyond the tradition of dramatic theatre" (his emphasis, 55).

It is time now to clarify the proposed relationship between contemporary British verbatim theatre and the postdramatic by pointing out, as Merle Tönnies does, that from the outset "'postdramatic theatre' with all its associations of formal experimentation [...] is seemingly far removed from 'real life'" (355) and, therefore, from the main concerns of a verbatim theatre practice. At the same time, and most pertinent to verbatim theatre, "the postdramatic finds many ways of pushing theatrical representation in the direction of greater authenticity, often destabilising the dichotomy of the represented and the 'real'" (Tönnies 357). Similarly, Liz Tomlin acknowledges the ambiguity of the relationship between verbatim theatre and postdramatic performance in her *Acts and Apparitions:*

9 This logic is reductively perceived under Lehmann's theorisation as conservative and logocentric as opposed to "a radical, poststructuralist postdramatic" (Tomlin, *Acts* 55). For Liz Tomlin, Lehmann's charge is in fact levelled at a subset of the dramatic output, namely "a discredited notion of the social-realist model of theatre" (*Acts* 57), according to his own definition of the dramatic.

> if verbatim practice shares some common ground with a postdramatic theatre practice in its rejection of the illusory, fictional world representations of the dramatic, it diverges significantly from the majority of new performance practices in the 1990s in its address to explicitly political and topical 'real' events. (117)[10]

Interestingly, Jérémy Mahut argues that "this type of theatre presents the peculiarity of undermining the theoretical foundations of dramatic, epic and postdramatic theatre" (1) whilst Jerome Carroll, Karen Jürs-Munby and Steve Giles include verbatim theatre in their study of postdramatic theatre, arguing that the strand "takes recourse to historical reality by using authentic historical documents or accounts, as it were *short-circuiting* the representative relationship" (5).[11] If a consensus cannot be reached, perhaps because of the elusiveness of today's verbatim theatre, it is clear that the notion of the postdramatic has broad applications for verbatim theatre, especially for the post-realist position I have observed. For now, let me attempt here to illuminate the formal elements and "aspects" in performance that presumably give shape to a postdramatic theatre practice and also consider their potential impact on verbatim theatre.

It appears, following the logic I set out, that not all verbatim theatre is post dramatic and it was consequently deemed necessary therefore to use the terms "postdramatic verbatim theatre" (Wilmer), "postdramatic reality theatre" (Garde and Mumford) or even "postdramatic documentary theatre" to signal a specific strand of performance practice. Put in another way, there is something within verbatim theatre that actively resists the terms of postdramatic theatre and it would seem that documentary realism is no stranger to this *a priori* resistance. It follows, as Jerome Carroll, Karen Jürs-Munby and Steve Giles jointly put forward, that in its postdramatic variant, verbatim theatre "do[es] not sustain a presumed authenticity in representation by means of the faithful and direct presentation of material, not least because such recourse to external, reliable, authentic reality is assumed to be unavailable" (26). In other words, verbatim theatre has aesthetically and epistemologically completely moved away from documentary realism in this new format, as its capacity to prompt strongly contrasting interpretations must illustrate.

Ulrike Garde and Meg Mumford provide the clearest definition of such an aggregation of concepts by stating that these performances: "create porous and

10 Others have remarked upon the ambiguity of verbatim theatre as regards the postdramatic discourse (see, for instance, Taylor, "Voice" 378–379).

11 My translation from the original French: "[c]e théâtre présente la particularité de bousculer les concepts théoriques qui définissent le théâtre dramatique, le théâtre épique et le théâtre postdramatique" (Mahut 1).

ambivalent worlds where real-life people, stories and places invade and are invaded by the frame of the stage" (148). They add that in these works "fiction and reality meet or even become indistinguishable, and fiction can seem to put us more in contact with truths than facts do" (148). Here, the audience witnesses a rupturing of the sealed nature of the verbatim performance that makes them "uncertain about how the people and events they are experiencing are connected to representation, reality and fiction" (Garde and Mumford 151). Furthermore, the area of extensive overlap between my conceptualisation of post-realist verbatim theatre and postdramatic theatre is made explicit through a focus on spectatorship. Typically in these performances, as to be seen, the spectator is exposed to a wider variety of stimuli (choreographic, audiovisual, musical, kinesthetic, physical, verbal, theatrical etc.). In other words, a tension is often created between the textual elements and the other stimuli in performance. For the purposes of this chapter, then, verbatim theatre can be said to enter the postdramatic frame and its operating terms when its verbatim text is no longer the unique or most important element in performance and becomes instead a non-hierarchical dramaturgical element against which the performance is constructed. For instance, a piece like Strawberry Blonde Curls Theatre Company's *Tanya* (2016) on asylum seekers mixes "spoken word, drama, dance, puppetry, live music, sign language and verbatim theatre."[12] In this context and in a more overt way than in new realist verbatim theatre, representation is no longer subverted to the mimetic order but to the performative and physical presence that takes precedence over conventional dramatic characterisation (see Lehmann 134–136).

In fact, for Lehmann, even some of the most innovative strategies such as the:

> use of choruses, narrators, interludes, plays-within-a-play, prologues and epilogues, asides, and a thousandfold more subtle openings of the dramatic cosmos – including in the end even the Brechtian repertoire of epic ways of playing – could all be incorporated and added to the drama without destroying the specific experience of dramatic theatre. Whether or not lyrical forms of language were effective within the dramatic texture, and to what degree epic dramaturgies were applied, made no difference in principle: the drama was able to incorporate all of these without losing its dramatic character. (Lehmann, *Postdramatic* 22)

In my view, postdramatic theatre and post-realist verbatim theatre align, not only through the use of common strategies that will be explored in the following

12 Quote taken from the official Vimeo account of the SBC Theatre Company here: https://vimeo.com/176805247.

chapters but, precisely, in the sense that what the postdramatic does to the dramatic is similar in intent to what post-realism does to documentary realist verbatim theatre. Now, to come back to my main concerns, the kind of shifts hitherto described and which have quickly been made into the principles of post-realist aesthetics may appear to be entirely governed by the postdramatic. As a result, this study could simply apply Lehmann's theory to the new corpus of verbatim performances and identify the same five aspects that are space, time, media, bodies and text. However, even though Lehmann is not overly prescriptive as regards what constitutes a postdramatic style of performance, he identifies a "palette of stylistic traits" that includes the following: "parataxis, simultaneity, play with the density of signs, musicalization, visual dramaturgy, physicality, irruption of the real, situation/event" (*Postdramatic* 86). The techniques listed in his palette do not necessarily match the strategies of what I call "post-realist" verbatim theatre even when they both suspend the meaning of the verbatim words through a practice of assemblage as opposed to montage.[13]

To understand "post-realism" as a new aesthetic attitude towards verbatim that interrupts the continuing encroachment of realism and further develop these considerations, one must at once turn to some of its practitioners' works. What follows may thus help to ground these somewhat abstract remarks in the particular verbatim performances with which this chapter is concerned.

13 Here I use these terms in the way Patrice Pavis has defined them: "*Montage* always happens in accordance with an underlying and meaningful logic: assemblage, on the other hand, is distinguished by its evident lack of organization and its art of incorporating the most disparate elements" (*Routledge* 14).

4.1 "Physical Verbatim Theatre": DV8's *To Be Straight with You* (2008)

In this chapter, I look at a kind of verbatim theatre that is perhaps most difficult to define due to its eclecticism in contemporary performance. From a historical standpoint, a case might well be made that trying to conjoin verbatim material and dance is something that was very much characteristic of the pre-1990s phase of British verbatim theatre. Documentary theatre-maker Rib Davis asserted in 1986 that "documentary drama [as opposed to what he calls – ironically for us in this context – "a straight play"] forces you into more imaginative solutions – perhaps mime, or puppetry, or an otherwise utterly non-naturalistic presentation, such as stylised dance across the dialogue" (10). What possibly necessitates further examination, however, is the way in which these "solutions" relate to and depart from the notion of stage realism.

These remarks aside, I intend to demonstrate here that today's physical verbatim theatre constitutes a significant departure from these earlier practices (that could arguably be framed in terms of new realism) and therefore deserves one's most serious attention.[14] In order to do so, one must turn at once to an important figure in this history that is identified by Lehmann as one of his eighty-four practitioners of the postdramatic (24) since his work with DV8 Physical Theatre is clearly set outside the parameters of the dramatic model and challenges the spectatorial habits of visual reception.

4.1.1 Lloyd Newson and DV8 Physical Theatre

DV8 Physical Theatre was formed in 1986 and, led by Lloyd Newson, quickly became "one of the pre-eminent UK-based dance theatre groups" (Balme 169), representing an enormous shift in both the content and process of creating theatre dance.[15] As they state in their artistic policy, DV8's creative approach is "on reinvesting dance with meaning, particularly where this has been lost through for-

14 Alternatively, this chapter also interrogates how this strand of verbatim theatre departs from contemporary verbatim pieces that simply include dance. For instance, Alecky Blythe's headphone-verbatim *Cruising* (2006) inserted some passages with dance that had no verbatim text in them. Here, the fact that the physical verbatim performances work with both at the same time changes the overall aesthetic experience and needs to be discussed in detail.

15 This term is used to designate "dance forms that are performed primarily in a theatrical context" (Balme 6).

malised techniques”.[16]More specifically, inspired by the work of Pina Bausch[17], *Tanztheater* and European dance theatre, director-choreographer Lloyd Newson, in the formidable 30 years of leading DV8 Physical Theatre, has always pursued what he calls “stylised naturalistic movement”, which he defines as a movement whose:

> origins are embedded with meaning – as were most dance forms until they were abstracted out of existence. If someone walks into this café, you have an immediate reaction about the way they look, how they hold their body, what their body’s telling you. As a creator, you become aware of that information and you find ways to reveal that formally, stylistically. But its source must always be clear. (Tushingham 46)[18]

Australian-born choreographer Newson – who was awarded an OBE for services to contemporary dance in 2013 – tried to find within a “high-energy performance aesthetic” (Harvie, “DV8” 69) a vocabulary of movements that holds a notion of body language that everyone understands and then he extends, heightens and exaggerates it in its conversion of lived experience into stage images. In other words, he explored how he could make highly stylistic movements talk about something important, about issues and ideas. This is to say that Newson stripped the performer of the codified gestural language of modern dance but also, to some extent, as to be seen later, the verbatim material of its referent.

Newson first worked with verbatim material in his *MSM*[19] (1993) – a performance for seven actors-dancers accompanied by a sound score composed by Jocelyn Pooks – as he conducted, with the help of six actors, sixty formal interviews with men on this topic (using especially the experiences of closeted men) and decided to keep nineteen of these for the final piece at the Royal Court Theatre.[20] He had also previously used tape-recorded interviews to inform his creative process, notably in his 1987 piece *My Body, Your Body* on the psychology of women who seek relationships with abusive men. In recent years, quite clearly,

---

16 This quote is taken directly from the artistic policy published on their website at the following address: https://www.dv8.co.uk/contact-dv8/artistic-policy.

17 Notably, Pina Bausch allowed her dancers to speak on stage and included techniques from other disciplines.

18 As I write these lines and, after the tremendous success of their latest physical verbatim piece (*John*), the company has been put on hold as the director is taking a break to reflect and think about the future.

19 “MSM” is a sociological term that is used to designate Men who have Sex with Men in public lavatories.

20 The piece was co-produced by the Royal Court and Nottingham Playhouse.

his London-based company DV8 Physical Theatre,[21] an independent dance collective, has been in the vanguard of the development of such hybrids with verbatim material in performance.[22] Their first proper foray into physical verbatim theatre can be said to have been in 2007 with *To Be Straight with You*, the piece I am about to discuss in more detail. After having seen Tamasha Theatre's *The Trouble with Asian Men* (2005), Newson started to closely collaborate with Louise Wallinger of Non-Fiction Theatre (see Chapter Five) as he became fascinated by the headphone-verbatim technique and verbatim theatre more broadly.[23] His contribution to the ever-growing field of verbatim theatre did not, however, end with this case study. In 2011, he created *Can We Talk About This?* – which won the prestigious Helpmann Award – a piece that focused on Islamic fundamentalism, multiculturalism in Western democracies as well as freedom of speech.[24] More significantly, perhaps, *John* (2014) – on men, love and sex – was their first verbatim performance to predominantly present a personal perspective instead of a collage of views.[25]

4.1.2 Definition

> There is actually no such thing as physical theatre. (McDermott 207)
>
> Physical Theatre is not codifiable. The term is applied to such a diverse range of work that it has become virtually undefinable. (Callery 5)

To add to Phelim McDermott's provocative statement and to Callery's oft-repeated warning above, the researcher, eager to make some sense of these unusual practices, encounters once again other terms such as "verbatim physical theatre," "documentary dance" (Wake, "Headphone" 331), "oral history-based performance about physicality" (Friedman 477), "dance-based verbatim theatre"

21 The name DV8 partly "refers to the 'deviant' homosexual identity Newson declares in his work" (Leask 228).

22 In Australia, Kate Champion, who has previously worked with DV8, now uses verbatim material in her dance pieces such as *The Age I'm in* (2008). Younger British companies such as those performing at the National Student Drama Festival have also been heavily inspired by DV8's approach to verbatim theatre.

23 Louise Wallinger advised Newson on the headphone-verbatim technique and ended up working as the sound editor of interviews for *To Be Straight with You.*

24 The piece attempted to portray an overall perspective through the contrasting testimony of 33 people with direct first-hand experiences.

25 It was also the first verbatim performance to be broadcast live to cinemas around the world via the National Theatre Live Scheme.

(Leung 2013) or even "verbatim dance theatre" that are sometimes used interchangeably to describe the kind of verbatim work under study in this chapter. To make matters worse, even when one agrees on the terminology, for Simon Murray and John Keefe, one should rather talk of "physical theatres" and not "physical theatre",[26] which inevitably complicates any attempt at a definition. The usual deluge of terms has also inescapably led to the existence of the "post-physical performance" (Chamberlain 2007) as, for some, the "physical theatre" appellation "no longer describes a movement of renewal in British theatre and performance [...] nor even a particularly useful critical term" (Chamberlain 120).[27]

On the occasion of the "Devoted and Disgruntled 11" events organised at the Birmingham Repertory Theatre in January 2016, and more particularly at the session focusing on "Physical Theatre and Verbatim Theatre – Creating Dynamic Performance", attendees rightly asked the following fundamental questions: "Is it possible to define the genre of 'physical theatre'? Isn't all theatre physical?."[28]The term "physical theatre" – that entered the mainstream thanks to the emergence of leading British experimental dance company DV8 Physical Theatre in 1986 (Murray and Keefe 14) – has now become "a catch-all phrase to describe touring theatre companies whose work has a strong visual dimension, companies who have developed a theatrical style which focuses attention on the physicality of the performers, and those defining themselves as 'new mime'" (Callery 6).[29]According to Helen Freshwater, what distinguishes physical theatre from the tradition of body-focused performance art that emerged at the same time is "the central importance of training" ("Physical" 173) whilst for Christopher B. Balme, "[p]hysical theatre could also be used to encompass older forms of primarily movement-based theatre, such as pantomime and/or mime, which have their roots in classical antiquity" (7). The spectrum of performance practice known as "physical theatre" has understandably generated competing definitions. For instance, Andrew Solway gives the following definition of physical theatre: "a theatre piece in which the physical aspects of performance

---

26 See their two 2007 volumes: *Physical Theatres: A Critical Introduction* and *Physical Theatres : A Critical Reader.*

27 For Chamberlain, "post-physical performance" is what comes after "physical theatre" in the works that are happening now.

28 See the published report on their website: http://www.devotedanddisgruntled.com/events/dd-11/reports/physical-theatre-verbatim-theatre-creating-dynamic/

29 DV8 Physical Theatre – which is not a permanent company – was founded by Lloyd Newson along with his collaborators Nigel Charnock, Michelle Richecoeur and Liz Rankin as an independent collective.

are at least as important as the dialogue, often more so" (Chambers 594). Interestingly, Ana Sanchez-Colberg strikes a different balance between these components by suggesting that the term characterises "an eclectic production" "which focuses on the unfolding of a narrative through physicalized events and which relegates verbal narrative – if at all present – to a subordinate position" (40). Patrice Pavis also adopts a similar view in his *Contemporary Mise en Scene* by considering that this type of performance "emphasises the actor's body, rather than their words or their minds" (182). However, looking at DV8's work over the years – especially since its growing incorporation of verbatim material –, it has seen a gradual importance given to the word, to the point that this appears to have become a *sine qua non* for all their productions. Newson himself uses the term "physical theatre" – almost out of necessity – to describe the particular work he does within his company since it seemingly has fewer restrictions and rules about what is admissible under its banner and can therefore encompass "pedestrian or naturalistic movements, circus skills, film, dance, song or text" (Newson "Promo"). As such, this is a category of practice that corresponds to Lehmann's conception of the postdramatic.

What can be gleaned from these successive definitions is that the term "physical theatre" is used to describe an unusual and complex aesthetic practice departing from modern dance and more attuned to devised theatre, a kind of theatre expressed through physical movements ranging from small looks and gestures to violent extremes of action. Most obviously, as it name suggests, it is also a hybrid form – based on the relationship between text and movement and how the two can support one another – that designates a specific performance practice in two main ways:

> 'Theatre' implied a connection with narrative development via action – as distinct from the conventional choreographic structures from which it was wanting to dissociate itself. It also reminded the public of the genre's strong commitment to a full exploration of the theatrical medium and the use of alternative methods of theatrical production which were by the 1980s fully accepted within avant-garde forms of theatre [...] but met great resistance in the general dance scene. However, the term 'physical' – separate from theatre – implied a particular approach to movement, and began to identify a particular style. (Sanchez-Colberg 48)

DV8's artistic policy perfectly embodies these ideas by stating that their work is "about breaking down the barriers between dance and theatre."[30] Clearly, as

30 This quote is taken from their own website: https://www.dv8.co.uk/contact-dv8/artistic-policy. However, one must acknowledge that DV8, in terms of organisation, functions very much like a conventional dance company as its actors train in ballet and contemporary dance.

Fiona Buckland argues, physical theatre is a way of working that draws on both disciplines whilst being "unconstrained by the conventional performance lexicons of either dance or theatre" (371). However, her arguments do not extend to appraising its particular relationship with verbatim and documentary theatre as physical verbatim theatre only reached prominence in the 2000s.

For the purposes of this study, physical verbatim theatre then points towards a more embodied theatrical re-enactment of the interviewees' stories. It indicates a kind of verbatim work characterised by a pronounced visibility of the performer's body that markedly disrupts the conventions of documentary realism and even departs from the terms of new realism studied in Chapter 3.

4.1.3 Methodology

In the case of DV8's Physical Theatre, the case study in this chapter, physical verbatim theatre precisely corresponds to a strand of performance practice that creates its choreographic material in response to some verbatim recordings and draws on these previously uttered words to make movements that have a clear intention and purpose. In other words, similar to the headphone-verbatim technique of Chapter 3.2, here a task such as "listen and repeat" becomes theatre.

Surprisingly perhaps, as far as the verbatim material is concerned, Newson's approach is on the purist side of the verbatim fence. Indeed, he only uses interviews with people that have had front line, first-hand experience of the issues being addressed in the piece and adds no extra words. More specifically, he edits out the interviewer and keeps the interviewees' words that then become mostly monologues in direct address to the audience on stage. His particular method is to encourage physical improvisations in rehearsals that can emulate the verbatim material. In practice, this is an adaptation of both the headphone-verbatim technique I described in Chapter 3.2 and of the Whelan Recording Technique.[31]

Once the DV8 group has collected a series of interviews, they then go into the studio and start to create the stylised movements – or what David Lane calls "interpretative dance" (65) – thereby turning the verbatim material into physical theatre. In practice, this means that the performers have an iPod and

31 The Whelan Recording Technique is based on the principle that actors always respond to the emotional content of a line. In a typical rehearsal using this technique, a recorder is used so as to capture a reading of the script done by the actors present in the room. Once this has been performed, actors are asked to physically act out the scenes to the playback of the recording that they have just made without using their lips.

listen to someone speaking and put movements to the interview so as to try and transform into that "character" as accurately as possible. As in the headphone-verbatim performance of Chapter 3.2, they do not only focus on the words but respond to the perceived emotion in the voice as well as the pauses and breathing patterns. In fact, in a way, their work corresponds to Blesser and Salter's understanding of listening as the "active attention or reaction to the meaning, emotions, and symbolism contained with sound" (5). More specifically, throughout the creative process, the performers listen to the original interviews and repeat them using iPods in the same manner as the headphone-verbatim method seen in Chapter 3.2, while they improvise physical tasks and others physically respond to them. Sometimes, Newson also gives them specific physical instructions as they practice the headphone-verbatim technique. Initially, Newson had in mind that the performers would keep the iPods in performance so that they could hear the interviews as they moved. This proved impossible as even the actors that have extensive experience with the technique – as seen in Chapter 3.2 – found themselves physically tied to the audio and could not really deviate from it (certainly not in the manner of a physical performance). In other words, here the performers' breathing patterns are very different from these of the interviewees, that is to say that their movements do not always reflect their delivery. In rehearsal, it is also very common for the DV8 group to record these improvisations via video technology which help them edit and shape the final piece. In the case of their latest piece, *John*, in 2014, "everything [was] filmed; from the sparks of ideas in workshops right through to performances 10 months later".[32]

What is more, sometimes the physical score is extremely close to the original interviews as one performer, Seeta Patel, explains in the context of their *Can We Talk about This?* (2011): "[t]hrough my improvisations Lloyd found movement he liked and we worked together intensely, in a fairly linear fashion with the text, to build up a sequence" (Bahl and Omi 5). In a scene from *Can We Talk about This?* that portrays an interview with Ann Cryer, a Labour politician, the performer playing her holds a teacup and saucer while a male performer (not in character) seemingly acts as an armchair.[33] Sometimes, on the other hand, it appears as a counterpointing commentary out of synch with the words that has its own

32 This quotation by David Grewcock (Assistant to the Director and DV8 Company Manager) is taken from page 5 of the NT programme's Education pack for students and teachers in relation to the production of DV8's *John* that was published in 2014 and can be consulted online here: https://www.dv8.co.uk/media/files/DV8JOHN%20Work%20Pack%20Final%20Small.pdf.

33 See the complete scene on the official YouTube account of DV8 Physical Theatre here: https://www.youtube.com/watch?v=vNVPumETpuA.

rhythm and meaning. Still in *Can We Talk about This?*, but in a different scene, this time a performer representing Flemming Rose – the editor of Jyllands Posten, the Danish newspaper that published the cartoons of the Prophet Mohammed – puts his trousers on and generally moves in a topsy-turvy manner to suggest that his whole world was turned upside down as a result of publishing those cartoons.

Finally, music scores are selected for each scene and ideas are tested for the set and costumes. There is also an important phase – like in any other high-profile production – that concerns itself with the lighting and sound design. As DV8 Physical Theatre is a touring company, each production changes in the course of its presentations: adjustments are physicalized, new scenes are created and older scenes are further developed – sometimes in response to recent events.[34] Perhaps this is one of the reasons why so few publications in the literature on verbatim have engaged with these physical verbatim performances. Last but not least, the work is also shared with the interviewees who agreed to tell their stories.

Another important aspect of their work – as exemplified by their name "Dance and Video 8" – is a strong commitment to visual media. The company always engages with video elements in their trademark multimedia sophistication. Taken together, these new elements in performance change the precious equilibrium of the verbatim edifice. In other words, what chiefly preoccupies Newson is the gestural and visual rhetoric over the verbal one as well as a concern with language embodiment. Of course, the verbatim words still matter, but they only comprise a small part of a much larger performance phenomenon. In this sense, these performances can be defined in terms of "postdramatic theatre" as they are as much interested (if not more) in the rhythms of speech (slowing down, emphasising certain points etc.) than in the content of the verbatim material.

As ever, DV8 Physical Theatre is illustrative of an approach to verbatim theatre and other groups have attempted to explore the potentiality of this innovative practice such as Theatre Delicatessen's *Pedal Pusher* (2009) on the Tour de France and the work of UK company The Paper Birds with *In a Thousand Pieces* (2008), their 2010 production of *Others* and their 2011 production of *Thirsty* which combined verbatim material, physical theatre and live music. Inspired by DV8, younger theatre collectives in Britain have also adopted this spectacular take on verbatim theatre and a new generation of students is currently devising

34 In the case of *Can We Talk about This?*, for example, changes were made to the production in keeping with the most recent events as the piece toured.

its first shows, following in their footsteps. To some extent, the fairly recent genre of verbatim circus-theatre with Flying Diplodocus's *Don't Mention the F Word* (2015) and Kristine Landon-Smith and Sita Brahmachari's *The Arrival* (2013) can also be considered as physical verbatim theatre as they combine aerial work and verbatim material.

Having just defined the initial parameters of DV8's physical verbatim performance technique, it is now time to look at *To Be Straight with You* more closely.

4.1.4 An Overview of DV8's *To Be Straight with You*

To Be Straight with You was produced in 2007 and marked their second collaboration with the National Theatre. It also marked a significant departure for DV8 Physical Theatre as well as the development of a closer relationship with the National Theatre, as since then all their shows have been performed there. The show opened on the Lyttelton stage on 30 October 2008 in the form of a 90-minute piece with an eight-strong cast and no interval. It also travelled internationally to France – where it won the "Grand Prix de Danse" –, Germany, Australia, Sweden, Switzerland and to the US. *To Be Straight with You* is largely a performance about homophobia and religious intolerance in Britain and around the world and it was inspired by Newson's own experience as a gay man as well as a Channel 4 documentary entitled "Gay Muslims."[35] More specifically, the piece looks at religious fundamentalism and homophobia, particularly within Britain's migrant communities. It is therefore about tolerance and intolerance. Throughout the performance, the theme of a line being drawn is often used, both literally and metaphorically.

For this particular production, Newson employed a full-time researcher and 85 people living in the UK were interviewed – including members of the clergy, the head of a gay police association, people opposed to homosexuality due to their religious beliefs, passers-by, people who are both religious and gay – and, as is common in verbatim theatre, "every word spoken on stage comes directly from the interviewees" (2008, 3). The performance also incorporates the words of human rights activist Peter Tatchell along with other public figures such as the late Nelson Mandela. Among the 85 people interviewed were immigrants who came to Britain to escape from violent homophobia and misogyny in their home country (often backed by the state). In one sense, this is also an in-

35 Newson concentrated his research on the countries from which the majority of British Muslims and Christians originate.

quiry into minority cultures in the UK, be they of an ethnic, sexual, political or social nature.

The multimedia piece productively intermingles verbatim texts with documentary material, dance, music, animation and film. As such, musical scores punctuate each scene and visual vignette.[36] The cast of eight – six men and two women – was largely multi-ethnic[37] and performers change characters all the time and accents including Pakistani, Rastafarian and Cockney English. The performance begins very powerfully with hate speech/"murder music" from Beenie Man and Buju Banton who are prominent Jamaican Reggae dancehall artists, having written songs advocating the killing, burning and shooting of homosexuals.[38] *To Be Straight with You* also includes scenes referring to the prosecuting and torturing of gay people in Uganda, Trinidad, Pakistan, Iraq, Zimbabwe, Nigeria – to name only six locations – and to hate crimes illegally committed in secular Britain. Despite this harrowing material, there are some uplifting dance scenes as well as some comic elements, as when a DJ reveals that his way to fight against the "murder music" is paradoxically to play it in gay clubs so that gay people can subvert it by dancing on it and having fun.

If the many reviews are to be believed, such a radical and extreme approach to verbatim theatre was entirely new at the time and we must at once attempt to understand the kind of contract *To Be Straight with You* is establishing with its audience.[39]

4.1.5 The Contract with the Audience

First of all, *To Be Straight with You* can be said to differ from the previous case studies in that it is aimed at a global, cosmopolitan audience. In this sense, what compels is that the technology, choreography, music and the verbatim words blend smoothly with one another so as to create a mainstream performance language that has a more global, perhaps even, younger, appeal. The audience can take pleasure in the recognition of international hits such as Justin Timberlake's *My Love* and Shakira's *Hips Don't Lie*. The dance lover can fully appreciate the

36 *To Be Straight with You* is in fact divided into segments or "tracks" rather than scenes.

37 Within DV8, there is one very devout Muslim and two other performers from a Muslim background.

38 Quite understandably, British gay rights activist, Peter Tatchell, coined the term "murder music" in the mid-1990s to describe the homophobic works of certain Jamaican musicians that encouraged physical violence as well as the murder of gay people.

39 For instance, Lyn Gardner in 2008 wrote that Newson "reinvents the rules with this piece."

prowess of the performers as well as the spectacular imagery. The gay and gay-friendly section of the audience can relate to some of the situations described such as the "coming out" phase. The verbatim "groupie" may rejoice in the discovery of a brand new style of performance. Responsible and conscientious citizens are also catered for with a series of selected facts and mini-lectures throughout the performance so that – in the words of Tiffany Jenkins – "they can reassure themselves that they are doing something." And of course, the right dosage of humour is provided for everyone's enjoyment.

More seriously, in one way, DV8's *To Be Straight with You* embraces the constraints of verbatim theatre since "every word spoken on stage comes directly from the interviewees."[40]This alone can be said to trigger the usual normative demands of documentary realism, endowing the whole performance with referential weight. However, I would suggest that, here, such a need to reference the original is not always coupled with the comfort of immediate visual and even sonic recognition. Indeed, there is more or less a constant distortion of the verbatim material throughout. Even when the original recording is preserved and replayed live, it is short-circuited by the visual aspect of the performance. One good example of this happens after we have heard Peter Tatchell talk about his stop "murder music" campaign. The abusive language he received as a response, such as "die fucking queer bastard," is then both played live and projected, each word being given a different size depending on its offensive value.

Most pertinent to the concerns of this chapter, however, is the fact that, here, the verbatim material appears in a changed light since the performance offers a physical supplement within a more experimental practice. In addition, contrary to both the documentary realist and the new realist verbatim performances we have seen that had kept a sense of linearity and unity through narrative, *To Be Straight with You* completely embraces fragmentation. For instance, after we have heard the story of a gay British man of Nigerian descent augmented with projections, one encounters performers wearing horse head puppets and dancing to the tune "Clap Your Hands and Say Yeah". A BBC radio recording immediately follows this, delivering the news whilst the performers – still wearing the horse head puppets – hold cards that read "Love Sinner Hate Sin". This brings us to the overall structure of the performance which appears to be, at first glance, somewhat incoherent, as is often the case in postdramatic theatre and in post-realism, as to be seen in the following pages.

One of the signatures of post-realism in verbatim theatre is that the terms of the contract with the audience are reconfigured so as to become a lot less pre-

40 This appears as a projection at the very beginning of the show.

scriptive, placing an emphasis on greater interpretative freedom. As Dymphna Callery argues, "physical theatre accentuates the audience's imaginative involvement and engagement with what is taking place on stage" (5). In *To Be Straight with You*, this indeed appears to be the case. The movements are conceived in such a manner that, even if one has no knowledge of dance, one will understand the emotional power of the body language and the visual imagery that is going on. Simply put, more questions linger as the verbatim text no longer determines everything and a performance engine is working underneath it. More than in the new realist performances of Chapter 3, here the performance no longer aims to reproduce an original but to create a new experience through a great variety of performative and stage design elements. As I have said, in *To Be Straight with You*, no main story seems to unfold and the audience experiences instead a series of (selected) fragmentary narratives on the subject of homophobia on several terrains at once. Arguably, these elements prompt within the spectator an actively critical attitude towards the verbatim material.

In addition, there is a deliberate blurring between the live and mediated, the digital and the physical, the aural and the visual that fosters a different way of looking at verbatim narratives and producing meaning. If the aesthetic of new realism had the potential to become a site of inquiry regarding the construction of narratives through its level of self-reflexivity, *To Be Straight with You* in comparison would seem to disintegrate or rather displace the authority of the verbal in a way that may appear irreconcilable. As a theatre strand that encourages this relationship, physical verbatim theatre is seen as a set of artistic choices, that is to say that the movements are not a verbatim re-transcription of the interviewees' words such as those to be found in the verbatim play *Black Watch* by Gregory Burke, an example of which follows:

> The Sergeant enters with a bundle of airmail letters (blueys). Stewarty notices him and takes the letters. He opens one and starts reading it, the words giving him comfort. Another soldier enters and takes the remaining letters. Stewarty creates a subconscious sign-language which expresses the content of his letter. One by one the soldiers enter, take the bundle of letters and, finding the one addressed to them, repeat the process for themselves. (39)

Instead, the movement material is sometimes no less than a radical intervention, as it insistently impresses itself on the verbatim narratives. The phrasing of the movements is sometimes more or less literal, in the sense that it is in tune with the rhythm of the verbatim speech. However, some movements are semi-abstract and therefore post-realist as regards to the verbatim material. Thus, connections and tensions are created as these elements engage in a performative relationship with the verbatim material.

A last change in the aforementioned contract has to do with the fact that the physical elements are constantly competing with the verbatim and theatrical ones. Susan Foster, in her *Choreographing Empathy* (2011), considers that there are three main components in the experience of watching a dance performance: the choreography, the kinaesthetic sensation – "a designated way of experiencing physicality and movement that, in turn, summons other bodies into a specific way of feeling towards it" (2) and the empathetic connection. By "choreographing empathy" in performance, she means the "construction and cultivation of a specific physicality whose kinesthetic experience guides our perception of and connection to what another is feeling" (2). Whilst the DV8 group is not one's typical dance performance, it would seem that the performers embody the principle of empathy as they physically attempt to feel as others.

To ascribe a particular "post-realist" effect on the reception of the verbatim material in performance, the creative team has therefore employed a wide array of strategies which I shall now unpick in more detail. The verbatim material is arguably always imbricated with the live body of the actor. Bodies on stage can be seen as sites of meta-messaging within a verbatim performance, or otherwise. As seen in the last two parts, verbatim theatre appeared particularly vulnerable to the figure of the actor and it remains to study how these considerations are taken onboard within a post-realist environment.

4.1.6 Acting

Within the realm of the post-realist, the performer has a crucial role to play.[41] In one sense, in physical theatre the role of the actor is even more prominent. Now the questions that preoccupy this section are the following: is there a particular "post-realist" acting process? And if so, what would this mean in practice? Does "post-realism" work against the "semiotic surplus" (Megson, "Act" 149) of the actor in the manner of documentary realism? Or conversely, can one entertain the idea that the actor is inherently a post-realist phenomenon? Before answering these questions, I want to talk more specifically about the kind of actors one is concerned with here.

Even though all the performers in DV8's productions are trained dancers with a particular athletic physicality and stamina, they are also extremely

41 Some "post-realist" documentary theatre-makers have created performances without actors such as *Tagfish* (2010) by the Flemish theatre group Berlin and *33 rounds and a few seconds* (2012) by the Lebanese artists Rabih Mroué and Lina Saneh. However, to my knowledge this has not yet happened in the UK (unless one takes into account VR performances).

good actors who are required to speak from the minute the curtain opens to the minute it falls. Clearly, actors in physical verbatim performances need a command of more skills than both the typical dancer and verbatim theatre actor in that they need to be able to talk and move at the same time. In effect, they are selected in relation to their capacity for physical improvisation, their overall dance technique and their delivery of text. Besides the necessity of possessing excellent skills in both dance and acting, performers are chosen based on their interest in the subject matter and general openness. But more than that, as Jeff Friedman wrote in relation to their earlier physical verbatim piece, *MSM*, "Newson insists on cast members' self-identification with the issues and themes that emerge during the data collection and creative process" (485). Arguably, this may be perceived as a complicating factor, one that could potentially affect how the performance engages with post-realism.

One of the defining features of the DV8 approach is that their performances are made in conjunction with the actors and the latter are substantially more involved in the creative process as a result. As one of the actors in *To Be Straight with You* explains:

> We put in weeks of work to produce five minutes of an 80-minute piece. That's why the performers were involved in almost five months of creation before *To Be Straight with You* was presented to an audience. (Bahl 5)

Another crucial difference, beyond their physical prowess, is that they are themselves to some extent the source material of the piece. As far as the physical improvisations are concerned, they rely heavily on each actor's intuition and analysis. In a sense, they are Stanislavskian because the performers relate the recorded material to their own experiences so as to lend the piece emotional weight and involvement: "[w]hen you know the inclinations of your own nature, it is not difficult to adapt them to imaginary circumstances" (Stanislavki 69). In other words, the basis of the preparation may be the headphone-verbatim technique but its execution in performance indicates a clear departure from these principles.[42] This process, therefore, betrays a very different attitude towards the verbatim material. If documentary realism privileged the erasure of all authorial traces, here they are self-evidently on display. Equally typical of a post-realist aesthetic is that the acting is not only based on the verbatim material, but on the images it conveys to the actors. Thus, to create characters, the team

---

42 As seen in Chapter 3.2, the acting process in the headphone-verbatim technique was exactly the opposite. Importantly, Newson also works from transcriptions of the interviews during the creative process.

behind *To Be Straight with You* does not solely rely on the words of the real people but uses physical expression to highlight a specific aspect of the interviewee's voiceprint, perceived personality or recounted experience. Put another way, the performers can be considered as contributors – albeit of the different kind from the interviewees – since the movements ultimately capture and recreate in real time their initial response as experienced. This is in keeping with Callery's understanding of physical theatre that conceives of the actor as a creator rather than a mere interpreter (5). Indeed, the concept of the creative performer seems to directly apply in this context for it is mostly the physical score that creates a sense of character in performance. Once again, this is a clear departure from both documentary realism and new realism that mostly based characterisation on the verbatim language.

In fact, one must not forget that there is also the key process of selecting and sequencing the movements. The company's creative process, in which the actors operate, is a very collaborative one, one that even alters the conventional roles of the technicians. As sound designer and engineer Adam Hooper has explained in an interview:

> The director, Lloyd Newson, gets quite heavily involved with all the technical aspects of the show, so as a technician for DV8, you're not just working with a designer for the lights or sound, you're dealing directly with the director and you have to translate what his ideas are into technical realities.[43]

More surprisingly – and this is a specificity of the verbatim subset of physical theatre –Hooper "was involved with the initial sound editing stages of the interviews, before any of the performers had been hired", instead of starting work after the show had been created. As such, the show is also created "around the capabilities of the technicians". To come back to the performers themselves, they are closely involved in the editing process of the verbatim material as well as the creation of the movements. Once they have downloaded the verbatim audio file into their iPods, the performers start improvising collectively in the context of long rehearsal periods and workshop exercises in which they compose movement responses and re-enact a theatrical mediation of what they hear. In other words, the material is filtered through the performers' imagination. How-

---

43 This quotation by Adam Cooper comes from page 6 of the NT programme's Education pack for schools in relation to the production of DV8's *To Be Straight with You* that was published in 2008 and can be consulted online here: https://www.dv8.co.uk/media/files/To%20Be%20Straight%20With%20You%20FINAL.pdf. All further quotations by Cooper will be from the same source.

ever, and maybe most decisively, the overall process strongly retains the author function and Newson remains responsible for the final product. Tellingly, performer Hannes Langolf – who has performed in their three latest physical verbatim performances – has expressed that "Lloyd makes the final decisions, in the end, it's his voice and vision" (Bahl and Omi 6).

One could argue that there may be something distinctly "post-realist" in this disparate process, notably in the way the verbatim material is filtered before it even reaches the stage. Like my concept of "post-realism", this gradual evaporation of authority still connects with what it attempts to outstrip. In other words, post-realism still operates within verbatim theatre, as DV8 does not completely do away with the author function.

Having discussed the complex process of creation of the DV8 group, this chapter now wants to consider how this might relate to what the audience sees on stage. First, one may well be thinking that, after all, the actor in verbatim performance is theoretically bound to either documentary realism or new realism (with a penchant for the latter). Indeed, one finds in *To Be Straight with You* strategies that are typical of the performances studied in Chapter 3. For instance, in one scene a man plays a woman who has given up her lesbianism for God. Furthermore, the choreographed movement phrases could be read as a mere mismatch between what one sees and what one hears. One could go on and force a "new realist" reading on *To Be Straight with You.* However, I would like to suggest that the arguments are not wholly convincing. Let me briefly explain why this is the case.

One key aspect that crucially tilts the balance in the direction of post-realism – and also the postdramatic in this case – is the curious presence on stage of performers who do not attempt to become characters. Even in new realism, for instance with the headphone-verbatim methodology, the actors were tuning in to the characters they were portraying. In one scene in *To Be Straight with You*, for example, a Christian character standing against a blackboard confesses that he has had homosexual inclinations whilst another performer draws images around him. The performer in question is clearly not a character and could be read as an actor – not in character – performing a task on stage. In another scene that one may choose to call "to be straight with you," one hears a recording where the interviewee is repeatedly using the words "to be straight with you," whilst on stage a performer – again not in character – makes some abstract movements in silence. More surprisingly perhaps, in a scene where the performer is supposedly portraying a female rabbi, the recording of the interviewee is played out as the silent performer spins and spins. In a way, this recalls choreographer Jonathan Burrows'*52 Portraits* (2016) in which, after an interview conducted with one person, a video is created in which a song and movements in a performance give a glimpse into that person, except that here we also have a process of dematerialisation of the interview. In other words, post-realism in physical verbatim performance manifests itself through a heightened physical theatricality of tasks and action.

Now to come back to my earlier point, stating that the actors self-identify with the issues could have an impact on the aesthetic of the performance, I would suggest that this detail – combined with the fact that the actors actually contribute to the material – participates in the creation of a situation of undecidability that betrays the imprint of post-realism. As ever, to understand "post-realism" more clearly, one must turn one's attention to other aspects of this production, especially since *To Be Straight with You* does not always rely on the actors to remediate the original recording. As seen earlier, *To Be Straight with You* required an extensive physical preparation. Arguably as it was a devised piece, such a preparation was in equal measure visual, sonic and verbal.

4.1.7 Staging

Already, a sense has been given of the way in which the aesthetic of post-realism crops up in contemporary British verbatim performances. In this section, I shall further elaborate on these ideas by looking at how the physical, visual and sonic elements are intertwined with the verbatim material in performance to create a structure of association and meaning that goes far beyond both documentary realism and new realism.

Writing a review in 2008 of the show for *The Guardian*, Michael Billington found evidence of what I call "post-realism" in verbatim theatre quite problematic in that "the prodigious inventiveness of Newson sometimes obscured the spoken material" and one experiences "sensory overload". The same year, Lyn Gardner in her review noted that the show was "visually over-busy", which she considered as being a problem. Several reviewers – including Billington and Gardner – also remarked that the music and sound effects made it particularly difficult to listen to the testimonies at times. For one thing, it would seem that post-realism is a curious aesthetic shift that one cannot easily miss. These performances appear to particularly like to accumulate gestural, sonic and visual signs that become vectors of intensity, heightening the experience beyond the terms of both documentary realism and new realism. In other words, these vectors militate against and open up new perspectives for the verbatim material in performance. They are disruptive in a number of ways that make them very distinctive from the works studied in Chapters 2 and 3 of this volume.

First, if one looks at the movements, one is encountering some that are directly drawn from the interviewees' verbatim words and, others, that can be said to multiply their meaning, thereby further implicating the audience in these, sometimes obscure, relations. In short, the non-verbal elements of the performance are no longer figurations that are reducible to their verbatim signifier. The movements are freed at last from the tyranny of the verbatim words in this post-realist formation. However, if the verbatim symbiosis is pronounced dead in this instantiation, this does not necessarily mean that the movements have become indecipherable. In a given moment, both verbatim language and movements can coexist without one overpowering the other. At other times – in a typical post-realist gesture – the movements take precedence and start to disintegrate the feeble causal link that connected it with the verbatim material.

If one chooses to focus on the set – and this may appear slightly odd, but I would suggest that these works encourage such a disjointed reading – it becomes clear that it is also an active participant in the verbatim performance.[44] Firstly, contrary to what one saw in the documentary and new realist verbatim performances, the set itself does not reference a specific location in the real world. Instead, the set resembles "a sound recording studio or a translation booth."[45] This collapsible set features a room with several points of entry, including doors and windows, a series of blackboards – that reminds one of a classroom setting – as well as movable walls. One may well make the case that here lies another manifestation of the post-realist. Clearly, the set is not documentary realist. But, is it new realist? I think not. As discussed earlier, new realism – even with the simplest of means – always references a place in the real world. Here, however, at times the set is completely abstract: the place is neither here nor there. Secondly, the set is augmented by a digital scenography that further complicates matters and even brings a comic distance. For instance, in one vignette, the British gay man of Nigerian descent, mentioned earlier, becomes all of a sudden framed by a digital comic strip that seems to more or less re-tell or rather co-tell his ongoing verbatim narrative. Post-realism is perhaps brought even more forcefully to mind when this performer directly interacts with the dig-

44 When the verbatim material enters a post-realist performance frame, the effect is so peculiar that by losing a sense of wholeness and narrative progression, the audience may find itself, at times, incapable of taking it all in, so to speak. In these moments, a member of the audience may focus on one or two elements instead or completely withdraw in introspection.

45 This quotation by Adam Cooper comes from page 6 of the NT programme's Education pack for schools in relation to the production of DV8's *To Be Straight with You* that was published in 2008 and can be consulted online here: https://www.dv8.co.uk/media/files/To%20Be%20Straight%20With%20You%20FINAL.pdf.

ital comic strip by virtually opening a door or by virtually clocking in for work whilst everything betrays – including the performer himself – that this ordinary gesture is empty.

Taken together, the scenes here are never documentary realist in that they do not attempt to fully recreate the original context of the interview, they are poetic and imagistic and they give a sense of incompleteness. As the DV8 company manager explains, "it might be a gesture, a particular line delivery or an interaction between 2 performers that ignites the concept for a scene."[46] Typical of what I would call a post-realist staging is therefore a practice of *détournement* of the principles of mainstream verbatim theatre combined with a taste for playfulness and code-switching – of which documentary realism is so ashamed – more pronounced than in the new realist performances of Chapter 3. A good example of this is to be found in the way in which the verbatim words are sometimes heard in the original voice through loudspeakers.

Arguably, verbatim theatre's very existence relies on an inevitable process of decontextualisation. It follows that documentary realism has attempted to mask this process by theoretically equating the new context with the original context. New realism, on the other hand, was adamant in declaring from the start that a decontextualisation has indeed taken place and that the new context is not theoretically equal to the original context. Now, post-realism also declares that a decontextualisation has taken place, but strangely, this is not followed by a full recontextualisation. Some elements in the performance appear almost out of place, as for instance, the start of a weather forecast right after the radio interview with Iris Robinson (a former MP) on homosexuality, telling the audience that "the weather's cloudy with patchy rain and drizzle and as we go through the morning and sunny spells [...]."

To come back to the earlier distinction I was trying to make between postdramatic theatre and post-realism, one must look again at the gestural movements in *To Be Straight with You*. As repeatedly claimed, these movements are of a different nature within the piece. Put simply, they can either clarify and support what is being said by adding a kind of subtext that was not necessarily there in the original interview or they can undermine the verbatim component. In both instances, one is in the domain of the postdramatic since the choreographic material is at least as important as the verbatim material. However, only in the second instance is one truly meeting the terms of post-realism as I have defined

---

46 This quotation by David Grewcock (Assistant to the Director and DV8 Company Manager) is taken from page 5 of the NT programme's Education pack for students and teachers in relation to the production of DV8's *John* that was published in 2014 and can be consulted online here: https://www.dv8.co.uk/media/files/DV8JOHN%20Work%20Pack%20Final%20Small.pdf.

them, that is to say, when the physical movements start subverting the verbatim words. One may recall at this point that some of the strategies seen in Chapter 3 proved to be particularly disruptive of the verbatim material. Whilst I may concede that this was indeed the case, these strategies simply called the material into question without really achieving any subversion.

Now that I have hypothesised a probable subversion on the basis of these observations of the post-realist phenomenon in physical verbatim theatre, and more particularly in DV8's *To Be Straight with You*, one last question must be examined: if post-realism has the capacity to sunder any of the relations that verbatim theatre strove to maintain at all costs despite the unavoidable slippages and excesses of performance, does this mean that it has lost any political potential?

4.1.8 Political Theatre

> I hope that through this work audiences will become more aware of the lives of people hidden under the veneer of a liberal and supposedly tolerant society. (Newson "Interview")

In the pages below, I set out to explore the idea of a post-realist political theatre based on the case study of DV8's *To Be Straight with You*. Arguably, the post-realist version of political theatre must be doing something relatively different from more conventional political performances. A key consideration in this section will be the one posited by Markus Wessendorf in the following:

> The major challenges for a postmodern drama intent on tackling the pressing issues of our time is how to combine and integrate a postmodern aesthetic that playfully explores dramatic form with a dramaturgy that critically engages with current events without falling into the trap of (re-) producing the master discourses of traditional political theatre. (332–333)

But before digging further into these problems, it is worth remarking that London-based DV8 (Dance and Video 8) has acquired a reputation as a performance group which creates "powerful socio-political pieces" (Lansdale 117) with "an oppositional agenda" (Peacock 209) that disturbingly and urgently attempts to shake people out of their complacency so as to provoke audiences to think and to incite re-engagement with certain real world issues including accountability, political correctness, liberal silence, cultural relativism and complicity.[47]Indeed, in their artistic policy, the DV8 group describes their practice as "determined to be radical" and, as *To Be Straight with You*'s actor Ankur Bahl has claimed, "you can't be in this show without tremendous grit, determination, persistence, and a steadfast commitment to the power of theatre to achieve social change" (5).[48] Similarly, *Can We Talk about This?* was dubbed as nothing less than "the riskiest show of the year" by Dominic Cavendish in an article published in *The Telegraph.* Therefore, it seems reasonable to suggest that there truly must be something at stake in their work, if one considers the fact that the television adaptation project for two of their physical verbatim theatre pieces – *To Be Straight with You* and *Can We Talk about This?* – by both the BBC and Channel 4 was inexplicably dropped. On their official website, the DV8 group makes it very clear that this was nothing less than a blatant act of censorship to prevent their work from reaching a larger audience:

> *Can We Talk About This?* is the company's most recent stage production. It is a verbatim theatre work dealing with freedom of speech, multiculturalism and Islam. From the 1989 book burnings of Salman Rushdie's *Satanic Verses*, to the murder of filmmaker Theo van Gogh and the controversy of the 'Muhammad cartoons,' the production examined how these events have reflected and influenced multicultural policies, press freedom and artistic censorship. . . . We felt that it would be an excellent production to adapt for film, and approached both the BBC and Channel 4. C4's commissioning editor for Arts, Tabitha Jackson, who had seen the live performance in London, championed the project, but it did not get approval by the Channel 4 bosses. BBC's commissioning editor for Music and Events, Jan Younghusband, who had commissioned *The Cost of Living* while at Channel 4, met with us, but informed us later that it was turned down for budget reasons. We offered to discuss the finances, but did not hear from her again. We subsequently wrote to Jan Younghusband asking her if she would consider filming our prior work *To Be Straight With You*, also a verbatim work, about Christianity, Judaism and Islam in relationship to homosexuality as she had expressly told the artistic director how important this work was. We heard nothing back. (DV8 2014, paras. 3–6)

To come back to *To Be Straight with You*, as Newson has expressed it himself, its power is that "it overrides sweeping statements: many of the stories don't fit nicely into politically correct, left-wing or right-wing boxes" ("Prejudice"). The vast array of opinions expressed – even when they are arguably radical or do not conform to political correctness – are serving the overall aim of educating, persuading, or transforming the spectators in attendance through an emotional engagement. In fact, there is also evidence that *To Be Straight with You* encour-

47 The term "Video 8" designates an earlier form of video recording format.
48 This quotation is taken from their artistic policy as published on their website here: https://www.dv8.co.uk/contact-dv8/artistic-policy

ages audiences to "feel a responsibility to challenge prejudice in communities that might not be their own" (Newson, "Prejudice"). Now, if one returns to my earlier point regarding the absence of a main story in *To Be Straight with You* in favour of a series of fragmentary narratives, I would suggest that it does not necessarily equate here to the absence of an underlying ideological narrative. From the outset, indeed, *To Be Straight with You* does not hide where its sympathies ultimately lie, as the words of Newson in the epigraph to this section make abundantly clear. More specifically, the piece has a series of interlinked targets. Firstly, as Liz Tomlin has cogently put it: "There was an explicit charge from the production that cowardice in the face of potential allegations of racism underpinned the refusal of the wider British public to interrogate homophobic attitudes within ethnic minority communities" ("Suspension" 170). Indeed, this is what Newson refers to as "a liberal blind spot, a lack of voices speaking up for some of our most basic freedoms," one that also extends to artists. Secondly, and perhaps more contentious, is the argument that seems to be stating that the fact that 25 % of the countries in which homosexuality is illegal today are former British colonies is no accident: "British rule may be responsible for institutionalised homophobia" (Shuttleworth).[49] Thirdly, there is also within *To Be Straight with You* a broader project of showing that the UK – despite its leading position as regards to the rights of gay people and its public image of tolerance, still has a long way to go when it comes to the daily life of this marginal group within society at large: "our interviews show how lesbians and gay men, if they choose to become visible, face intimidation or physical abuse" (Newson 2008, 3). In other words, Newson wants people to know and understand what the gay community goes through, the psychological distress, the fact that they are demonised and even killed, that they are scared of being who they are because of a society that still judges them.[50] For instance, in *To Be Straight with You* a casual bus ride or a meal in a restaurant assume a completely different meaning.

These systematic persecutions, discriminations, bigotry, internalised prejudices and homophobic crimes should not be glossed over and Newson's political message is one of tolerance and diversity in multicultural Britain. As British journalist Owen Jones has recently claimed on behalf of the LGBT community "we all grow up in a society that still treats us as if we are inferior: we have all repeatedly encountered homophobic abuse, the stress of coming out repeatedly, or the

---

49 This quotation come from page 3 of the NT programme's Education pack for schools in relation to the production of DV8's *Can We Talk about This ?* that was published in 2012 and can be consulted online here: https://www.dv8.co.uk/media/files/CWTAT_DV8_Workpack.pdf.

50 Some of the interviewees, for fear of being recognised, refused to have their voice used in the show.

fear of holding hands with a partner in public." Unsurprisingly, the new mayor of London, Sadiq Khan, who is often described as one of the UK's "most prominent pro-LGBT Muslim politicians", received death threats because of his public views. After seeing the show, Ian Shuttleworth confessed that "[i]t is difficult to listen to so much hatred and ignorance" and Phillip Hammond wrote that "[he] beg[a]n to wonder just how far we have moved in terms of liberal, civilised and intelligent thought right here at home."

As such, *To Be Straight with You* fulfils one of the primary targets of a political verbatim practice if understood as "seek[ing] not only to challenge the master narratives, but to do this specifically by enabling the voices that such narratives silence, oppress or marginalize, to speak out on their behalf" (Tomlin, "Suspension" 168). This target – that Tomlin calls "the political imperative" of verbatim theatre ("Suspension" 169) is never threatened by the aforementioned plurality of voices, as for Tomlin, "the voices of opposition serv[e] [...] effectively to undermine their own positions and consequently strengthen the preferred ideological narrative of the piece" ("Suspension" 169). Quite clearly, the contextualisation of the fundamentalist religious perspective of Iris Robinson on homosexuality with constant interruptions and performers wearing horse head puppets seeks to undermine and ridicule her views.[51] Newson's intentions from this point onwards are fairly militant regarding sexual politics and, more generally speaking, about what is considered a "schism" within the Anglican Church as well as religious fundamentalism more broadly.[52] This in itself would seem to indicate a departure from the postdramatic as theorised by Lehmann with regard to his articulation of the political, inasmuch as the postdramatic provides an ideological critique without the presumed pitfall of reaching a particular ideological conclusion (*Postdramatic* 185). Also important here is how to define the political project of *To Be Straight with You* in the light of the above considerations. Astonishingly, one may be persuaded that *To Be Straight with You* achieves the same political intervention as the documentary realist performances of Chapter 2,

51 This 2008 on-air interview sparked outrage at the time – a 27-year-old man had been badly injured in a homophobic attack in Newtownabbey and Mrs Robinson's husband had just been elected as the Northern Ireland Assembly's First Minister – as she proclaimed that homosexuality was "an abomination" and provocatively offered to refer gay people to "a lovely psychiatrist who works with [her] in [her] offices and his Christian background is that he tries to help homosexuals, trying to turn them away from what they are engaged in." As Democratic Unionist Party's health spokeswoman in the Stormont Assembly, Mrs Robinson had to abide by equality laws and should have refrained from condemning homosexuality on a live radio phone-in show.
52 The term "schism" refers to the potential split within the Anglican Church over the question of homosexuality.

but from fundamentally different end of the verbatim spectrum. Put differently, they are to be understood as manifestations of the same impulse in the sense that all the elements of the performance are held fairly firmly in the service of one specific construction of reality that attempts to pass for the real. By exploring such questions, the overall premise is that the use of post-realist strategies may be seen here as simply an attempt to cover its agenda.

Having discussed in broad terms how DV8 and *To Be Straight with You* may be said to initially engage with the category of political theatre in the sense that their politics is made explicit, I now would like to refine this claim by looking at how post-realism fares in such a scenario. As was seen, the numerous post-realist strategies encountered in this chapter had quite a corrosive impact on the verbatim material. One of the outcomes has been that the verbatim words are amplified just enough to be audible, forcing the audience to listen more attentively. However, on the whole, the political effectiveness of these discomforting strategies, as well as specific corporeal practices, is difficult to ascertain, unless one invokes neuroscience. In effect, juxtaposing the verbatim material with a strong emphasis on the presence of bodies may potentially have a deeper impact on audiences in the light of recent research on mirror neurones applied to performance studies:

> The discovery of mirror neurones adds support to the theory of theatre as a mirror to those watching and observing, firing neurones in empathy with the characters as they play out the drama. If the same sets of cells are firing up for the observer as the actor, then the notion of empathy takes on a deeper dimension – we could actually be experiencing vicariously the world and life of the character onstage. Watching an actor onstage is more than just a visual event; theatre is a transaction, a sharing of ideas and a site of reflection. (McCutcheon 147)

However, the fact that post-realism has eroded the dramatic notion of characters and that the movements are also of a different nature in the case of the physical verbatim performances under analysis in this chapter, can be said to present a strong challenge to this reading, however fascinating this suggestion might be.

On a very different level, I would concede that DV8 politicises the body by revealing through movements its internalised restraints. As Judith Butler has argued in *Undoing Gender* in relation to femininity and masculinity – and this could be extended to other social constructions – these movements are largely cultural and not natural as some would have it:

> The very attribution of femininity to female bodies as if it were a natural or necessary property takes place within a normative framework in which the assignment of feminity to femaleness is one mechanism for the production of gender itself. Terms such as 'masculine'

> and 'feminine' are notoriously changeable; there are social histories for each term; their meanings change radically depending upon geopolitical boundaries and cultural constraints on who is imagining whom, and for what purpose. (10)

In this sense, the disjunction between the verbatim words and the body of the actor can be seen as a radical intervention. In one scene, for instance, one of the interviewees pretends to be straight whilst the movements of the performer – unrelated to the real movements of the interviewees – betray what is really going on for him: his limbs are straight but stiff and awkward in their attempt to avoid at all costs being perceived as "bent." In another section, a performer playing a British-born Muslim woman of Pakistani origin – who was disowned by her family for being a lesbian – pulls her own shirt harder and harder so that it tightens round her body, reflecting the constant and increased pressure to conform to the impossible cultural demands of her environment. These post-realist gestures put on a visceral level the daily surveillance, the fear, the violence under which some members of society are living. In other words, they express what it is to live a life that is so constrained by an unsympathetic, unsafe and dysfunctional society, what this does to the body. DV8 gives us here a window on a world in which the body is a victim of social exclusion, showing how this operates on an experiential day-to-day level.

Arguably, the distortion of the verbatim words engendered by this post-realist setting may have an additional effect. Besides the content of these harrowing stories, one can see or rather feel a pattern emerging. Clearly, these verbatim monologues are as much about questioning, second-guessing and suppressing and, at the level of form, the audience experiences for 90 minutes an intense and claustrophobic universe emblematic of the violence wrought by men upon one another. A good example to illustrate this point is the final image in which a gay imam, after having given his testimony, seems to physically look for an escape but to no avail. He then starts being surrounded by darkness and pounding sound, and desperately moves his fists in the air until he becomes completely swallowed up. Furthermore, as I have argued elsewhere, the use of technology in *To Be Straight with You* is also a crucial component that may have some radical potential in relation to the ways in which it overlaps and intersects with the performers' bodies "so that they ultimately become out of sync and reveal cracks in the totalising realist narrative which in turn suggests possible resistance to dominant ideologies" (Garson and Gonzalez 2015).

Last but not least, and concomitant with the post-realist frame in which the verbatim material has been placed, I would suggest that, beyond an attempt to dispel the myths and correct the record, *To Be Straight with You* wrestles with the very notion of representation itself. These post-realist strategies of storytelling have also the power to change minds and hearts as they ultimately reveal the vulnerability with which the verbatim theatre-maker works, whilst nonetheless keeping the desire to intervene in the public sphere alive. Spectators here are not addressed as uninformed about the issues discussed in the performance, but as sharing the same vulnerability by being open to hitherto unconsidered possibilities. Perhaps in this way, the audience of verbatim theatre can fulfil what Jill Dolan describes in her *Utopia in Performance:* "being passionately and profoundly stirred in performance can be a transformative experience useful in other realms of social life" (15).

4.1.9 Conclusion

Whilst this chapter has focused exclusively on exploring physical verbatim theatre, much of the discussion anticipates ideas which recur over the course of this chapter. As I compare and contrast the "post-realism" of *To Be Straight with You* with both documentary realism and new realism, one sees that it envelops the whole edifice of verbatim theatre in a way that makes it less recognisable as verbatim theatre. In this new arrangement, verbatim theatre's inner hierarchy has been inverted, its usual markers have been replaced, the words themselves do not sound like they should, the actor is no longer what s/he was, realism is no longer possible. With respect to these notions, DV8's *Can We Talk About This?* was perhaps even more experimental as it involved a semblance of audience participation as well as computerised modifications of the verbatim recordings.[53]

Through this study of DV8's *To Be Straight with You*, we came across four main principles that seem to underpin what I am calling "post-realism":

1. A constant state of uncertainty
2. A fragmented postmodernist aesthetic
3. A (self)decentring of the verbatim material
4. The possibility of social intervention

This last point is important. If these "post-realist" and intermedial strategies certainly disrupt the tradition of verbatim storytelling that seemed to be the core of its political power – especially within the documentary realist performances in

53 At the beginning of the performance, for instance, the audience was asked whether they felt superior to the Taliban – a question quoted from the words of writer Martin Amis at London's ICA in 2007 – which prompted some of its members to raise their hands.

Chapter 2 – verbatim theatre in this configuration has nevertheless not abandoned its political project. In fact, through the medium of physical theatre, one can see an important gesture rehabilitating the power of affects. For psychologist and researcher Diana Fosha, in her study of the affective model of change, "the power of affect to transform is enormous. Unlike other change processes, it is not gradual and cumulative, but intense and rapid" (4).

Through this chapter, we discovered a series of dematerialising strategies unknown to verbatim theatre (as conceived in Chapters 2 and 3 of this volume) that has some radical implications for these performances. On the whole, *To Be Straight with You*, and by extension, physical verbatim theatre, has been a fascinating site for exploring some of the tenets of what I have called "post-realism." Before I can pursue these remarks any further, one also needs to look at other manifestations of post-realism in verbatim theatre: those manifestations that do not revolve around physical presence as such. The next chapter will thus investigate notions of musicality that potentially transcend both documentary realism and new realism as defined in Chapters 2 and 3 of this monograph.

4.2 The "Verbatim Musical": Alecky Blythe and Adam Cork's *London Road* (2011)

This case study takes us back to the Royal National Theatre and to a play that has slipped effortlessly into the contemporary canon. To carry out research on *London Road* is immediately to encounter a milestone in British verbatim theatre history, a "groundbreaking" (Spencer, "London Road" 408) production that lifted verbatim theatre to a whole new level, making a claim for attention that was to rival both the genre of new writing and that of the musical. Indeed, Alecky Blythe and Adam Cork's *London Road* (2011) was "a theatrical triumph" (Brown, "London Road" 411), "immensely successful" (Cantrell, *Acting* 142) by today's standards and generally considered as both "the theatrical highlight of the year" (Brown 411) and "one of the most original productions the National Theatre has ever staged" (Rosenthal 814).[54] For critic Kate Basset, *London Road* was "the most excitingly innovative musical [she's] heard or seen" whilst, for critic Gary Nunn, it was nothing less than "the boldest, most innovative production [he's] seen." For the purposes of this volume, it may therefore be desirable in the first instance to clear away the eulogistic chorus, and to focus instead on what makes it a significant departure from both "documentary realism" (Chapter 2) and "new realism" (Chapter 3) as well as "an important moment in the history of verbatim theatre in the UK" (Forsyth and Megson x). In this chapter I shall therefore ask whether or not the presumed "post-realist" aesthetic of *London Road,* and, by extension, that of the verbatim musical, has stretched verbatim theatre to breaking point. Before answering these pressing questions, let me for now provide a brief overview of what I mean by "verbatim musical."

4.2.1 Definition

As seen in each chapter of this study, the provided definition of verbatim theatre is always provisional and it seems that *London Road* has once more extended one's understanding of verbatim theatre still further. First of all, one would sure-

54 It received five star reviews from *Time Out, The Independent, The Financial Times, The Sunday Times, The Evening Standard, The Sunday Express, The Mail on Sunday, Metro* and *The Guardian* and its first run at the National Theatre was extended to accommodate the high demand for tickets.

ly argue that the term "verbatim musical" signals a paradox of sorts.[55] As Stuart Young has pointed out in his review of the performance, "verbatim and musical theatre make improbable bedfellows" (101). Indeed, verbatim theatre's integrity and seriousness seem very far removed from the joyous lightness and buoyancy of the musical that is all glamour, entertainment and charm.[56] Verbatim theatre is essentially interested in recreating the level of the everyday, whereas, by contrast, the musical requires exuberance as well as a higher and enchanting level of energy. In other words, they reflect quite different attitudes towards life, and for our purposes, this generally pertains to opposing aesthetic considerations. That is to say: the documentary realism of verbatim language cannot be more opposed to the sheer artificiality of songs and musicals.

Additionally, from a technical standpoint, it would seem virtually impossible to reconcile the formal structure of music with the verbatim meanderings and inarticulacy of everyday conversational language. By way of example, verbatim language tends to be characterised by various changes of subjects midstream, as the words of one of the characters in *London Road* can attest:

> RON. The more publicity we get – around here – the less likely it is – I think – fer de (*Beat.*) ne'er-do-wells ta start creepin' back. (*Beat.*) Get the – uh (*Pause.*) I mustn't keep sayin' this I keep tellin' (*Beat.*) every meetin' we hold 'we gotta get this street tarted **up**'. (*Laughs heartily.*) An' they always say 'it's not the right choice a words'. Yeah. (8)

On the other hand, songs tend to have a more structured dynamic overall. Further, the sequential placement of different oral verbatim material does not necessarily match the sequential placement of musical sections that can include repetitions (sometimes with alterations).

In the context of *London Road*, another difficulty arose due to the unsettling subject matter – the 2006 Ipswich serial murders – and the delicacy of the material that would seem, from the outset, not to lend itself well to such a treatment, becoming an insurmountable ethical and political challenge.[57] In fact, be-

---

55 Different terms have been used to describe this technical innovation in the field of theatre such as "verbatim-theatre musical," but the term "verbatim musical" seems to have been used more consistently in the literature (see Haydon, "2000s" 48; Julien; Roesner 213).

56 I am very much aware that these are generalisations and the spectrum of the musical genre cannot be equated to these clichés. However, more often than not, musicals are understood as "a light-hearted story on a contemporary theme performed as music theatre, and using solo songs, duets, trios, choruses and straight dialogue to convey the dramatic dialogue" (Nzewi 181).

57 The residents still live on this road, the parents of the victims are still grieving and for many people at the time, the project – however respectful it may be – appeared to come too soon after the events.

fore *London Road* even opened, the media tried to stir up controversy. *The Daily Mail*, for instance, titled one of its articles "Theatre Show about the Suffolk Strangler's Killing Spree to be Staged" (Daily Mail Reporter) and subtly implied that this could potentially be a traumatising experience for the people affected by the tragedy: "some of the families might be offended." Similarly, *The Eastern Daily Press* called its article on the subject "Former Norwich Pub Landlord's Prostitute Murders to Be Turned into a Musical" (Staff Reporter), with all the alarming implications that such a title could trigger in the general public. As for the BBC – perhaps as a result of the aforementioned *Daily Mail* article – they revealed that "Suffolk Murders Play Angered Tania Nicol's Mother" (BBC) and reported that Ipswich MP Ben Gummer had claimed that "[i]t [was] a pretty sick-making feeling to make a musical, or musical documentary, out of something which [he] th[ought] most people in Ipswich would like to be left alone" (BBC).[58]Taken together, these reactions would seem to indicate that the term "musical" is in itself heavily loaded.

Unavoidably, defining the term "musical" is not an easy task. Interestingly, Millie Taylor defines musical theatre as "a combination of song, visual spectacle and verbal text" (1). Hence, one can assume that a verbatim musical is a combination of these things, with the exception that both the verbal text and songs are drawn from interview material. One question that springs to mind, however, is whether the verbatim musical is simply a species of musical theatre manqué and, perhaps also, an illegitimate type of verbatim performance. Another one has to do with the function given to the songs in the verbatim performance. Do they serve as a verbatim narrative or do they clash jarringly with the unsung spoken words, brimming with the pleasures and celebratory tones of the spectacle? Finally, is the song material comparable to the movements in the physical verbatim performances of Chapter 4.1 inasmuch as it would signal an engagement with what I have termed "post-realism"?

Faced with a completely new type of verbatim performance and musical theatre, critics have tended to use a rich variety of expressions to encompass the phenomenon of the verbatim musical. The National Theatre itself initially called *London Road* a "documentary play with music." Other appellations by critics have included (but are not restricted to): "a highly unusual slice-of-life musical drama" (Calhoun), "a poem laced together with melodies" (Hitchings), "an extraordinary new form of sung drama" (Lawson "Musicals"), "a quasi-operatic libret-

---

58 As a result, Alecky Blythe wrote to all the families of the victims to explain in more detail what the show was about and what it was not about, thematically as well as aesthetically, so as to allay some of their fears. She also explained that their daughters and themselves were not going to be represented in the performance.

to" (Brown "Choreographer") or even "an uncategorisable oratorio/musical/documentary/singspiel" (Sutcliffe 2). One thing is certain: the verbatim musical significantly departs from conventional understandings of both verbatim and musical theatre. Consequently, it appears almost counterproductive to attempt to define a strand of performance largely in its infancy. Suffice it to say that, over the course of this chapter, a verbatim musical is understood as a verbatim performance which prominently displays its musical and sung material and, in so doing, it inevitably disturbs the conventions of mainstream British verbatim theatre (especially the conventions of documentary realism discussed in Chapter 2 of this study). Conversely, a verbatim musical is understood as a break from mainstream musical theatre since both the music and the songs are created through a verbatim methodology of some sort. Last but not least, the verbatim musical, contrary to a given verbatim show with songs and music such as Robin Soans's *Life after Scandal* (2007), often blurs the lines between verbatim speech and songs.

As ever, one can also find examples of similar explorations in earlier periods of British verbatim history, albeit with a different name and intent such as the "documusical," "documentary musical" or "musical documentary." Music is an affecting and effecting device that was prominent during the first wave of British verbatim theatre. Verbatim pioneer Peter Cheeseman even considered music as "a bloodstream that pulses through [documentary]" (Nevitt, "Peter" 8) and Derek Paget wrote, of the inspirational production in 1963 of Theatre Workshop's *Oh What a Lovely War* that arguably "started the documentary ball rolling" ("Chilton" 1), that it established the musical documentary method of "juxtaposing popular songs with evocation of historical reality" ("Chilton" 3). More surprisingly, back in 1977, Gary Yershon created within *Cheshire Voices* the song "Henry Farrell's Mystery Tour" that "can lay claim to being a verbatim *song*" as "the lyric was taken directly from a taped interview" (Paget, "Oral History" 336). However, the purpose of music in a "verbatim musical" like *London Road* can be said to largely exceed the aforementioned uses, which more often than not served as a link between scenes as in Sheila Rowbotham's documentary piece, *Friends of Alice Wheeldon* (1980) and therefore stood "apart from the action" (Favorini and Elvgren 185).

A simple example may serve us well: Roy Nevitt's *All Change* (1977) is typical of earlier forms of documentary and community theatre creatively engaging with music. In the piece, original songs were "performed by a mixed vocal and instrumental group" (vii) and could remind one of the genre of the ballad "in a robust traditional folk style" (Winkler 307). In the pre-1990s phase of verbatim theatre, the songs themselves could be seen as documentary in the sense that they tended to be found through archival research, if not written in response to the "original" verbatim material. Today, this strand of practice has survived but is almost never seen in mainstream venues.[59]Despite this heritage, I would contend that the verbatim musical is novel in that it entirely relies on the use of technology and its main activity is the musical treatment of the verbatim material in performance, rather than the mere addition of musical elements. In other words, the verbatim speech is here set to music and songs are created. For instance, in *London Road*, Alecky Blythe's interview with the prostitutes in which she asks them about their profession and they reply that they "used to be" and that they have all henceforth stopped is turned into the song "We've All Stopped" (56). However, as opposed to the songs we tend to find in musicals that comprise a kind of language that has been cleaned up, in a verbatim musical the "lyrics" are made using the "uhms and erms" of spontaneous speech, as in the following:

> ALL. *Obviously you gotta lotta people goin' out f' th' weekend*
> *th' gonna be a bit worried*
> *So erm we're handing out personal alarms to erm.* (13)[60]

Arguably, these shows tend to be enormously expensive in comparison to the other strands of performance encountered in this volume and, as such, they represent one of the smallest sections of the contemporary verbatim output.[61] Yet, since *London Road*, their numbers appear to have grown steadily. More recently, there have been various cases of what one could call a "verbatim opera", notably *Sex Workers's Opera* (2016) by the theatre company Experimental Experience.[62] Most strikingly, Victoria Brittain's *The Meaning of Waiting* (2010) had also two characters performed by opera singers and this process transformed the verbatim material into aria forms.[63] The intersection of experimental music and verba-

59 The work of Banner Theatre with their video ballad productions such as *Burning Issues: The Miners, 1984–2004* (2004) is a good example. Indeed it belongs to a tradition of "documentary musicals" in Britain that "aims to present some aspects of working-class history to a local audience familiar with the setting" (Winkler 307).

60 As legend would have it, Alecky Blythe allowed Adam Cork to add one "erm" in the piece, that was not in the original recording, for musical purposes.

61 These kinds of shows require a larger cast, a longer rehearsal process and sometimes a band as was the case in *London Road*. Their development process also spans many years and their experimental nature makes them a particularly risky artistic and financial venture.

62 To some extent, the National Theatre's *Jerry Springer – The Opera* (2003) based on the outrageous 1990s global television phenomenon, *The Jerry Springer Show*, could enter this category.

63 *The Meaning of Waiting* is a "verbatim music theatre" piece that gives an insight into what the War on Terror has done to individual families, from a female perspective.

tim theatre has also given birth to some fascinating post-realist fusions such as Born Mad's *Sister* (2016) which interlaced verbatim material with songs and live electronics. Other verbatim musicals include Anna Braithwaite and Michael Betteridge's *In Their Own Words* (2016) on loss and survival, *Land of the Three Towers* (2016) made by activists from Focus E15 Campaign about their occupation of Carpenters Estate, and Leeds-based 203 Theatre's *Untold Wars* (2015).[64]

4.2.2 Methodology

In one sense, *London Road* is similar to *To Be Straight with You* as there was no blueprint to start with. If DV8's innovation in verbatim theatre was through the dancing of actual speech, *London Road*'s is in its singing of it. Indeed, the musical elements can be said to follow natural speech and sound patterns in the manner adopted by Lloyd Newson from the DV8 group who composed the more abstract and stylised physical score in direct response to the detail of verbatim material. As British composer Adam Cork has acknowledged, he specifically tried with *London Road* to "discover responses suggested by the material on a moment-by-moment basis" (viii). However, unlike verbatim physical theatre that tends to set all the selected interviews for the show to movements, here only some of the interviews in the piece were set to music. More specifically, the project of making a "verbatim musical" initially started in the spring of 2007[65] through an experimental workshop programme in the National Theatre Studio[66] that randomly paired up Alecky Blythe with British composer and sound designer Adam Cork. By this time, Blythe was already considered as "an old hand at documentary drama" (Spencer, "London Road" 408) and Adam Cork had under his belt – among other music projects for the theatre – the experience of having written the music and 47 songs in the context of an allegedly unmanageable translation of Brecht's *The Caucasian Chalk Circle* in 2002.

Developed as a further articulation of the headphone-verbatim methodology, this time it saw actors leaving the headphones behind in performance, not un-

64 At the time of writing, the National Theatre was about to open a new musical, *A Pacifist Guide to the War on Cancer*, by Bryony Kimmings and Complicite based on the stories of real cancer patients.

65 At the time, Steve Wright, the presumed killer, was on trial.

66 Since autumn 2015, the NT Studio, which can be considered as the experimental wing of the National Theatre, has become the New Work Department, incorporating also the Literary Department.

like Anna Deavere Smith who created the technique.[67] In other words, this was the very first time Blythe departed from the headphone-verbatim theatre methodology explored in Chapter 3.2. This departure, prompted by artistic decisions at the National Theatre, necessitated a complete revision of the way in which Blythe had worked since *Come Out Eli* (2003). To implement this deviation (with as little compromise as possible), Blythe organised a gradual rehearsal schedule: "[w]e had six weeks of rehearsals, and the first three weeks, they worked with earphones, and then the second three weeks, the earphones came off" (Blythe qtd. in O'Malley).[68] Technically, however, the process is initially very similar to what was seen in Chapter 3.2: that is to say that Alecky Blythe proceeds in the manner of an investigative journalist, as she spends a very long time with a given community. This "technique" allows her to obtain material of a different nature from that which a traditional journalist would normally collect. Over time – and very often years – Blythe builds a relationship of trust with her interviewees. By the same token, Blythe followed her usual and refined research methodology, that is to say that she records the oral material, listens to it and logs it.[69]The transcription thus happens much later, usually to coincide with the publication of the playtext. However, here, Blythe had to change the way she usually works because of the musical component of the production. In other words, she had to transcribe more material than usual so that it could be turned into songs.

As just seen, perhaps the best way to define the verbatim musical is by seeing it as an aesthetic compromise between mainstream verbatim theatre and the genre of musical theatre. In the case of *London Road*, due to the subject matter and the close and constant scrutiny of the press, it is very possible that the balance tilted more in favour of verbatim theatre.[70]In other words, a different verbatim musical based on less controversial material might have different aesthetic priorities tending towards an all-dancing and all-singing spectacle. In effect, the main priority here seems to have been to avoid distracting the audience from listening to the verbatim material. As such, the methodology adopted by composer Adam Cork was to attempt to transfer Blythe's strict headphone-verba-

67 Blythe also experimented a little with the headphone-verbatim technique in *Cruising* (2006) by adding some ballroom dancing but, to date, none of her productions has stretched her methodology as much as *London Road* did.

68 There was also one week of music, so it was effectively seven weeks of rehearsals in all.

69 Alecky Blythe always uses a logbook to help her organise the impressive amount of material that gets condensed into her under two-hour shows.

70 In addition, as is the case in most verbatim theatre productions, the real people behind these stories always come to see the show.

tim methodology – as laid out in Chapter 3.2 – to the medium of songs and music. Technically, this means that Cork transposed every spoken sentence and half sentence as if they were a tune. In other words, he focused on the rise and fall of conversational speech in the musical line and was forensically faithful to the rhythm and pitch of the real people, to their 'voiceprint'.[71] Very skilfully, he moulded the structure of the songs around the verbatim testimonies, down to the minor details of pauses in everyday language.[72] For instance, in the second song in the show entitled "Everyone is Very Very Nervous," the Radio Techie sings the following:

> *I mean the best person to talk to is the bloke*
> *by the gen – by the erm lamp-post over there.* (11)

Viewed as a whole, this new methodology of verbatim theatre-making marks a distinct break in Blythe's practice, moving her for the very first time into the realm of music as she became a co-lyricist, writing songs on this particular production (that is to say, turning individual lines into choruses, repeating some of the verbatim lines, and so on). More specifically, her decision to leave the headphones behind and to push the verbatim material further – with the help of Adam Cork's "slightly freer hand" (ix) – would seem to indicate a significant departure from what I called "new realism" in Chapter 3 of this study.

4.2.3 An Overview of *London Road*

London Road was first performed – with a cast of 11 and a six-piece band – on 14 April 2011 at the Cottesloe auditorium of the National Theatre, London, in a production directed by Rufus Norris and designed by Katrina Lindsay that played to packed houses. The production also raised £25,000 for the Iceni Project in Ipswich (a drug rehabilitation organisation) through audience collections after

71 As described in Chapter 3.2, this is a kind of vocal fingerprint. Arguably, when we speak, we hit particular pitches, to which we tend to return to when we repeat rhythm in our speech. Here, in this particular context, this corresponds to the verbatim detail of the voice in a musical sense.

72 As Cork has explained, the "fairly solid musical structures" he created to contain the "anarchic" verbatim material were mostly "built out of key elements of the transcribed voice, translated into harmonic progressions, or rhythms in the accompaniment" (ix–x). On other occasions, Cork used a slightly different process whereby he composed "a 'container' unconnected with the musical surface of the words, but inspired by their literal content, or the tone in which they are spoken, or the mood of the situation in which they were uttered, or that of the situation which they describe" (x).

each performance.[73] Following its success, including the reception of Critics' Circle prize for Best Musical and Show of the Year from *Time Out London* as well as critical rave reviews, it was then revived the following year for the much bigger space at the National Theatre, the Olivier, on 28 July 2012. Similarly to David Hare's *Stuff Happens* (Chapter 3.1), an extract from *London Road* was revived for a special performance on the occasion of the National Theatre's 50th anniversary in November 2013. Due to its success, *London Road* has also been revived throughout England in amateur and student productions such as one by Sedos in summer 2016 at the Bridewell Theatre in London and, by the Bristol Old Vic School in June 2014. It also received a North American premiere that was produced at the Bluma Appel Theatre in Toronto. More recently, *London Road* was adapted for the big screen and the script was again written by Alecky Blythe, with added material that had not been used in the theatre show.[74]This was the first time the National Theatre was premiering a film and also the first time a verbatim theatre production was adapted for the cinema. In a sense, this experimental kind of verbatim theatre can be said to have made an impact on cinema itself. At the time of its release, critic Peter Bradshaw struggled to find a term to qualify this new type of cinema production and finally settled for "cineopera in a reportage verbatim style" ("Musical"), before changing his mind again and calling it "a film oratorio" ("Musical").

London Road is a "'social crisis' play" (Wake, "Headphone" 325) that focuses on the lives of the Ipswich inhabitants on the eponymous road in 2006–2007, at a time when the whole community was at the height of its fear after having experienced a brutal series of murders committed on young prostitutes working in the area, by a man who was to be known as the "Suffolk Strangler" and which made national headlines. Instead of focusing on the victims or the family of the victims like the media coverage did at the time, Blythe turned her attention to the community going through (in a relatively short space of time) a state of rising panic on the way to a sort of post-traumatic regeneration, that is to say a community healing itself. As one of the characters says in the play, Blythe was "about the only person that **has** – actually come down [t]here [London Road in Ipswich], and **asked** the residents how they feel" (9).[75] In other words, she fo-

73 The orchestra had three woodwinds, a guitar, a keyboard and percussion.

74 The film was also directed by Rufus Norris with music by Adam Cork and kept the entire original cast whilst notably adding to the ensemble Olivia Colman and Tom Hardy. The camerawork was by Danny Cohen.

75 To some extent, *London Road* also enabled the residents to counter some of the media representations that sensationalistically depicted this particular area in Ipswich as a "red-light district".

cused on the "**other** victims [...] the people that lived [t]here and [...] put up with it" (9), which is something very reminiscent of her *Come Out Eli* discussed in Chapter 3.2 of this volume. Throughout the process, Blythe also interviewed police officers, the surviving sex workers in the area and, perhaps in what one would call a self-referential manner, members of the media working on the case.[76] In other words, the play is not a record of the startling event itself but, instead, lays emphasis on the panic response of the London Road community to that event. In particular, the piece narrates how the community strove to blot it out so that ordinary life could resume.

The two-act performance comprises fifteen musical numbers in total and is framed by an annual, local flower and hanging garden competition entitled "London Road in Bloom" – designed by keen gardener and Neighbourhood Watch Events Organiser Julie – who has attempted "to make the area look pretty [...] an' getting people – involved – an' getting interested in their road and their homes again" (7) after it was tarnished by the tragedy. As such, the piece begins at the Neighbourhood Watch AGM with its chairman Ron Alder and his wife Rosemary as well as other residents (where we are first introduced to the contest) and ends with the prize-giving ceremony, thus creating a positive *denouement*. The performance spans multiple locations from the streets of Ipswich to some London Road sitting rooms and its timeframe corresponds to a period stretching from around September 2006 to February 2009. Act I mainly concentrates its attention on the community's reaction after the five murders and before Steve Wright's arrest and it closes during his trial, which was to find him guilty of the five counts of murder and sentence him to life imprisonment. Act II concerns itself with the end of the trial leading up to the verdict and how the community reacts in real time to each instalment of the story. In the meanwhile, the "London Road in Bloom" competition continues to attract more people each year and it would seem that the prostitution problem has finally been removed from London Road.

Most of those who have written on *London Road* to date have stressed its strangeness as well as its puzzling engagement with both verbatim theatre and the genre of the musical. Despite its arising from the rather grim context of the Ipswich murders and aftermath, it was perceived as a seismic shock – a

---

76 Blythe was in fact at the Royal Court Theatre making *The Girlfriend Experience* – mentioned in Chapter Five – when the events in Ipswich occurred. The first series of interviews she conducted happened at the time when the murderer was still on the loose and five women had been murdered. These were vox pop interviews and it was not until the arrest of Steve Wright that Blythe would start knocking on doors and meet Ron, the chairman of the Neighbourhood Watch, and his wife Rosemary.

groundbreaking moment in British theatre history – that radically challenges an audience's habit of attending both musicals and verbatim theatre. In the following sections, I would also like to suggest that the meeting point between these improbable allies that are verbatim theatre and the genre of the musical have some bearings on documentary realism and may consequently shed light on my conceptualisation of "post-realism."

4.2.4 The Contract with the Audience

> At the performance I saw back in 2011, this laboriously practised spontaneity from the performers went unrewarded by at least one audience member, who, after an especially inarticulate riff from a character who started a phrase several times, lamented to heri neighbour: '[t]he problem with coming to previews is that the actors sometimes haven't learnt their lines'. (Lawson "Musicals")

> It's also possible to notice some effects of the verbatim song forms we've developed, whilst confessing that many of them were unforeseen. (Cork qtd. in *London Road* x)

If one applies Jauss's concept of "horizon of expectations" – that is to say "the familiarity of previous aesthetic experience" (25) with which we always see new works – it would seem that none of these past experiences could have possibly helped in framing and understanding *London Road* from an aesthetic standpoint. It is firstly in this sense that *London Road* can be considered as post-realist. If verbatim theatre automatically assumes a contract that stands apart so that the audience is more willing to accept an unconventional format, its post-realist instantiation can be said to violate that code in the sense that it is a journey of discovery for both theatre-makers and audience alike. The term "verbatim musical" certainly did not make much sense to the audience at the time. One can also assume that many members of the audience did not know what verbatim theatre really was (as we have seen, it can mean such a variety of things depending on the theatre-maker), but they certainly knew what the term "musical" entailed. In other words, their expectations were almost entirely based on what the word "musical" can mean. In the epigraph, Lawson describes the experience of one such audience member unfamiliar with the conventions of verbatim theatre, mistakenly confusing the actors' delivery with a lack of professionalism or skill. In fact, as will be seen later, the work of the actors is quite the opposite.

Arguably, the terms of the contract with an audience typically start even before the audience has entered the auditorium. Before buying a ticket to see the show, the audience would have at least some prior knowledge about some of these elements: the publicised genre of the performance (here, "a verbatim musical"), the location of the performance (National Theatre), the subject (the 2006 series of murders in the quiet rural town of Ipswich in England), the author(s) (Alecky Blythe and Adam Cork), the director (Rufus Norris), what it looks like visually (some official photographs), perhaps even what it feels like to a critic (published reviews or "tweets" on Twitter) or to a member of the general public (social networks), and finally how the show might be staged (publication of the performance text). As crucial as these parameters may be in shaping the reception of a given audience member, this section will look specifically at the contract in performance, that is to say the moment at which someone has reached the auditorium and tacitly accepted to become an audience member for Alecky Blythe and Adam Cork's *London Road* at the National Theatre.

I would contend that the first minutes and musical notes of *London Road* set the terms of a contract that goes well beyond what we have been accustomed to in both documentary realist and new realist verbatim performances. As the audience enters at a low level, a recording is played of the "prelude" of the original Neighbourhood Watch AGM that took place in May 2008 at the Church Hall, just off London Road. Some members of the audience may recognise Alecky Blythe's voice on this recording and therefore be surprised not to see her on stage.[77] In a typical "post-realist" moment, reality and fiction mingle with each other. The mixture of voices from the original recording strangely echoes both what the actors are saying to each other – as they seemingly get ready for the performance of *London Road* – and what some audience members are saying or whispering to one another as they try to find their seats in the auditorium. In fact, the spectators have become AGM attendees without their knowing, as three levels of reality intermingle without clear sound boundaries between them.

Here a first breach of both documentary realism and new realism is taking place at the level of sound. As seen in Chapter 2 of this book, documentary realism is a self-contained, closed off universe that attempts to recreate what happened in the distant or not so distant past with absolute exactitude. In other words, it abhors any interference, be it theatrical, performative or simply linked to the accidental and unpredictable nature of a live event. New realism in Chapter 3 of this volume, on the other hand, was more playful with its "authentic" verbatim material, but always within the strict terms elaborated by its authors. Allusions to the "making of" process were always present – the earphones in Blythe's *Come Out Eli* and the character of "the author" in David Hare's *Stuff Happens* – but, they never offset the investment in a certain version of realism

77 In fact the real Alecky Blythe appears in this production much later, but she is barely recognisable this time as she plays a cameo (one of the newsreaders).

because, essentially, these opposing strategies were more or less kept separate.[78] Put in another way, if, before, the gesture of the headphone-verbatim or massaged verbatim performer placed the verbatim material in quotation marks as a separate entity from the overall theatrical experience of the performance, the post-realist verbatim pieces seem to be founded on a different logical basis.[79] Now to continue with the study of the very first moments of *London Road*, at the visual level, the audience can observe some actors stacking chairs, not decidedly knowing whether or not this is part of the piece. The actors from the company also make some tea and drink it on stage. Soon, however, it becomes clear that this is for the staged scene of the AGM. They start hearing "[t]*he original audio recording of* RON'*s opening speech*" (5) before it "*fades out as* RON *starts to sing*" (5), accompanied by some music played live by an orchestra that cannot be seen. This is the second breach in the contract, as referred to in the second epigraph opening this section.

Compared to the original recording where Ron (the Chairman of the Neighbourhood Watch) said:

> *Good evening.* (Beat) *Welcome.* (Beat) *We've a bit short on residents tonight. This really is our first AGM after we reconstituted in 2006 and then all the awful events happened and we became stronger and stronger. Erm* (Beat) *aft' so after our er reconstitution we made a lot – we made a lot of progress in regenerating this street. We've put new signs up, thanks to Ken. Hope-hopefully the problem with the girls has disappeared. We don't see them now. I believe they are still a few round in Hanford Road but er* (Beat) *we really can't concern ourselves with them.*[80]

the sung version of the verbatim material is an important alteration. "*Good evening. (Beat.) Welcome. (Beat.)*" (5) whilst still recognisable – and oddly identifiable as a faithful reproduction of Ron's speech patterns – is repeated many times unlike the original recording. The actor playing Ron even comes to shake some hands in the audience as he repeats "*Good evening.* (Beat.) *Welcome*" (5). The low

78 As seen in Chapter 3.1, David Hare adopted a specific visual strategy to separate the verbatim from the non-verbatim material. As for the headphone-verbatim pieces, once the contract had been established, it tended to gradually erode any potential for distancing effects as the audience simply accepted the new conventions and focused on what the actors were saying.

79 The use of music in verbatim performance does not necessarily create this phenomenon in practice. For instance, Robin Soans added some songs in his *Life after Scandal* (2007) where the characters start singing. However, as soon as the song was over, the action would simply start again. As such, I therefore consider the aforementioned logical basis as typical of the verbatim musical.

80 These are my notes from the recording of a Cottesloe performance of *London Road* available at the National Theatre Archives.

level of light in the Cottesloe auditorium means that the moments of direct address appear to be aimed at a specific member of the audience (eye to eye). The following words "[t]*his really is our first AGM*" (5) are then repeated by all the actors singing chorally, significantly departing from the original. A few seconds later, this is repeated again. The audience is being introduced here to what a verbatim song, and by extension a verbatim musical, sounds like.

A few minutes later, the contract would seem to have changed again in a typical post-realist gesture. This is reflected by the song "London Road in Bloom" (9) performed at the very end of the AGM, after Julie has announced all the forthcoming events for the London Road community and Neighbourhood Watch. This time, the verbatim song is clearly announced as a song. As soon as Julie says "[t]hat's about it" (6), we start hearing some music, the stage lights change – including a bright green light to represent the gardens – in a way that may remind one of musicals. The song itself is musically more elaborate and ample, alternating between canon and choral singing. Music starts to overtake the verbal material as the physical movements did in DV8's *To Be Straight with You* studied in the previous chapter. Furthermore, conditions of sound volume change throughout the song, distorting even more the verbatim material. Halfway through the song, the non-sung verbatim material resumes. However, this time the audience is no longer cast as the London Road community attending the AGM, but as Alecky Blythe:

> ALFIE. I won the first prize. So – it'll be in the papers on Thursday. Yes. Well I had people coming all week here ya know what I mean – everybody. [...] Today the er st-*Evening Star*'s come down and yerself – who's gonna be next? (8)

Interestingly, this unsung verbatim material is undercut by a continuous musical accompaniment that makes it sound strange, the rules of documentary realism having been "mercifully" broken again:

> *Underscored:*
> JUNE. I don't know a weed from a plant I don't. But I did do a-a pot.
> TERRY. Yes we got er June's pot./ An' these are double petunias. (7)

The last three verbatim testimonies one hears in this section seem to have been juxtaposed together for their dramatic and musical qualities, bearing no resemblance to the real context of the interviews. Indeed, the three characters end up laughing at the end of their respective testimonies for unrelated reasons. Music is also allowed at times to completely dominate the verbatim material, as when, at the end of the song "London Road in Bloom", the words have ceased to be uttered but the music remains for a few seconds.

Having said that, it would seem, however, that the contract of *London Road* leans more towards verbatim theatre than the genre of the musical. Indeed, one is told in the "Note on the Text" that the "[i]*nconsistences in spelling and grammar are deliberate and indicate idiosyncrasies in the speech and delivery of the characters*" (4) and the audience understands from the start that these are the words of real people, which arguably changes everything in terms of reception. Moreover, audiences are very unlikely to clap their hands in a show of appreciation after one of the verbatim songs, even after hearing its most catchy song "London Road in Bloom" (9). Similarly, this particular and typical moment when a character is about to break into song is completely exempt from the verbatim musical apparatus and the singing voice of the performers sounds more "spoken" than is usually the case. Furthermore, here, even the songs are affected by verbatim theatre and most of them are audience-directed – a musical equivalent of direct address.

Nonetheless, visually, one can see a smoke/haze effect throughout that makes the general atmosphere slightly unreal, as does the sound of the orchestra, the hanging flower baskets being flown to the ceiling, the lighting and the repetitions of the verbatim words. In other words, this would seem to indicate a postdramatic approach to verbatim theatre. Yet, *London Road* does not fulfil the postdramatic promise of doing away with characters and an overarching narrative. Clearly, one finds an exposition scene at the start of the piece which makes clear to the audience what has gone before on London Road and what the present situation is for the community. As expected, the tension increases in response to the continuous threat and there is a real closure: something good has happened in the community and the girls have been kept off the street and drugs. Perhaps, this is where "post-realism" in contemporary verbatim theatre distinguishes itself from Lehmann's theorisation of the postdramatic.

To sum up, one could define post-realism in contemporary verbatim theatre as an aesthetic contract essentially characterised by instability and uncertainty, perpetually conveying contradictory messages. To add further complexity to the aforementioned post-realist contract, it is relatively difficult on many occasions to know precisely whether we are hearing a singing or a speaking voice. In the published playtext, even though one has specific codes – "*Songs and sung text are indicated in italics*" (4) – in performance, they are arguably a lot harder to differentiate and, therefore, no clear boundary between them can be drawn. It almost seems as if the actors go seamlessly from speech to choric moments and ensemble textures and then back again into speech or what sounds like speech. Here, the spectator constantly experiences a sense of not knowing what organised combination of verbatim sounds might be coming next. Their relationship with the original interview, as seen in the last chapter on the occasion of

DV8's *To Be Straight with You,* can support or deny the verbatim component. Indeed the re-sequencing of the original speech patterns into songs allows the verbatim material to acquire a different energy and strength as well as a metaphorical meaning. In other words, similar to physical theatre, music possesses a communicative meaning that can potentially oppose the corollary of the verbatim language, having the capacity to add dramatic intensity, emphasise the state of mind of a character and/or facilitate empathy more broadly. Additionally, as seen earlier, it may even become dominant following the post-realist impulse, creating a prominent subtext other than that inferred by the verbatim words. Furthermore, the alteration in the timing of speech delivery may have a bearing on how the audience apprehends the verbatim content. As the character of David Hare argues in Hare's verbatim play *The Power of Yes*, pace affects understanding in relation "to the speed of waves in the brain" (23). Whether or not one agrees with this facile scientific explanation, it is clear that, in normal everyday conversation, pauses tend to relate to something meaningful and emotional (see Chapter 3.2).[81]

As discussed earlier, perhaps the only element of *London Road* that makes it recognisable as verbatim theatre is language. Yet, it would seem that the post-realist onslaught on verbatim signifiers has not even spared its most cherished trait. In order to fully assess the observed destabilisation of verbatim theatre, one must at once turn one's attention to language itself.

4.2.5 Language

At a time when it appears more and more difficult to distinguish verbatim theatre from the genre of new writing on the British stage, it would seem that language has remained its unique and irreducible signature. In addition, language – even in the case of the new realist verbatim productions of Chapter 3 – has always been what encourages a reading of verbatim theatre in terms of stage realism. Crucially, Kate Dorney in her study *The Changing Language of Modern English Drama* considers that contemporary verbatim theatre marks the passage into a new realm whereby "[l]ifelike-ese and speech have finally melded" (225). Linguistically speaking, the presence of a unique "idiolect", that is to say one's "own distinct and individual version of language" (Coulthard 431) is generally what defines a non-verbatim playwright. Even if, theoretically, a playwright

81 Having said that, I am aware that the way the interview is set up suggests that there is a certain level of performance involved.

can choose any word at any time, s/he eventually tends to make typical choices of preferred words. If one follows this reasoning, the proportion of unmodified verbatim material in a given performance text is key to its categorisation and recognition as verbatim theatre. In this scenario, integrating verbatim speech with song is a post-realist act of trespass *par excellence*. However, it is in fact a far more complex phenomenon since the selection itself of the verbatim material, as a form of artistic expression by one author that reproduces actual speech, can also betray him or her. In other words, the way s/he puts together the verbatim words on a page, the choice of certain conventions and techniques and, particularly, the limitation of his or her artistic expression to those very few qualities are always revealing. On a given page, the so-called "voice" of the author may not be as evident in the context of verbatim theatre. Nonetheless, one would certainly not confuse a verbatim play by Alecky Blythe with one by David Hare or Chris Goode.

Let me now consider again briefly Blythe's use of verbatim language in her headphone-verbatim pieces. In Chapter 3.2, we saw that what characterised "new realism" was a specific textualisation of the material. In other words, Blythe attempted to fit the verbatim material into a conventional dramatic frame. The display of such an attempt within the piece itself was a step beyond documentary realism. It follows, therefore, that "post-realism" is doing something else with verbatim language that goes beyond what I recognised as typical new realist strategies. For example, in Robin Soans' *Life after Scandal*, the opening verbatim speeches in Act I "*are not related to the characters in the play*" (27), which marks a departure from the conventions of mainstream verbatim theatre in Britain. My surmise here is that the musicalisation of verbatim speech may foster such a departure. Martin Julien validates this novelty by stating in a recent publication that "a lengthy monograph could be essayed to analyze and celebrate Cork's innovative musical composition alone" (164), a full exploration of which is, however, well beyond the scope of this chapter. Even though Blythe writes that "Adam [Cork] was able to set these words to music by following the cadence and rhythms of the original speech patterns so accurately that their verisimilitude was not diminished" (Blythe, "Singing" 6), I would suggest that this argument is particularly difficult to sustain when the audience witnesses, for instance, the character of Ron say "*Good evening.* (*Beat.*) *Welcome*" (5) no less than twenty two times and slide into song. As seen earlier, Blythe reduced Cork's musical interventions to a minimum but, ultimately, he had to versify spoken language at times, leading Blythe to modify the forensic accuracy with which she usually works.[82] Inevitably, going from unversified, spontaneous speech to songs heightens and formalises it.

What one thus encounters in the performance text are competing kinds of verbatim language.[83]On the page, one notices three kinds of language: the spoken, which remains unmarked, the sung, presented in italics, and the emphatic/ stressed presented in bold. Although the headphone-verbatim technique of Chapter 3.2 required very precise transcripts, the entry of verbatim theatre into the musical realm has exacerbated this tendency to produce a script similar to a musical score:

> FEMALE REPORTER (*and the two other reporters*). *Well this is*
> *the third time that Steve Wright has ap –*
> *peared in Court but it's the* **first time** *that he's*
> *formally entered a plea he arrived here at*
> *Ipswich Crown Court.* (42)

This forensic transcription is clearly faithful to the rhythms and pitches of the "real" people, yet it is not exactly like conventional poetry, as it does not rhyme. Within the sung, the chorus – the main body of a song – clearly stands out as it constitutes the most distinctive departure from the aesthetic conventions of verbatim theatre. For instance, the song "Everyone is Very Very Nervous" in Act I, Section Two – that is interspersed by a singing Santa statue on a plinth interpreting Edward Pola and George Wyle's *It's the Most Wonderful Time of the Year* – ends up being performed by the full cast as follows:

> ALL. *Obviously you gotta lotta people goin' out f'th' weekend*
> *Th' gonna be a bit worried*
> *So erm we're handing out personal alarms to erm*
> *A t" sort of ladies out shopping today*
> *So that, so that's why we're here.*
> *Everyone is very very nervous.* (15)

82 Following Blythe's methodology, Cork did not versify too significantly the verbatim material so as not to lose its original quality. Nevertheless, his intervention, however constrained, is meaningful enough to impel us to write a whole new chapter in this verbatim theatre history in the making.

83 Similarly to what was noted in Chapter 3.2 on the headphone-verbatim strand and in Chapter 4.1 on physical verbatim theatre, there are also here some extracts from the original recordings that are played at times, thereby creating a fourth kind of verbatim language. There is additionally a fifth kind of language: the surtitles for the audience to read.

It also deviates even more from its original context at the end of the song, arguably in such a manner that the published text can no longer account for it. Indeed, everything accelerates, the music overtakes the words, a police siren joins in, and the word "very" is repeated within a cacophonous medley of sounds that completely upsets the remnants of realism. In another song entitled "The Verdict" in Act II, Section Three (59 – 62), this time more passages are pure music, the verbatim material having in effect turned into a spectre looming above the piece. In other words, in several musical instances the verbatim material has been stretched beyond itself, to the point where its original meaning has been lost and new meaning cannot yet be created.

Adam Cork's determinedly "post-realist" interventions into verbatim language can be divided into two main levels. First, the verbatim line is simply sung instead of spoken, whilst keeping the same pitches and rhythms of the "original" voiceprints. That is to say that the origin of these songs in recorded speech is always audible. As discussed earlier, Cork was extremely faithful to the tune of the words, retaining the ups and downs, the speech impediments, if there was any. This level is as close as possible to the documentary realist rendering of speech patterns whilst no longer being documentary realist. For example, in the song "Everybody Smile" in Act II, Section Four, that takes place in a garden on London Road, Carol (one of their local Neighbourhood Councillors) – as one of the judges – announces the results of the "London Road in Bloom" contest in a manner that is hardly recognisable as song:

> WARD COUNCILLOR CAROL. *Uhm. Hi everybody. This is*
> *the second*
> *Annual er- traditional party in Dodge's garden,*
> *After er-judging the best gardens.*
> *And they just get better an' better every year.* (74)

The "speech-singing" naturally invites us to focus on the words themselves, onto the way they are said.[84]Paradoxically, one cannot overlook the fact that the material is sung, the music being continuous throughout. In fact, at this point the "song" is so oddly close to the "real" rhythm of speech that the audience reaches a level of uncertainty between two opposite aesthetic experiences, one of them being the unattainable documentary realism. Yet, this focus on doubt is consistent enough throughout, which suggests that even at this level the experience is

---

84 This style of verbatim composition may remind one of the *sprechgesang*, that is a particularly recitative manner of singing in which the pitches are sung, but the delivery is rapid and loose like naturally occurring speech.

in keeping with the post-realist project of verbatim theatre. Secondly, there are more openly "post-realist" gestures when music and sound are already present. In these pieces, the repetitions are overpowering, turning into duets, trios, choruses, and sometimes even canons that twine themselves around the "original" verbatim line until it musically implodes into itself. A good example of this lyrical leap can be found in the song "It Could Be Him" whereby the chorus "[y]*ou automatically think it could be him*" (22) outstrips the whole material in a rhythmically consistent manner. If the headphone-verbatim technique relied exclusively on "locked" sounds that contained some emotions to be reactivated by the live voice of a performer, this time, each verbatim sound is openly allowed to be changed in response to the musical environment.

Looked at in this light, the sung version of the verbatim score appears to attempt to capture the perceived essence of the London Road community in a similar way to the physical movements in DV8's *To Be Straight with You* that strove to emotionally convey each individual testimony to an audience. The song "That's When it All Kicked Off" in Act I, Section Four is a clear illustration of this. The piece starts with Ron retrospectively telling us how he experienced Steve Wright's arrest:

> RON. *I got up and all I could see was police cars* (Beat.)
> *goin' up the road. Course – ya know – I –*
> *we knew obviously in charge –*
> *ya know – somethin' to do with these poor girls.* (26)

As such, the musicalisation of verbatim speech serves a typical diegetic function, generally increasing the sense of flow. The same words will then be sung by "various" (27) in vocal harmonies and then distorted through a kind of echo-singing technique, creating the sense of a whole community being brought together by these tragic episodes. Furthermore, one of the key effects of setting verbatim language to music and taking it beyond its customary social significance is that it makes the audience look at the unsung verbatim material in a different way. Indeed, all of a sudden, the spoken material sounds more sung, one notices the symphonic structures of spoken language like never before. If the vernacular is often associated with negative connotations of some kind, here this new perspective elevates these verbatim words into another register. In addition, I should note, the staging of this kind of language is inevitably "meta", in the sense that it makes anglophone audiences reflect on their own language. As repeatedly argued throughout this study, verbatim theatre is particularly sensitive to the concept of voice and how it locates one geographically, socially and even psychologically. In a sense, one is witnessing here – for the first time in this volume – what a true post-realist verbatim language may be like in practice. Arguably, even the headphone-verbatim practice saw its initial disrupting effect being ultimately annihilated by the forces of realism. Here, on the contrary, realism can never quite fully envelop each destabilisation. In other words, even though some of the strategies discussed here heavily rely on repetitions, they never become repeatable conventions themselves as they are constantly mutating.

But what one also needs to take into account in the assessment of these aesthetic innovations are the ways in which the verbatim language is physically embodied by the performers in *London Road*. And one is led to another crucial element of the performance that may set one on the route to clarification or, rather, ambiguity, in keeping with the tenets of "post-realism." I thus turn, as one must, to the unusual figure of the verbatim musical performer.

4.2.6 Acting

The first thing one should note is that the activities the verbatim musical performer engages in appear to be of a different kind from any of the strands hitherto discussed in this volume. In the same way as physical theatre, the musical arguably demands a larger array of skills from its actors. Indeed, typically, the performer in musical theatre is expected to conjoin acting, singing and dancing at any given moment.[85]More specifically, the specific kind of singing in the verbatim musical sketched out earlier, for example, challenges from the outset any attempt to approach the actor's practice in terms of either documentary realism or new realism. In this light, it may be of interest in the first instance to look more deeply into these practices, as one should always be wary of exaggerated claims for the originality and singularity of any verbatim practitioner.

One might thus wish to consider, at this point, the fact that the acting process was so unusual that the performers had to prepare for the performance in a completely different way. In particular, the singing in the performance was also very unusual and bizarre as it attempted to retain the quality of speech and therefore differed from both conventional operatic and musical theatre practice. In the words of composer Adam Cork, he created "a speech-like way of singing" (qtd. in Bosworth). Consequently, the full company of *London Road* initially started working with the Music Director (David Shrubsole) for a few weeks before the rehearsals properly started with the whole team (as well as the technical rehear-

85 The process was even more laborious and intensive here as the actors used the earphones in rehearsals for the first three weeks and then, gradually, took one earphone off at a time.

sal process). This way of working was so complex and formally challenging that the actors had to have one-to-one sessions with him so as to learn the score. When the actors then started rehearsing for *London Road*, the creative team did not know at the time whether or not it would be possible to sing in this manner, as it was a completely new way of making musical theatre. For actors, learning the very detailed audio was considerably challenging, especially once they had to drop the earpieces. Indeed, not only did they have to learn the exact words and non-sequiturs of real people, they had to sing them exactly as they were composed by Adam Cork with strict adherence to each unpredictable and "labyrinthine" tune. In other words, there was technically no room for improvisations or variations despite the fact that the headphones had disappeared from the stage. As Anita Dobson – an actress who was not originally involved in the stage production but appears in the film adaptation – has explained:

> It is a song, but it's not a song, because you're learning dialogue but not dialogue as we learn it normally in a script: it's got every cough, every hiccup, every mispronunciation... every pause – that's all sung! (qtd. in Brazier)

In order to achieve such prowess, the cast of 11 – each playing one of the London Road residents as well as nearly 60 roles between them – rigorously policed themselves and invented what they called the "anti-piping rule".[86] Piping up means in this context an embellishment/alteration of the audio as a result of an acting process akin to a sense of ownership that could lead to caricature.

Another crucial component in most musical theatre is dance. Here, however, there is no such thing as dance or physical theatre movements. For all that, there is a stylised choreography of some sort that is not as easy to discern as the music and songs in the performance. Faced with the double challenge of working with verbatim material whilst attempting to create a musical piece of theatre, movement director Javier De Frutos had to change the way he usually choreographs a piece. He therefore invented a new style of choreography that was essentially made of – what he calls – "verbatim movements." In this respect, he worked from the principle that if he knew how someone spoke, he would know how they sat, got up and moved more generally. Actors were therefore encouraged to improvise movements in response to the audio material – not dissimilar to the way in which Lloyd Newson initially worked to create *To Be Straight with*

---

86 This is explained by the director Rufus Norris in an account of the rehearsal process of *London Road* in the context of a National Theatre Platform event with Dan Rebellato, Alecky Blythe and Adam Cork recorded in April 2011 and available online here: https://soundcloud.com/nationaltheatre/london-road-platform.

You – and De Frutos filmed these movements. In this manner, actors behaved in an incredibly natural way within the character they had been creating. The last stage in this process was for the company to eventually learn these subtle and complicated movements. To some extent, this is a movement version of the way Cork worked on the musical component, since it was "faithful to the material but also manage[d] [...] to elevate it out of the literalness" (Blythe "Interview").

Now, the question that arrests us is the following: how does this innovative acting process relate to "post-realism"? Typical of the post-realist aesthetic, as seen in the last chapter, is a disruption of the conventional rules of characterisation. As Cork has argued, "hearing the natural speech patterns sung in this way can have the effect of distancing the audience from the 'character', and even the 'story'" (x). Interestingly – and without using the term "post-realism" – Cork describes this distortion of the verbatim speech as having "the potential to explode the thought of a moment into slow motion [...] [and allowing] us more deeply to contemplate what's being expressed" (*x*). Some questions remain here. Is one still in the realm of "new realism" or is there something more to it that would trigger an entry into what I have called "post-realism"? Or rather, could one envision that "post-realism" is simply an extension of "new realism," that is to say an intensification of some of its practices?

To answer these questions, I must point out again briefly how *London Road* departs from the new realism of the headphone-verbatim technique. One of the main differences with the latter is that the actors here are no longer engaged in the act of listening, they would seem therefore to have completely moved into the act of telling. From this perspective, it is surely clear that by doing this, the actors also ceased to kinaesthetically experience the emotional potency of recorded verbatim material and its breathing patterns. And yet, the striking preponderance of the act of singing throughout makes particularly difficult any theorisation of the performance. If one looks at the song "Cellular Material" in Act II, Section Two, on the news report following the start of Steve Wright's trial, it is obvious that its effects on the performance aesthetic are extensive in proportion to the new realist verbatim performances of Chapter 3.2. Throughout the song, the following "scene" is repeated three times.

> HELEN. *I think he probably did do it.*
> GORDON. *He must have done it really.*
> JUNE. *I think he did. Definitely did.*
> DODGE. *I think he'll be found guilty.*
> RON. *I think he did but he's gonna get away with it.*
> ROSEMARY. *It's only circumstantial evidence.*
> TIM. *I couldn't sit there an' say 'Yes he dunnit'.*

JAN. *Really?*
JULIE. *He's gonna get away with it.* (50)

This highly theatrical exchange – despite the singing – could hardly pass as documentary realism as it implicitly suggests both the act of editing and representation of verbatim testimonies. Interestingly, Lib Taylor has argued in relation to *London Road* that such performance processes introduce "deferrals of the realist matching of voice to real person, and offer the audience a layered experience in which authenticity is refracted" ("Voice" 376). As my general line of argument on headphone-verbatim theatre has repeatedly shown in Chapter 3.2, the presence of the earphones/headphones on stage also introduced similar deferrals, even more so when the actors were openly cast against type. Arguably, in *London Road* it strikes one that such deferrals are continuous throughout and can never be fully absorbed or remedied. It is for this reason – among many others – that the verbatim musical embraces the concept of "post-realism" as charted in this study. In other words, the distortion of the verbatim material through voice and music has an impact comparable to the physical movements of DV8's *To Be Straight with You* studied in the previous chapter.

In addition, the movements of the actors induced by the singing, especially as they all sing chorally in all the songs, are also a clear break from both documentary realism and new realism. As seen in the previous chapters – most notably in the last one on DV8 Physical Theatre – the gestures in verbatim performance are typically seen as an extension of the verbatim methodology. In effect, there is a "verbatim gestural vocabulary" being created in each production. In documentary realism, these gestures attempted to avoid drawing attention to themselves by mimicking the real gestures on which they were based. In new realism, it was a combination of two kinds of gestures: some documentary realist gestures such as when David Hare *Stuff Happens* recreates some real-life events and "fictional" gestures such as when the character of "An Actor" makes some disruptive interventions. And in post-realism, then, it would seem therefore that new parasitic gestures enter the realm of the performance. As Rib Davis has rightly remarked, these movements give "the impression of scenes" ("London Road" 126). Indeed, some of these scenes are neither recreation of the original interviews – Blythe could not record a whole community on her Dictaphone – nor conventional scenes.

In arguing that the acting processes in *London Road* significantly contribute to the creation of post-realism, another crucial locus of aesthetics deserves mention: the staging. Given that both language and the figure of the actor strongly challenge the possibility of documentary realism, have the staging decisions attempted to limit their implications for the reception of the verbatim material by an audience?

4.2.7 Staging

To define the staging of *London Road* is to attempt to articulate its competing intentions between largeness and constraint as well as between verbatim performance and musical theatre. In many ways, post-realism can be seen as a kind of aesthetic that gives more artistic license to the verbatim theatre-makers in terms of distortion and imaginative leaps. In other words, the verbatim material is removed from its original context and more openly heightened in this new context. Likewise, I would like to suggest that – besides a manifestation of post-realism through a distortion of verbatim language by both actors and music – the verbatim material undergoes a further distortion through the use of visual metaphors that are comparable to the physical score in DV8's *To Be Straight with You.*

As has been seen from earlier sections, the staging of a verbatim piece is always a complicated affair but *London Road* set the bar rather high, as no one, not even Blythe, had worked in this manner before and new rules were urgently needed to get the show up and running. On stage this meant that a series of restrictive rules, mostly linked to verbatim theatre, were created to shape the staging of the *idea* of what a verbatim musical ought to look and sound like. The most important rule was, as said earlier, that the performance had to avoid – almost at all cost – doing anything that would make the audience stop listening to the verbatim material. As such, and contrary to what one would expect of a musical, *London Road* did not include the typical song and dance numbers in its frame. In essence, as Blythe said herself, she wanted to have her verbatim cake and eat it too, in the sense that she longed to preserve the audio quality of the headphone-verbatim technique "in a way which incorporates music without losing its honesty" (v). In a similar fashion, Cork confessed that he wished "the musical score [...] [to] be like a time capsule inside which the speech rhythms would be captured and contained, frozen and fossilised in music just as they have a fixed existence on Alecky's recordings" (ix). In other words, what Cork calls "the ideals of verbatim documentary theatre" (ix), something to which he strove to remain faithful, is exactly what I have termed "documentary realism" throughout this book. Ironically, and despite the fact that *London Road* was nothing like the documentary realist performances of Chapter 2, the reproduction of the verbatim words on stage were sufficient to provoke, on one occasion, the following incident:

> When Julie said of the prostitutes, 'They were a complete pain in the neck. Ya know – they – they're better off 10ft under... That's a horrible thing to say isn't it?... But I'd love to shake his hand and say, 'Thank you very much for getting rid of them,' a heckler shouted, 'Shame on you!' and Fleetwood, still in character, kept repeating, 'It's just what I said,' until he shut up. (Calkin)

Even though this may be an isolated case, this phenomenon has some fascinating implications for this theoretical discussion. Is one wrong in thinking that documentary realism can ever lose its hold on verbatim theatre? Does this mean that no matter what happens on stage and no matter the proportion of actual verbatim words, documentary realism will always have its way as long as a given performance is framed as verbatim theatre? I may even take this point further by suggesting that documentary realism is in fact no more than a cultural construction. Consequently, and within a contemporary British context, the term "verbatim theatre" may be considered as so overcoded as the narrative of documentary realism discussed in Chapter 2, that even "post-realism" cannot do anything to upset this curious state of affairs. Can one thus entertain the idea that even the most post-realist staging of verbatim material, in which everything refutes verbatim theatre, could yet potentially be subjugated to documentary realism?

Having said that, the creators of *London Road* seemed to think otherwise and considered that they had to tread carefully on such treacherous aesthetic ground. As a consequence, any artistic decision departing from documentary realism was considered as a potential threat to the overall verbatim project. As I have already outlined, above all they felt the need to tame the musical components of the piece. At the same time – and perhaps paradoxically – they aimed for an inventive staging. This kind of contradiction can be seen most clearly in the way the various strings of verbatim and musical material were pieced together. In *London Road*, there is obvious dramatic cohesion, which is perhaps once again what sets a post-realist verbatim performance apart from postdramatic theatre. For instance, one of the most unsettling sentences in the piece in which the character of Julie – the originator of the "London Road in Bloom" competition – confesses the following:

> Ya know, I feel sorry for the families but not them [the murdered prostitutes]. Ya know it was just a pain in the arse. They were a complete pain in the neck. Ya know – they (*Beat.*) they're betta off ten foot under. (*Beat.*) That's a horrible thing to say isn't it? But. What's happened's happened but I'm not sad. (*Beat.*) Ya know. (*Beat.*) I'd still shake his hand [Steve Wright's]. I'd love to just shake his hand an' say 'Thank you very much for getting rid of them.' (65)

is placed at a strategic and dramatic moment.[87] The same treatment applies to the humorous and comic passages in the performance. Furthermore, Blythe did not scrupulously follow the chronology of the interviews and the performance used flashbacks.[88] However, it is clear that the chronology of the events in *London Road* is dramatically coherent, except perhaps on one revealing occasion. Indeed, the last scene is a reprise of the song "*London Road in Bloom*" (78) – one of the high points of the show – heard in Act One, Section One.[89] One could assume here that musical coherence prevailed or that this was a way to end on a positive note what is essentially a grim story and/or it was seen as important to strike a balance between the genre of the musical and verbatim theatre and therefore occasionally have "musical-like" moments.[90] Also, typical of the musical genre is the presence of self-introduction "songs" such as the following:

> JASON. *Me, I'm uh, uh – I'm a freelance, photographer* (*Beat.*)
> *but I'm – working for – some a the nationals, today.* (*Beat.*) (30)

Now if one looks at staging decisions proper, despite the fact that the headphones have disappeared, they are far from endorsing documentary realism by any stretch of the imagination: hanging flower baskets are repeatedly flown, the lighting design is made obvious and does not always serve realism, and objects like the Santa in Act One, Section Two have their own stylised entrance.[91]Another particular moment in the performance stays in the memory, the one in which the real television news reporter Simon Newton is shown on a mon-

87 Until that part in Act II, Julie had been introduced to the audience as a sympathetic character and this revelation towards the end of the piece was clearly intended to produce a momentary and unexpected emotional shock.

88 For instance Section One in Act I mixes a Neighbourhood Watch AGM that took place in May 2008 with the London Road in Bloom announcement that was in fact made in July 2007. Section Two then flashes back to December 2006, a recording that was made in "*London Road sitting rooms*" (9) after the five murders, but before the arrest of Steve Wright.

89 There is also a "prologue" – as discussed in earlier sections – in which we hear Alecky Blythe. In some sense, there is also an epilogue in which we hear the voices of the remaining prostitutes in the area. Both of them, as "pre-show" and "post-show" material, are not mentioned in the published playtext.

90 It is very common for a musical to end on a big production number with choruses of singers, performers in vivid formations, splendid costumes and colourful lighting as well as settings.

91 An interesting comparison can be made with the snowman in Alecky Blythe's *Come Out Eli* that is simply there on stage. By contrast, in *London Road* the singing Santa enters from upstage centre and runs down stage in a straight line.

itor. This filmed TV newsreel footage is then juxtaposed with the musical recreation that the report actor, Michael Shaeffer, gives to a live camera:

> SIMON NEWTON. (*on TV*). For a second day DNA expert doctor Peter Hau from the forensic science service arrived at Ipswich Crown Court in a van. Today, the jury heard more about the pair of gardening gloves. They were found inside a high-vis jacket recovered from Steve Wright's house at 79 London Road. [...]
>
> SIMON NEWTON. *Using his hands to demonstrate Peter Wright explained how forensic scientists had extracted Steve Wright's DNA from other cellular material inside the thumbs of both gloves. Doctor Hau said that within that cellular material he'd found the incomplete profile matching the fifth victim. The chance of it not being hers 'one in a billion' he said.* (52–53)

This particular report is then sung by all the actors on stage, closing on the following words by an Anglia newsreader on TV: "*Simon thank you*" (53). It is evident from this extract alone that we are dealing with a highly stylised staging – albeit in direct address – unparalleled by anything before or since. Even more disrupting to documentary realism, perhaps, is the use of blue and white police incident tape creating a web on stage in some of the scenes. It fully becomes a post-realist gesture in the piece when it appears around the residents' sofas, an artistic decision that neither the documentary realist nor the new realist verbatim theatre-makers would have happily considered. This metamorphosis of the London Road sitting rooms – where the majority of the scenes take place – is a visual metaphor for what the residents experienced in the midst of those tragic events. On stage, one sees the residents trying to get on with their lives whilst being constantly reminded of the incidents. The actors standing in for the London Road residents are literally caught in these incidents, physically trying to circumvent the police tape.

These staging strategies are not altogether surprising when one considers the impact the music has on the verbatim material in *London Road*. As seen earlier, on the one hand, the songs are extremely tied to the verbatim language and there is a very smooth transition between the speaking and singing parts. On the other, as the main argument in this chapter made clear, the musical elements of the piece on many occasions completely absorbed the verbatim material. Most tellingly, they succeeded in dismantling the perennial documentary realism of contemporary British verbatim theatre. All of this does suggest, as argued previously, that this entails a shift at the core of verbatim theatre and opens up new and unforeseen aesthetic relations.

Having studied how a post-realist aesthetic manifests itself in various ways in *London Road*, it remains to see whether these have any broader political aims than self-consciously questioning the capacity of verbatim theatre to represent the world.

4.2.8 Political Theatre

In one sense, post-realism does not sit very well with the notion of political theatre. The mere idea of treating the verbatim material as "clay for [the verbatim theatre-makers] to experiment with" (Blythe qtd. in Lombardi) would seem to suggest that political theatre is certainly not the aim and, if such an artistic manipulation leads to some effects on the world outside the theatre, they can only be accidental. Or rather, it does not sit very well with conventional understandings of political performance, for it is a theatre in which politics are made less explicit. For one thing, even if the cast of *London Road* held a collection to raise money for the Iceni Project, a drug rehabilitation charity in Ipswich, the piece itself is less interested in a frank and authoritative presentation of the verbatim words, but in a semi-controlled aesthetic chaos. Rich-textured aesthetic surroundings recreate some of the excitement of the musical, using abstraction and distortion to break up the verbatim environment into near-fantastic fragments. Considered together, these may appear as a relaxation of the play's political intentions as Lib Taylor has suggested in one article: "[its] important musical dimension [...] may function to dislodge the popular perception of earnest sincerity of documentary theatre thereby making it more entertaining" ("Voice" 369–370).

Following this argument, one may wish to look for another and less evident understanding of political theatre. Once such understanding can be found in the way Josephine Machon conceives political performance – in reference to immersive theatres – as an "intervention [that] can work more subtly via resonant engagement with what it is *to be human*. Through this, what we as humans are capable of questioning and how we might shift attitudes and sensibilities during and subsequent to a performance event, in and of itself marks its own political action" (*Immersive* 287–288). Indeed, much of the power of *London Road* comes from its recreation of a microcosm of community life and its fault lines. More specifically, the show is about community life in Britain, Ipswich being generally seen as a peaceful and quiet town. To some extent, London Road is the central character in the story and the title *London Road* can be seen as synecdochic as there are many London Roads in England. In addition, the prostitute situation at the heart of the narrative can also represent any other issue that may affect a community and disturb the pattern of the very controlled and necessarily ordered lives in which most of us in the Western world live.

At the same time, *London Road* is more openly political than a piece like Blythe's *Come Out Eli* because the post-realist interventions of the creative team make clearer some of the points in the piece, as Blythe has explained in an interview: "with the music we could [...] make a bit more of a statement with it [the verbatim material], and the music could also emotionally raise the temperature a bit" (Blythe qtd. in Butler "Alecky").[92] A good example of this can be seen in the representation of the media in the performance. In *London Road*, the media has an overpowering presence that affects the everyday life of the residents. In other words, their lives are punctuated by the "24[-hour] live coverage" (38) of those events and, in many scenes, we watch the residents watching the news. The media – perhaps even society to some extent – has crushed them and, in the performance, they are often standing on the top level as opposed to the residents who remain on stage level (in their sitting rooms). The stage itself is repeatedly peopled by a fury of reporters and other media, adding to the general sense of panic: "[*t*]*he sound of laptops being packed away, chairs being moved as the media pack leaves the room in a frenzy*" (59). At the end of the performance, the residents finally get their street back, away from the media turmoil, and the only media presence left is represented by Gemma, an innocuous photographer for a local newspaper with a very soft voice. In one important sense, therefore, *London Road* harbours a severe criticism of the media, or at least a critique of the media coverage in the course of those events. The parodic scene in Act II, Section Two entitled "cellular material" is perhaps similar in intent to the scene in DV8's *To Be Straight with You* where Mrs Robinson was expressing some extreme views towards homosexuality. Both scenes can indeed be considered as textbook examples of the post-realist act of distortion of the verbatim material in performance. In *London Road*, the musicalisation and repetition of Simon Newton's ordeal in trying to report the latest instalment in the story for the six o'clock news heavily accentuate its potential for comedy:

> SIMON NEWTON (*to camera*). *Using his hands to*
> *demonstrate Peter Wright explained how forensic scientists*
> *had extracted Peter Wright's – Peter Wright – Steve Wright.*
> *This is the problem with this bloody trial!*
>
> SEB. *Too many Wrights.* (50)

It also offers a strident and acerbic criticism of the media, reducing their activity to what often seems derisory.

At the same time, at the level of form, the performance is filled with overlapping interferences – such as those coming from the sextet and the ensemble

92 Of course the editing hand of Blythe certainly enabled some of her points to come through, but she never allowed herself to repeat a word or to change anything from the tape in order to make these points clearer.

pieces – in the sense that the audience does not always quite know which sound they should be listening to. As such, *London Road* is very close to DV8's *To Be Straight with You*. It follows that *London Road*'s political efficacy may lie within a postdramatic reconfiguration of political theatre. All of this would suggest that the performance aims to challenge an audience through a "politics of perception" (Lehmann, *Postdramatic* 185) or, at least, a multiplicity of political readings. These interventions, as Lib Taylor proposes, "may provide a dynamic strategy for the destabilisation of the real to promote critical insight" ("Voice" 370). To come back to my last point, regarding the treatment of the media in *London Road*, one is at all times presented with several constructions of reality including that of the media. Hence, the "cellular material" scene I discussed in the last paragraph clearly accounts for the mediatised simulation of reality as we are shown the construction of the report that differs importantly from the version that will be broadcast on the residents' TV sets a few minutes later.

Finally, perhaps the strongest political point in the piece is made through the absence or quasi-absence of the voice of prostitution, which is almost entirely referred to in the third person.[93]Contrary to the headphone-verbatim pieces of Chapter 3.2 that always embodied the interviews, this time the only interview with the surviving prostitutes is played on one occasion without being remediated by any actor in a way that recalls DV8's *To Be Straight with You*. Towards the end of the interview, however, we start hearing Adam Cork's music in the background until the music begins to completely dominate the interview, silencing again the voices of the prostitutes. Such aesthetic intervention is arguably problematic and seems to indicate the general hypocrisy of society. As such, it equally challenges an audience to think more actively about the politics of *London Road*, to problematise or even question the relationship the performance seems to entertain with its verbatim sources and, by extension, with the media version of those events.

Crucially, the audience throughout *London Road* is often caught in a dilemma, in a constant state of not knowing what sort of reactions exactly is expected of them. For instance, many times the audience wonders whether they are laughing at or with the real people represented on stage. They may also wonder whether this is a kind of laughter of recognition, or even more problematic is the question of how one can reconcile the fact that the verbatim song is aesthetically pleasing to the ear and yet it depicts the atrocities in relation to the five murders. These questions offer no easy answers. Yet, it seems that they are typical of post-realism, in the sense that they viscerally engage an audience, addressing their

93 The only exception to this is the song "We've All Stopped" in Act II, Section Two (56–58).

preconceptions, what they identify with and what they feel passionately about. A good example of such moments, as has been previously argued, is *London Road*'s most controversial passage pronounced by Julie (one of the main characters in the story), claiming that she'd "love to just shake his hand an' say "Thank you very much for getting rid of them.'" (65). Admittedly, this passage is constructed in strong association with the comedy genre since its placement within the narrative whole is one of surprise and incongruity for the audience. Another example of such dilemmas happens in the course of the song "The Five Counts of Murder" that refers to the real name of the murdered prostitutes.[94]One particular passage springs to mind, the one in which an unfortunate rhyme that helps build a certain musical harmony with the verbatim material coincides with the horror and brutality suffered by one of the victims:

> *Gemma Adams*
> *Her body of course was the*
> *first to be found that was*
> *fa-found on the third of December.*
> *Anneli Alderton*
> *Whose body was found in Nacton.* (43)

Even though the music is certainly designed to inspire respect and compassion for the victims as it clearly marks a break from the previous musical passage by being more solemn in tone and pace, one cannot help question one's own response. In other words, the active hypothesising process of the audience is once again put to test.

It is possible to conclude, then, that *London Road* is not intended as a radical performance insofar as it does not explicitly seek to effect change, as was the case with DV8's *To Be Straight with You*. Furthermore, as is well known, Blythe never articulates her verbatim works in these terms. However, it is my contention that post-realism has triggered some unforeseen radical possibilities that complicate the nature of the relationship between aesthetic experimentation and the possibility of political theatre.

4.2.9 Conclusion

As I began to articulate to what extent *London Road* may exemplify an inherent post-realism in the verbatim musical, this chapter has made clear that the verba-

94 In the film adaptation, their real names have been changed.

tim musical integrates music into the verbatim material more completely than in any of the other strands of performance exposed to our critical gaze during this research. As a result, the rules of verbatim theatre appear looser than in both documentary realism and new realism and what we see and hear on stage is no longer exclusively dependent on what supposedly happened at that time. As one studies the growing field of "post-realist" verbatim theatre, the dislocation of documentary realism has appeared in a variety of ways in the first two case studies subjected to a detailed scrutiny. The technique of contrasting the verbatim material with non-textual elements is, it would seem, an enduring one in these performances. As demonstrated throughout this chapter, *London Road*'s contribution to the wider British mainstream theatre sector has been colossal and has opened the valve for more "post-realist" experimentations.

The exploration of Alecky Blythe and Adam Cork's *London Road* in the above sections has foregrounded a number of important features that are useful in this attempt to identify a "post-realist" tendency in contemporary British verbatim theatre. Firstly, the verbatim musical is distinct from both documentary and new realism in that the songs – whilst keeping faithful to the exact words and rhythm of real people – defamilarise our conventional relationship with the verbatim material in performance; nor is there an indication that priority must be given to the unsung verbatim components. In a sense, more than in the headphone-verbatim productions of Chapter 3.2, verbatim theatre is finally allowed to completely shift focus onto the cadence of real speech, relegating the "original" meaning of the words to a second, and sometimes even third position. In other words, musical logic here dominates both narrative and visual logic, disregarding the documentary realist bind that always attempts to preserve the verbatim referent. Secondly, as is the case in DV8's *To Be Straight with You*, there is, here, a significant emphasis on non-textual aspects of performance that are more stylised than in the verbatim performances encountered in Chapters 2 and 3 of this study. Here, as Lib Taylor has appropriately remarked, the performance is crucially much more invested in postdramaticity as it questions "authenticity, realism and the function of the text" ("Voice" 369). Thirdly, one is witnessing in *London Road* a dislocation of both verbatim language and realism: arguably two of the core principles of contemporary British verbatim theatre following my main argument in this study.

Admittedly, the addition of music and more crucially, the musicalisation of the verbatim material automatically break up the possibility of a documentary realist aesthetic apparatus in the minds of (most) audiences. Whilst it is beyond the scope of this volume to analyse the film version of *London Road*, it may be tempting to ask whether the concept of "post-realism" is adaptable to another medium – provided one is dealing with a verbatim film. Surprisingly or rather unsurprisingly, the film adaptation that premiered on 12 June 2015 through the National Theatre Live scheme had a far more important aesthetic battle to wage than the stage version. Indeed in the filmic version, the chronology was rearranged along realist lines, the characters that populate *London Road* were given more substantially realist patterns and some of the ambiguities and uncertainties of the theatre production were clarified.[95] Put in another way, despite the disruption unleashed by the moments of music and singing, it was reduced to manageable and inoffensive bubbles, making the film safer for a broader mainstream consumption. And this difference is of major importance in characterising "post-realism" as a significant departure from "new realism" in verbatim works. That is to say that the mere presence of "out-of-sync elements" is not necessarily enough to create post-realism in a given production, if they do not entirely short-circuit the narrative expectations of documentary realism. In short: these elements must be sufficient in number and/or force to provoke an overall off-cadence and open-ended structure that is both truculent and endemic in that it affects everything the audience experiences.

Now, to further test this idea, I want to consider another "post-realist" verbatim performance, one of another fractious and estranging kind that instates space as a verbatim landscape that carries its own specificity in the creative and receptive process: site-specific verbatim theatre.

95 This of course does not preclude the possibility of a "post-realist" verbatim film in the commercial sector. However, this may mean that for such a "post-realist" verbatim experiment to exist in practice, different and more medium-specific aesthetic strategies are required.

4.3 "Site-Specific Verbatim Theatre": Jonathan Holmes' *Fallujah* (2007)

Now something quite new has entered the post-realist picture of verbatim theatre that puts an even greater emphasis on the affective, corporeal, physical and visceral effects of this type of performance, what industry terminology often refers to as "immersive,"[96] "environmental"[97] or "interactive"[98] in order to point to its "here and now" and relational quality. It is my contention in this chapter that site-specific parameters reconfigure, trouble and complicate the familiar operations and understandings of verbatim theatre in a way that requires immediate attention, notably in relation to its spatial signifiers. In order to reduce the scope of the investigation, I will focus mainly on Jonathan Holmes' *Fallujah* (2007), but references from other site-specific verbatim performances will periodically appear when they serve the overall analysis and/or add to my main argument.

---

96 Gareth White defines "immersive theatre" as "a trend for performances which use installations and expansive environments, which have mobile audiences, and which invite audience participation" (221). By contrast, "site-specific theatre" does not necessarily imply that the audience is mobile but can still be considered as an immersive experience.

97 This term was coined before "site-specific performance" but is nonetheless still used today. Richard Schechner in his *Environmental Theater* (1973) explains that he borrowed the term from "Allan Kaprow whose book, *Assemblages, Environments, and Happenings* had a big impact on [him]" (ix). Before *Environmental Theater*, he had first theorised these practices in a 1968 essay entitled "Six Axioms for Environmental Theater". The meaning he gives to these works – "one in which *all the elements or parts* making up the performance are recognized as alive" (x), that is to say that they "change, develop, transform [...] have needs and desires; even potentially [...] acquire, express and use consciousness" (x) – can be said to differ from current site-specific performances (even if Schechner simply says that environmental theatre has "earned a new name: 'site specific' performances" (ix)) and to the ones to be discussed in this chapter.

98 Interactive theatre is often employed in conjunction with the use of technology in a performance that creates a specific experience perceived as a significant departure from the conventions of theatre: "[it is] a new type of media that introduces new digital interaction methods into theatres. In a typical experience of interactive theatres, people enter cyberspace and enjoy the development of a story in a non-linear manner by interacting with the characters in the story. Therefore, in contrast to conventional theatre which presents predetermined scenes and story settings unilaterally, interactive theatre makes it possible for the viewer to actually take part in the plays and enjoy a first person experience" (Sood and Vasilakos 423). While site-specific performances overlap with this kind of performance, they tend to lay more emphasis on the physicality and materiality of space.

4.3.1 On Space and Verbatim Theatre

Space has been mostly ignored and overlooked in scholarship, even though the localist dimension of verbatim theatre is paramount in most productions. Even a mainstream verbatim piece such as David Hare's *The Permanent Way* (2003) was staged in a meaningful context and location by opening at the York Theatre Royal in York, an English railway town. At the heart of verbatim theatre, that is to say, the way it was originally conceived in Britain, the regional or community aspect of the production played an integral – if not dominant – part. In Derek Paget's oft-cited 1987 article, playwright Rony Robinson mentions "region" ("Oral History" 317) before all else in his definition of verbatim theatre:

> it is a form of theatre firmly predicated upon the taping and subsequent transcription of interviews with 'ordinary people', done in the context of research into a particular region, subject area, issue, event, or combination of these things. ("Oral History" 317)

Robinson then goes on to explain the importance of feeding back into the communities ("Oral History" 317). Furthermore, in terms of staging, one can argue that there is a particularly mainstream verbatim way of using space that reminds one of Peter Brook's famous claim, according to which theatre can be reduced to this:

> I can take any empty space and call it a bare stage. A man walks across this empty space whilst someone else is watching him, and this is all that is needed for an act of theatre to be engaged. (7)

Arguably, in the case of verbatim theatre, this is envisionable in its post-realist incarnation as long as the actor is specifically reproducing verbatim what happened. More often than not, it would seem that an act of verbatim theatre needs the following in order to be considered as such: I can take any empty space and call it a bare stage. A man walks across this empty space as he says some words in direct address to someone else who is watching and listening. These words are the exact words of someone real, perhaps the person who is watching, and this is all that is needed for an act of verbatim theatre to be engaged. In other words, verbatim theatre tends to create intimate spaces as a general rule, especially when it aims for a documentary realist aesthetic experience based on the original circumstances in which the words were uttered. Space in verbatim theatre is thus often expressed through physical proximity. For example, in Robin Soans' *The Arab Israeli Cookbook* (2006), one of the actors "offered a bowl of olives to the audience" (Jeffers, "Refugee" 4) and in *London Road*, as seen in the last chapter, the actor playing Ron shook the hand of some audience members at the start of the performance. The use of space may therefore be seen as creatively very restrictive, the design being verbatim itself in aspiration. As aesthetic considerations in relation to verbatim theatre do not always seem to go hand in hand in the minds of some critics, it may be the case that its spatial strategies also failed to be taken seriously. However, it is my contention that space is a crucial component of verbatim theatre that can make or break the documentary realism of a given production. Put another way, space might very well foster in the audience a realist reading of the verbatim material, unique individuals becoming types, as Marvin Carlson has recently and cogently argued:

> The normal image of a theatrical performance is one that takes place inside a space particularly created for such an activity. Thus, although the actor and his language come in from outside, the space within which he performs is most often constructed as part of a theatre apparatus, even when it imitates extra-theatrical spaces. (60)

More specifically, the transferral of the exact words of "real" people to a theatre space means that everything has to be scaled up and one may almost consider that one witnesses in verbatim theatre an experience of displacement. Interestingly, quite the reverse happens to time and verbatim theatre's unique temporality, which seem to have attracted a lot of attention in comparison to space.[99]Due to the relative novelty of site-specific verbatim performances and the impossibility of archiving the transient experience of being there, very little literature yet exists on the subject.[100]One might infer that this is perhaps one of the reasons why site-specific theatre is so difficult to write about once it has vanished. One cannot help but feel underequipped for the fullest demands of these works. Most profoundly, such complications are exacerbated by the fact that readers may only experience here my limited description and biased response

99 Indeed, verbatim theatre typically attempts to recreate an impression of "real-time" as if the performer were the original interviewee and s/he were now talking to us and only us, something that almost becomes possible through the headphone-verbatim technique of Chapter 3.2. In other words, verbatim is playing with our sense of lived time, especially when it updates its performance text to match what has happened outside the theatre, or when its staging is made to coincide with or be close in time to some events in the real world. One could say that verbatim theatre endlessly strives to capture time *verbatim*.

100 To my knowledge, one of the only publications on the specific interaction of verbatim and the site-specific is Joanna Ostrowska's 2013 book chapter entitled "Site-specific Theatre as a Form of Documentary Theatre". If it will always be impossible to really record the experience of a site-specific performance, it may be possible in the future to experience one of these shows from home. As I write these lines, the British group Punchdrunk is currently working on a promising virtual version of their show *Sleep No More*.

to what possibly goes on in these pieces. Since it appears impossible to completely escape this, one may as well declare it from the outset. Josephine Machon in her study of Punchdrunk – one of the most prominent site-specific performance groups in the UK – explains that her analysis "is intentionally personal, a feature that results from the decidedly individual nature of audience reception that arises from immersive practice in general" ("Punchdrunk" 255).[101]Now to come back to the remarks on space in verbatim theatre: if space, as I am suggesting, is a defining component of verbatim theatre, it is rendered even more explicit and far-reaching in the performances I am about to discuss which create these containing and moving structures that appear to operate on their own terms, as well as play themselves out so as to renegotiate their meaning with every performance and successive audience. Before considering the relationship between location and locution and how space and verbatim theatre renegotiate each other in site-specific productions, let me now attempt to define their main characteristics.

4.3.2 A Definition

> [W]hat's *specific* about a site-specific show is the audience, not the site. (Holmes qtd. in Megson, "Conversation", 68)

First and foremost, it must be acknowledged that the term "site-specific" itself "did not enter the theatrical vocabulary until the early 1980s"[102] (Carlson 72) and the term I propose, "site-specific verbatim theatre", is largely in its infancy.[103] Indeed, one of the latest publications of the Writers' Guild of Great Britain,

101 I am basing this study on my own experience of these kinds of works. As it was not possible for me to experience *Fallujah* first-hand at the time in 2007, I benefited from a DVD copy of the production, courtesy of Jonathan Holmes. I have had the chance, however, to be a momentary spectator at the National Theatre of Scotland's *Enquirer* in 2012 and at the Royal Court's *Who Cares* in 2015. I have also greatly benefited from a long conversation with Rebecca Hillman on her own site-specific verbatim piece *The Pact* (2011) and two short interviews I conducted with Jonathan Holmes and Mimi Poskitt.

102 Mike Pearson who is the leading British expert in site-specific performance, usually claims that it came much later in the late 1990s through visual art practice (for instance in land art).

103 Understandably, very few critical publications use the term and, when they do, it is frequently in a tongue-in-cheek manner as is the case with Mark Fisher in his *How to Write about Theatre* when he discusses the act of classification in theatre and ends his list of genres with the term "site-specific verbatim comedy" (37). However, British theatre companies them-

The Working Playwright: Engaging with Theatres, co-edited by David Edgar and Amanda Whittington, concedes that "[s]ite-specific productions are a relatively recent phenomenon in Britain on their current scale" (23). British actor, Simon Callow goes as far as calling the first decade of the new millennium, "the decade of the 'site-specific show'" (2012) whilst Marvin Carlson prefers to call it the "'immersive theatre' vogue" (124). For Olivia Turnbull, this is nothing less than "arguably the biggest single shift in the theatre world (since the beginning of the new millennium)" (151); therefore it is no wonder that this shift should eventually reach verbatim theatre productions and dramatically change them as a result.[104]

If the matter and purpose of site-specific performances[105]are subject to constant definition and redefinition at the present time,[106]it may be worth spending some time discerning its salient features or, rather, its characteristic processes in many of its differing guises – beyond its rupture of fourth-wall realism – whatever these might be. Reputedly, site-specific theatre or "[s]ite-specific devised theatre" (Oddey 125) is indeed a contested and "slippery concept" (Filmer 128) and one is consequently not surprised to find no working definition in Mike Pearson's seminal *Site-specific Performance* (2010). Pearson, the former associate artistic director of Brith Gof,[107] purposely resists making definitive statements and, instead, provides a series of 18 provisional distinctions between the auditorium and site work, such as the following:[108]

> The auditorium is cloistered. At site, bounds and perimeters may be extant or installed.
> In the auditorium environmental At site, environmental conditions may change and need
> conditions are stable. to be accepted or actively countered.

---

selves have described their work in terms of "site-specific verbatim performance", hence our decision to retain the term regardless.

104 This quotation is taken directly from the V&A app "Played in Britain: Modern Theatre in 100 Plays" in a video entitled "2000–2010: The State We're in" recorded on 25 September 2002.

105 For practical purposes, I have purposely avoided discussing the many terms that have derived from the notion of "site-specific performances" such as "site-generic", "site-responsive," "site-oriented," "site-related", "site-sympathetic" etc.

106 Crucially, site-specific and immersive theatre are not a genre with clearly defined codes and conventions (see Machon *Immersive*, xvi and Pearson *Site-specific*). I do not consider them as synonymous either, as the notion of the "site-specific" clearly lays more emphasis on the site itself.

107 Founded in 1981, Brith Gof was a Welsh theatre company co-directed by Clifford McLucas that was renowned for its large-scale site-specific works and innovative approach to site more generally.

108 Funnily enough, Josephine Machon makes a similar table for the concept of "immersive theatres", comparing them to "[a] 'traditional' theatre experience" (*Immersive* 54).

> The auditorium is dark and quiet. Site is only dark or quiet if chosen for such qualities or rendered so. (*Site-specific* 16)

If one looks at a close term such as "immersive theatre" for elucidation, one is again facing the very same problem: "[it] is impossible to define as a genre, with fixed and determinate codes and conventions, because *it is not one*" (Machon, *Immersive* xvi).

As a starting point for the discussion in the pages to come, I would therefore like to offer the following definition by Charles London: "theatre designed for a specific space or location that is not itself designed for theatrical use" (Chambers 693). In other words, and this is an obvious enough thing to say, a site-specific piece typically lies outside what is sometimes called "building-based theatre." Whether or not one agrees with this conception of "site-specific performance," when such a term is deployed in conjunction with verbatim theatre, it generally refers to a collaborative performance characterised by non-conventional and formally open strategies of theatrical presentation taking place outside a permanent theatre building (warehouses, mobile and stationary vehicles, disused buildings, purpose-built structures, railway arches, shipping containers, etc.). A site-specific verbatim performance is therefore a theatre piece based on a text assembled from interviews with real-life people in real-life situations "conceived on the basis of a place in the real world" (Pavis, *Dictionary* 337). Put more simply, a site-specific verbatim piece can be understood as the juxtaposition and interplay of "found texts" with "found spaces" (Pearson & Shanks 23) to create a verbatim performance outside traditional theatre buildings.[109] Chiefly, the site used for the performance may suggest the dramaturgical structure of the verbatim piece to be devised. In Jonathan Holmes' site-specific verbatim piece, *Katrina* (2009), the performance text is divided into four storeys instead of scenes or acts.

A site-specific verbatim performance is a live piece of theatre at the crossroads between installation and performance art, its designated driver being symptomatic of wider trends in British theatre that traffic across a variety of apparently incompatible disciplines beyond the authorial model. The verbatim words are thus framed by the site, by a place one can never return to, and the performance attempts to embed the audience in the worlds evoked. The main difference with "site-specific performances" is that the verbatim subset of these works admittedly does not place the audience within fictive worlds in the man-

109 David Hare in the essay "Obedience, Struggle & Revolt" compares the process of making verbatim theatre to sculpture: "[y]ou find the driftwood on the beach, but you carve the wood and paint it to make it art" (29).

ner of groups such as Punchdrunk.[110]I might pursue this attempt at a definition one step further by stating that "promenade performances"[111]are not necessarily site-specific. Indeed, Michael Wynn's *Who Cares* (2015) on the NHS was originally conceived and promoted as a site-specific piece but, because of the impossibility of finding an adequate site, it eventually became a promenade piece staged at the Royal Court Theatre.[112]

Now, another key feature of site-specific works must be highlighted here and I must refer to the quotation that opened this very section. To fully grasp the concept of the site-specific, one must focus on the issue of spectatorship. In this new "post-realist" setting, breaking away from conventional playgoing routines, verbatim theatre showcases a particular concern for audience receptivity, active participation, productivity, responsibility and agency, suggesting a more subtle, fluid and unpredictable interplay between audience and performers in four dimensions, what might be termed a new "art of spectating" insofar as it is seemingly not conceptually predetermined.[113]In this new configuration or "mode of ambulatory spectatorship" (Drees 101) that allows a more limited number of audience members than usual, then, the audience is invited to cross the fourth wall (and any such "separation" between performers and spectators), to be subjected to "the gaze of other spectators" (Alston, "Funding" 197), to have an experience and consequently become fully implicated in the collective act of transgressive creation.[114]In other words, the performance is precisely this whole situation of a unique, live, unscripted and more fully developed encounter between perform-

110 Within a UK and international context, the British company Punchdrunk has become emblematic of the enunciated mainstream site-specific development. A typical Punchdrunk production uses a large-scale site and adapts an existing fictional text so as to create an immersive environment in which the masked audience "freely" circulates.

111 A promenade performance typically implies that the audience goes on a physical journey – sometimes across several locations – as opposed to being static throughout and watching a performance.

112 *Who Cares* was staged in and around the Jerwood Theatre Upstairs as well as other spaces throughout the building of the Royal Court Theatre.

113 Following Dave Beech's distinction, site-specific verbatim theatre enables the audience to become participants and not collaborators who are perceived "to share authorial rights over the artwork that permit them, among other things, to make fundamental decisions about the key structural features of the work" ("Include" 3). Such an active participation usually concerns the following sensory acts: "touching and being touched, tasting, smelling and moving" (Alston, "Audience" 129). Pearson equally mentions that the spectator can expect "to feel a performer's breath, to feel a performer sweating" (Morgan 55).

114 For instance, a site-specific verbatim show like Look Left Look Right's *Caravan* (2008–2009) only welcomed eight audience members at a time and performed five times a day.

ers and audience, the audience being perceived as part of the verbatim work or even "at the centre of the experience" (Machon, "Punchdrunk" 261). Importantly, the interactions between spectators is also valued and the audience is constantly free to choose its positioning in the course of the performance; thus one witnesses a kind of unfinished quality in relation to more traditional staging practices.

Alongside these considerations, it is important to highlight that site-specific and verbatim performance share some common features inasmuch as they engage with "the real" in performance. If verbatim theatre relies on the real words of real people to challenge theatrical conventions, site-specific works tend to operate in a "real world" context: "real trees, real oil drums, real cars [...] introduc[ing] into the heart of the theatrical discourse, a whole array of questions about reality and pretence" (McLucas, qtd in Morgan 46). This multidimensional medium that arises from a fusion of site-specificity, immersivity and verbatim theatre entails key questions. What is the relationship between the verbatim material and the site in these works? Does the site function as a verbatim text (documentary realism) or does it fully compromise the verbatim component of the performance (post-realism)? Conversely, one may ask: what does the verbatim technique do to the strand known as site-specific performance?[115]

As is the case with each chapter, I see Jonathan Holmes' *Fallujah* as representative of this whole spectrum of practice I have just discussed. According to Stephan Koplowitz – whose particular focus is on site-specific dance – there are four main categories of site-specific performances. The fourth category is what he calls "reframing the known" to designate "the experience of taking something like a work of art that was created, and then [that] ends up being placed in a particular site."[116]This, in itself, poses a threat to the divide between site-specific and non-site-specific works and, to my knowledge, very few verbatim texts in Britain have yet been exposed to such a treatment, except perhaps in the context of a touring production or in exceptional circumstances when, for instance, the tribunal play *Half the Picture* (see Chapter One) was performed at the Houses of Parliament in 1994. Koplowitz's third category, "reframing from studio to site," identifies the practice of artists that "intentionally take a work of theirs and decide to put it in a specific location." In the context of verbatim theatre,

---

115 As fascinating as this last question may be, it falls largely beyond the key aims of this book and will therefore not be treated in this part. For now, my position is that – as site-specific performance is not a genre – it can easily accommodate verbatim strategies within its "walls."

116 This quote and all subsequent ones by this author are taken from a transcript of an online course "Creating Site-Specific Dance and Performance Works" from the California Institute of the Arts made available freely on the web through the website Coursera: https://www.coursera.org/learn/site-dance/lecture/n4O2U/defining-four-categories-of-site-specificity-part-1

again this is relatively rare and therefore does not constitute the bulk of the practices of concern in this chapter. His next two categories, however, are very meaningful for the verbatim works under discussion here. His second category, "site-adaptive," corresponds to "a work that is actually wholly site-specific, meaning that the artist does not make any decisions about making a work until they have encountered the site." Crucially, he adds that "the site is not unique, the site is not one of a kind. It is a site that exists in multiple locations. And so this work that is made for this one specific place can actually be adapted and placed in other places once it's made". His first category, "site-specific," is used for a piece that "is completely, and in essence, one of a kind [...] this is work that cannot be repeated anywhere else." Typically, I would suggest that these site-specific verbatim performances do not wholeheartedly endorse site-specificity in the strictest sense of the term (category one). In effect, they are very much part of the mobility turn in site-specific works identified by Fiona Wilkie, that shifts "from performance that inhabits a place to performance that moves through spaces" (90), that is to say category two. This tendency, outlined by Wilkie, thence goes directly against the longstanding idea behind the concept of the "site-specific," that is to say "a work of art created for and theoretically fully understandable only within that location" (Carlson 76). By way of exemplification, in Jonathan Holmes' *Katrina* (2009), another site-specific verbatim piece, the playtext specifically indicates an openness to new sites that would multiply its possibilities of staging:

> *Reflecting the play's first site-specific performance by the Jericho House at the Bargehouse in London, the play is divided into 'storeys' rather than 'scenes'. For subsequent incarnations in other spaces these terms can naturally be removed, though it may be handy to retain the broad structural transitions they implicitly mark.* (7)

In the same way, *Caravan*, created in 2008 by one of the most high profile site-specific verbatim companies, Look Left Look Right, was particularly mobile, as its name indicates, and was temporarily parked outside the Royal Court Theatre in Sloane Square in February 2009. In short, these works therefore appear to be closer to the idea of the "situational," that is to say that they are repeatedly interventionist across context, time and place, as an event-in-progress, rather than by virtue of a fixed and stable notion of the site-specific.

To throw in one or two last preliminary comments on these very ambiguous activities, I want to point out again that not all site-specific performances embrace my concept of "post-realism", although it seems that they have a particularly strong post-realist inclination. Frame by frame, it is a matter of how the following components are put together and experienced by an audience: verbatim text, acting, space, design, movements, sound, music, lighting and also video in the case of this representative site-specific verbatim performance, *Fallujah*. Perhaps, one of the strangest site-specific verbatim performances that employs this concept of "post-realism" at its fullest, is one that turns out not to have been staged in a site but in a theatre space. Cressida Brown's *re-Home* (2016) on social cleansing is the sequel of a more "traditional" site-specific verbatim performance called *Home* that was staged in 2006.[117]*re-Home*, produced with her company, (appropriately named) Offstage, maintained an active engagement with site as it sought to revisit a community ten years later at a time when the site – a dilapidated tower block in the Beaumont Estate in Leyton, East London – no longer existed.[118]In a sense *re-Home* applies the principles of post-realism to site-specific verbatim itself by turning the theatre into a site. In addition, the piece critically questions the right of a theatre-maker to make such shows, or at least shows like *Home* which, ironically, can be seen as another type of gentrification. Strikingly, in the performance, the intermittent act of actors reading verbatim a physical copy of the interview transcripts – however unassuming in its theatrical ambition and execution – further questioned the ingrained editing process as regards to what verbatim theatre-makers think is important. It also completely ripped apart any remnants of documentary realism.

As we have just seen, site-specific verbatim theatre is perhaps less of a genre than might be the tribunal plays of Chapter 2.2 or the headphone-verbatim pieces of Chapter 3.2. It is rather an aesthetic trajectory and in order to fully comprehend the spectrum of practices that falls within its remit, I must attempt here – as I did in Chapter 3.2 – to provide a short history of its development since the late 1990s, especially as it has not been the object of much documentation in the literature despite the increased visibility of both verbatim theatre and site-specific performance.

4.3.3 A Short History of Site-specific Verbatim Theatre

Site-specificity has infiltrated contemporary British verbatim theatre in many different ways throughout its history. At the purest point of this sits a perfect matching of verbatim material and site and, at the other end of the spectrum, the ver-

117 The piece was also an instance of "minority theatre" as it was made with the community, for the community and in a community setting.

118 The site that was in an area notorious for gang violence and crack houses was demolished as part of a regeneration scheme in the lead-up to the London 2012 Olympics and the rehousing process took over six years.

batim occupation of a site. For Chris Goode, here lies the key distinction between the "genuinely site-specific work" and the not-so-genuine site-specific performance "which is merely presented in sites not normally used as locations for performance" (*Forest* 75), the former being a lot less frequent than the latter. Put another way, there is a significant difference to be acknowledged in relation to the specific level of engagement of a verbatim piece with a given site.

Now, it may be very controversial to suggest that site-specific verbatim theatre was born in the late 1990s. Indeed, verbatim theatre's new arrangement is in truth not as new as one would at first think. If one focuses on the promenade strand alone, it appears that verbatim theatre can legitimately make a claim of kinship. According to Max Stafford-Clark, the first promenade piece in the UK was in fact verbatim.[119] *The Speakers* (with the late William Gaskill), based on a book of interviews of the same name by Heathcote Williams, by Joint Stock in 1974 was indeed a verbatim promenade recreation of Speakers' Corner in Hyde Park. Delving more deeply into the concept of site-specific verbatim performance, there have likewise been some substantial explorations and early experiments with space and documentary material at the time of the alternative theatre movement in the UK.[120] In contrast to the British companies of that era and political ilk, the site-specific verbatim performances of today seem to have a very different agenda. But, let me for now shed light on these little-known contemporary activities.

The very first performance that can clearly be considered as such dates back to 1999/ 2000 and was undertaken by Non-Fiction Theatre, the performance group discussed in Chapter 3.2.[121] Thus, it was a combination of the headphone-verbatim technique and site-specificity and the performance was entirely created from verbatim interviews gathered by the actors themselves – Jo Harper, Louise Wallinger and Lexi Strauss – and recorded on mini-discs under the directorship of Mark Wing-Davey. The piece was crucially site-specific to the Museum of Me (Oxo Tower, London) and the actors performed as living exhibits wearing

119 This information came to light in the context of Danielle Merahi's interview with Max Stafford-Clark and Stella Feehily on 12 December 2001 that is reproduced in note form in her *Theâtres du Réel.*

120 Depending on the adopted definition of site-specific verbatim performance, earlier examples can be found. However, the contemporary British practitioners I interviewed were largely unaware of prior examples of such practices and tended to consider themselves to be pioneers.

121 More research into this may bring to light earlier examples that I am not aware of. This short historical overview does not have the pretension of being exhaustive and only aims at introducing the reader to an under-researched phenomenon in contemporary verbatim theatre, thereby situating *Fallujah* within a larger strand of contemporary practice in the UK.

headphones and repeating the verbatim recordings. In the same "genre," Alecky Blythe created *Voices from the Mosque* as part of Headlong's collaborative site-specific piece *Decade* that marked the tenth anniversary of 9/11 and ran from 8 September to 15 October 2011 in a disused office building in St Katharine Docks, a marina just west of the Tower of London. Two other notable examples of site-specific verbatim performance – are London Bubble's 2006 production of *My Home* – directed by Karen Tomlin and exploring the immigrant experience, the hopes and aspirations of the latter through multisensory installations performed in houses across London – and Rebecca Hillman's *The Pact* (2011), performed in a disused pub called *The Coopers Arms* in the town centre of Reading that asked how neoliberal ideologies and capitalist systems affect people on a day-to-day basis and how to resist their negative impacts. Besides Jonathan Holmes' Jericho House, founded in 2007, that is the specific focus of this chapter, Look Left Look Right Theatre Company, co-created by BBC researchers Ben Freedman and Mimi Poskitt in 2005, can be said to have spearheaded the development of these types of works in Britain with pieces such as *The Caravan* (2008–2009), *Counted* (2010) and *NOLA* (2012).[122]

The National Theatre of Scotland itself was born out of the idea of an art institution without walls and the kinds of political imperative and radical politics that can be nurtured in this context. Vicky Featherstone, its first artistic director, said in an interview conducted in 2012 that the NTS model, springing from a "determination not to have a building" (Keidan and Mitchell 63), went hand in hand with having "no barriers, no boundaries" (Keidan and Mitchell 63). As a statement of intent, the NTS decided in 2006 to call its inaugural production *Home*, a site-specific piece that was simultaneously performed in ten different locations throughout Scotland. Their own breed of mobile and fluid verbatim and non-verbatim pieces is generally perceived as a direct response to the London mainstream and, in particular, to the National Theatre, so as to become integral to its very identity, as Vicky Featherstone seems to suggest in the following: "the site-specific adventures that morphed unlikely spaces into magic portals, so that an airport or a car park, a high-rise tower block or forest, became not just a theatre, but the National Theatre of Scotland" (Keidan and Mitchell 69).[123] Other site-specific works have added to the marketing story of the NTS

122 The company also created one non-site-specific verbatim piece, *Yesterday Was a Weird Day* (2005) on the 7/7 London bombings.

123 It should be noted, however, that the National Theatre building in London is also site-sensitive: "When Denys Lasdun designed the building in the 1960s, he wanted a building that was not a monument, standing apart from society and embodying a rarefied set of artistic values – he wanted a building that was literally and metaphorically continuous with the society around

such as the site-specific verbatim piece *Enquirer* (2012) – in coproduction with the *London Review of Books* – based on the exact words of 43 leading UK journalists channelled verbatim by a cast of six actors and staged on the empty floor of an office block overlooking the Clyde in Glasgow.[124] The performance dramatising a day in a newspaper office brings the audience into close contact with British journalists at a time of crisis.[125]The resonance with an audience was truly pushed to the fore as the performers had complete licence to incorporate anything that happened in the news on the performance day.[126]

In comparison, Wales has an even richer tradition of site-specific works and it is therefore no surprise that in 2009 the newly formed National Theatre Wales included site-specific works in its first programme under the directorship of John McGrath. More specifically, in 2010, the first season of National Theatre Wales was perhaps even more engaged with site-specificity with no less than twelve works in twelve different locations. They also produced *Ricochets of Verbatim Tales* in 2014 that was "an immersive, site-specific, verbatim theatre experience" (National Theatre Wales website). Having briefly surveyed a rather alluring body of verbatim works that may potentially engage with the tropes of "post-realism", a turn to *Fallujah* itself will certainly facilitate such an exploration.

4.3.4 An Overview of *Fallujah* and Jonathan Holmes (Jericho House)

In May 2007, Jonathan Holmes' *Fallujah* opened in a partnership with the ICA in London, three years after the sieges of Fallujah (April and November 2004) conducted by the US and the UK as well as other international coalition forces in Iraq. The city of Fallujah – located in the Sunni belt north east of Baghdad and inhabited by 300,000 people – was by far the most troubled province in Iraq during the war and these sieges were, according to Amy Goodman and Denis Moynihan, "among the most violent and devastating attacks on a civilian

it. The National Theatre, when it eventually opened in the mid-1970s, had no single point of entry; instead its terraces and walkways multiply connected it with the rest of the South Bank Centre and the streets around it" (Rebellato "Introduction" 4).

124 The performance I saw in London (under the auspices of the Barbican) was recreated in an old toy factory in Clerkenwell that felt very much like Old Fleet Street.

125 This crisis is both economic – the industry is changing rapidly and its business model appears somehow outmoded, archaic even – and moral (the sense of purpose has been overwhelmingly replaced by commercial instincts).

126 This was timely as the Scottish performances coincided with the Leveson Inquiry, at the time when Rupert Murdoch was giving evidence live.

population in recent decades" (xi). The verbatim text was drawn from interviews with former residents, witnesses, soldiers, politicians, journalists and diplomats, exposing multiple human rights violations not justified by military necessity – allegedly breaching over 70 articles of the Geneva Conventions – including the bombing of hospitals, health centres and schools as well as the use of chemical weapons, the killing of unarmed civilians holding white flags, the prevention of civilians from leaving the city and relief organisations from entering. The piece was site-specific as it was originally conceived for a disused brewery in London's East End. However, *Fallujah* was not confined to one specific site and was very much a portable format adaptable to other sites. As such, it was characteristic of the aforementioned mobility turn in site-specific performance. Indeed, *Fallujah* travelled to Berlin and Prague in a very stripped-down form and had other parallel lives in the US – it was on the university syllabus for a time and there were a number of amateur productions and protest readings – which departed from the format of the London production, depending on the focus given to music (which was continuous and written by multi-award winning composer Nitin Sawhney) or the design (which was by the renowned Paris-based artists Lucy and Jorge Orta). It was also remounted in Paris, Tel Aviv, Amsterdam and Glasgow.

Fallujah is not Holmes' first site-specific experiment. In June 2005, for example, following the rediscovery of the scores of several songs by the English poet John Donne, a performance was held at St. Paul's Cathedral by the sopranos Dame Emma Kirkby and Carolyn Sampson and the choir The Sixteen led by Sir Harry Christophers. The following year, Holmes mounted another site-specific and devised piece, *Unlawful Assembly*, that was performed throughout the town of Chipping Norton in the Cotswolds. Holmes' ongoing exploration of site-specific performance was initially sparked in 1998 by *Gargantua*, a piece by Ben Harrison he saw at the Edinburgh Festival Fringe that was part of a series of works Harrison made with the company Grid Iron in the late 1990s and early 2000s.[127] Significantly, in 2007 Holmes then founded the company Jericho House, whose main focus revolves around hospitality, found spaces and cross-media collaborations as opposed to more regular theatres and, *Fallujah* was to become his first foray into site-specific verbatim theatre. The same year, Holmes won the Guardian/Unltd Social Entrepreneurship Award for his work on *Fallujah* (as writer, director and producer).

Since 2004, Holmes has been an Artist in Residence at Peace Direct, a charity founded by triple Nobel Prize nominee Dr Scilla Elworthy and actively in-

127 The piece in question was based on Rabelais's *Gargantua* and was performed in an old bank by Scotland's leading site-specific company.

volved in conflict prevention projects.[128] In this capacity, he took part in their 2005 seminar on "Learning from Fallujah" and, following the Chatham House Rule, was allowed to take down what was said without attributing it to the speakers.[129]He also subsequently conducted extensive research and a series of interviews, not always in the flesh, with people that had been in Fallujah at the time of the atrocities. Among the 40 verbatim testimonies from independent sources he collected, three figures particularly stand out: Jo Wilding – a human rights activist from Bristol – who heroically took a small circus to Iraq to work with traumatised children, helped wounded civilians get out of Fallujah, gave them medical supplies and blankets, writing down their stories and putting them on her blog; Dahr Jamail, an independent US journalist notably writing for *The Guardian* and *The Independent* in the UK who was a tireless unembedded chronicler in occupied Iraq, and Iraqi activist Rana Al-Aiouby who had been risking her life to deliver essential medicines to the wounded in Fallujah.

Within the space of four acts and 32 scenes lasting a little over 90 minutes, *Fallujah* – which uses the preferred British spelling in its title – depicts a very complex and "atrocity-producing situation" in a chronological manner from October 2001 to 2005.[130] On 31 March 2004, over a year after the invasion and occupation of Iraq, "[t]he discovery of the burned, brutalized and mutilated bodies of four American civilian contractors" belonging to Blackwater (160) precipitated the first siege, also known as "Operation Vigilant Resolve".[131]On 4 April 2004, a full attack was thus launched on Fallujah so as to take complete control of it. The second siege – dubbed by the US as "Operation Phantom Fury" – began on 8 November 2004 and was the largest military operation during the occupation. It was essentially a symbolic operation that destroyed 70 % of the city and killed about 5000 people (the vast majority were civilians). The performance of

---

128 She also provided the basic material for Max Stafford-Clark's verbatim production of *Talking to Terrorists* (by Robin Soans) mentioned in Chapter 2.3.

129 This verbatim material eventually ended up being dramatised in the fifth scene of *Fallujah*.

130 An exception was made in the epilogue that is in fact drawn from a Sunday school class on 4 August 2002 at the National Presbyterian Church in Washington with Condoleezza Rice. The term "atrocity-producing situation" was initially coined by leading American psychiatrist Robert Jay Lifton in the context of the Vietnam War and is now widely used in relation to other conflicts. Lifton defines the term as "a situation so structured, militarily and psychologically, that ordinary people are capable, when entering it, of committing atrocities" (qtd. in Goodman), which is sadly what happened in Fallujah.

131 This was a public mutilation and the bodies were dragged through the city and hung on a bridge. These private contractors were in fact mercenaries who did not belong to the US army. Private paramilitary contractors were often accused at the time of coming into the city to aggravate the situation, carrying out assassinations and generally seeking profit out of war.

Holmes' piece begins and ends with Condoleezza Rice. Opening on a television interview with *Al Jazeera* on a raised stage, *Fallujah* closes on the same raised stage, but this time the atmosphere is not quite the same as Rice is in a Midwestern Presbyterian Church. Principally, the site-specific verbatim piece follows scene by scene the incidents as well as the courageous and honourable daily efforts of aid workers including Jo Wilding, a Christian Iraqi cleric, translators and journalists who remained in the city during the assault despite the overwhelming atmosphere of devastation and death.

In *Fallujah*, one confronts the unconfrontable/unstageable as one is drawn into the world of the eponymous Iraqi city that news agencies were not allowed to visit at the time of these two separate months of siege warfare. One is told by the actors, standing in for the besieged and bereaved Fallujans, that there is an ongoing dire humanitarian crisis happening beyond view and outside the care of the international community as indiscriminate, despicable and illegal military tactics are being systematically used, thousands of homes, shops and mosques are being devastated, infrastructures are in a shambles, air strikes have destroyed Fallujah's main hospital, aid vehicles are being turned away, ambulances are being shot at, major Iraqi institutions, such as the police force, have been dismantled, citizens are trapped without electricity, clean water and food as well as other basic necessities, US snipers target anything that moves including children, civilian casualties abound and bodies lie in the streets for days, being gnawed at by dogs in a bloodthirsty and lawless atmosphere. In light of the steady decline of the situation in Iraq as well as these harrowing accounts of chaos, violence and suffering, no one could still believe that the war was for democracy, human rights, development or freedom.[132]

In its London incarnation, *Fallujah* was performed 40 times by a cast of seven actors in the converted industrial interior of a brewery in London's Brick Lane with a roving minimal stage light and a space crammed with anti-contamination suits, stylised corpses, hospital beds and equipment, clothes and shoes, stretchers, water canisters, a Red Cross ambulance, along with television screens hung from the walls and some folding chairs for the audience. A soundscape punctuates the narrative and the audience is implicitly required to follow the performers around the set of the old industrial space.

In a nutshell, *Fallujah* is a politically motivated and explicitly anti-war verbatim piece that attempts to address the failure of the West to condemn Ameri-

132 By that time, the WMD narrative – so crucial to the verbatim accounts of *Justifying War* (Chapter 2.2) and *Stuff Happens* (Chapter 3.1) – had been officially dismissed and swiftly replaced by another one. According to the Bush administration, they were there to liberate the Iraqi people.

can behaviour in Fallujah that had conveniently slipped under the mainstream media radar. The verbatim production was covered by every English broadsheet, as well as the BBC, *The New York Times* and *Al-Jazeera*. Some of the responses mainly took issue with *Fallujah*'s stance that they summed up as "anti-US" (O' Reilly 42), missing its complexity and nuances. The latter will be the very object of this chapter, as I attempt to articulate how the aesthetic of post-realism manifests and impacts on the verbatim material. Interestingly, one of the responses to the performance from a journalist on the *New York Times* caused quite a stir at the time and "set off a mini-storm of e-mails" (Hoyt) when the journalist in question claimed that: "[t]he denunciations of the United States are severe, particularly in the scenes that deal with the use of napalm in Falluja, an allegation made by left-wing critics of the war but never substantiated" (Perlez). This, combined with the death threats Jonathan Holmes received, only confirm that the information *Fallujah* contained was still very controversial and potentially subversive.

Having briefly surveyed the difficulty of the kind of material to be found in the piece, it is now time to explore the specific terms of its contract with the audience so as to perhaps test Michael Rothberg's proposal that "[t]he problematics of representing and coming to terms with an extreme historical event push the realist project [...] to its limits" (99).

4.3.5 The Contract with the Audience

As the audience enters the warehouse space, they can hear several layers of ambient noise and music, the profoundly disturbing sounds emanating from various undefined locations in the space that may or may not have occurred in the besieged city of Fallujah. One moment they are at a press briefing in Bagdad (160), the next they are in Fallujah (163). In the introduction to the play, Holmes writes that "[s]pectatorship [...] takes place on two levels: the reception of the performance on the level of art, and that of the script on the level of immediate testimony" (142). While it is useful to remind us, as Holmes does, that these two modes of spectatorship are of a different order, I would suggest instead that within the frame of a verbatim performance – whether documentary realist, new realist or post-realist – the audience, being endowed with critical awareness, can never reach the level of immediate testimony, even when the "real people" are also the performers in the piece. What verbatim theatre can do effectively, however, is through a series of strategies – belonging to documentary realism, new realism or post-realism – make the audience intermittently invest in this fiction and potentially shake them to the core as a result. As ever, the exact outcome of this is incalculable and the audience can obviously refuse or rewrite the roles assigned such as "theatre spectator, tourist, game player, partygoer, voyeur, connoisseur, witness, scientific observer, detective" (Turner 25). Acknowledging this draws attention to the relationship with an audience, one that is always fundamentally part of the performance and perhaps even more so in the context of verbatim theatre. In its site-specific formation, I would like to argue that verbatim theatre's audience relationship completely shifts to the forefront. Now, without further ado, let us examine exactly how this works in site-specific verbatim performances and to what extent this contract may be typical of what I have defined as "post-realism."

As seen throughout this study, the default contract for a contemporary British verbatim performance is documentary realism. New realism in Chapter 3 was, notwithstanding, about including certain things apparently destructive of documentary realism as we go along. From the outset, I want to posit that, in essence, a post-realist contract – as exemplified by both *To Be Straight with You* and *London Road* – is characterised by a growing uncertainty and ambivalence towards its own terms, as opposed to a certain sense of mastery still prevailing amongst the new realist verbatim works. In sum, each contract betrays a very different attitude towards the verbatim material in performance. A common thread running through the various contracts covered in this study (from Chapter 2.1 to Chapter 4.2) is that, in verbatim theatre, the contract tends to assert itself from the very start. Jonathan Holmes' *Fallujah* in this regard is no exception. As usual, the screens read: "[e]verything you are about to see and hear is drawn directly from the testimony of those involved in the sieges of Fallujah in 2004," whilst at the same time the audience is immediately aware that – as Holmes has put it – "[t]his play differs from most other plays of the recent verbatim boom" (144). Indeed, the site-specific terms have so much entered the contract that the scenography cannot help but distract attention to itself. In addition, more than in any other verbatim performances covered in this volume, the spectator is aware of herself as part of an audience. The premises of the relationship between performers and audiences are established in a way that puts a premium on corporeality and the liveness of the performance. In this new configuration, the scenography is always transparent and the audience never forgets that they are in a disused brewery. The scenography is also as big as the space, there are no separate spaces of actors and spectators and, here, the performers often step through the absent fourth wall into where some audience members have gathered, creating very dynamic scenes in the crowd. For instance, in a stage direction in the third scene of the first act, Holmes categorically writes that "*[a]t a suitable juncture RANA addresses us directly. There should be no acknowledgement of a fourth wall – she is talking to us*" (161). In another one in Act

III, Scene 23, the performers "*mingle with the audience as they talk*" (195). In the case of *Fallujah*, this phenomenon is admittedly exacerbated in comparison to most site-specific verbatim performance as the set is both a stand-alone exhibition and a component of the performance. The set is both a kind of verbatim design that seemingly recreates a sense of what it might have been like to be in Fallujah during the 2004 sieges and a visual and sensual response to the verbatim material in the manner of the physical movements of DV8 in Chapter 4.1 and the songs of *London Road* in Chapter 4.2.

In other words, *Fallujah* hovers between two aesthetic positions. On the one hand, the characters are all based on real people, the play is still monologue heavy, the verbatim material has not been "augmented" (141) in the manner of the massaged pieces of Chapter 3.1, and it attempts to give us the most accurate insight into what happened in Fallujah as well as recreate the initial intimacy of the interviews.[133]More than this, the created characters all appear passionate and embody what the author felt when he heard these people. This is exactly what Suzanne Little identifies – in the context of Jonathan Holmes' second site-specific verbatim performance *Katrina* in 2009 – as "[a]n attempt at [...] literal repetition" ("Repetition" 47) so as to "match the emotional tenor of the testimony and/or to re-create the original traumatizing event" ("Repetition" 47). On the other, it fully acknowledges the impossibility of such an enterprise. In fact, in the performance text itself – that differs from most verbatim texts in that it gives more sense of how it will exist physically in relation to an audience, that is to say that it re-anchors the verbatim words in the physical world – Holmes confesses that the performance "cannot be made to look like a besieged city, however gifted the designer, never mind smell like one [...] [d]ocumentary realism is not possible in this case" (142). Accordingly, this new contract creates a very different experience and one no longer expects to be within the familiar territory of documentary realism as described in Chapter 2, that is to say where the performance would strive mimetically to recreate the events of the sieges with as much accuracy as possible. Legitimately, one could say that Holmes is subtly subverting the genre he has invoked. For one, even the usual disclaimer found at the start is tweaked by being repeated at the very end of the performance. Second and more trenchantly, the use of multimedia technology affects any possible documentary realist reading as sound moves between speakers and images between screens. Third, the spectator finds herself mobile, witnessing the scenes being

133 Holmes explains in the manner of the documentary realist verbatim dramatists that he has scrupulously followed the applicable rules of practice in this context "with the exception of occasional greetings at the start of a scene" (141) and the character of Sasha who is "a composite figure, built from testimony given by different journalists" (143).

performed by actors and being re-mediated live on screens at the same time. Between the simulated and "the real" (screens), the audience can choose to be on the outside looking in or be more implicated in the ongoing narratives. Fourth, as Adam Alston has argued (via the philosopher Robert Nozick), in site-specific/immersive performances, "audiences enter 'experience machines'. Experience machines are enclosed and other-worldly spaces in which all the various cogs and pulleys of performance – scenography, choreography, dramaturgy and so on – coalesce around a central aim: to place audience members in a thematically cohesive environment that resources their sensuous, imaginative and explorative capabilities as productive and involving aspects of a theatre aesthetic" (*Beyond* 2).[134]The end result is that the overall experience is heightened and more immediate than the conventional verbatim performances essentially focused on words. Clearly post-realism, as seen in both *London Road* (a musical experience) and *To Be Straight with You* (an emphasis on corporeality), has a unique inflection in verbatim theatre because it is more connected with what Janelle Reinelt perceives, in the context of documentary theatre, as a "phenomenological engagement" ("Promise" 7). In *Fallujah*, the accent is placed on the senses, on the fullness of the audience encounter with the situation in the besieged city, on the interaction with the real in performance, on an awareness of how we position ourselves in space and over time. In other words, this is the phenomenological condition *par excellence* if understood as:

> a human being, thrown into a world not of their choosing, encountering the environment with the body, attempting to act in or against that environment, reflecting on the situation, and revising future actions in light of what has already taken place. (Fortier 31)

Furthermore, as Josephine Machon argues in the context of "immersive theatres", there is also, here, to some extent, "some kind of "contract for participation'" ("Watching" 35), a new clause in the contract of verbatim spectatorship. Arguably, even though the site-specific verbatim works identified are minimal in terms of participation, compared to the most daring immersive ones in the period under study, it is my contention that, in this case, the imperative to participate is at its most irresistible. The performance challenges the spectators to act. Of course, the audience can leave the performance at any moment deemed suitable or withdraw from the centre of the action "with chairs available if they want to sit down" (147). At the same time, there is an acknowledgement and absorp-

---

134 Scott Magelssen prefers to use a different vocabulary and he describes the simulated, immersive, performative environment in which audiences meander, "simming" (2014), a term he has borrowed from online gaming.

tion – within the frame of the performance – of the presence and even "agency" of the audience, their real bodies being in motion and the actors really looking at them in the eyes when they talk. In this sense, *Fallujah* is also remarkably contingent on each spectator's physical actions and positions. That is to say, the implied terms of the site-specific verbatim performance hold open an invitation for the audience to see their real-time decisions and choices as ultimately completing the work.

If new realism was a makeshift bricolage approach to gently sidestep the critique of documentary realism, post-realism goes a step further and deepens its aesthetic predicament. It is as if *Fallujah* forces one to experience the edge of its own endeavour, as if it were expressing the central fear in verbatim theatre, that impulse that inevitably leads to its self-destruction and yet, perhaps surprisingly, trusting one to sustain the paradox of a permeable boundary despite its immense fragility. For all one knows, thus fuelled, it may be at this precise moment that verbatim theatre reveals its full strength, that is to say its capacity to move one that may have been ironically partially suppressed by both documentary and new realism. Let me illustrate these tentative points by suggesting that, at this juncture, post-realism springs from the impressive range of modes of encountering the verbatim material, a range that encompasses mimicry, narration, the extra-textual such as the embodiment by the performers of a reaching outwards towards an encounter with the other, the physical crossing of boundaries of experience, musical and sound stimulus as well as an emphasis on the mutually constitutive relationship between audience and performance and intrapersonal events that enrich our experience. Crucially, the piece never encourages one to privilege one of these modes over the other whilst also subtly offering a shattering and acerbic critique of language itself that I shall explore in more detail in the fifth section of this chapter.

Some consequences of such a change in perspective that I call "post-realism" will be suggested in the now familiar categories of investigation in what follows, with particular attention to site-specific verbatim performance through the case study. One is thus now in a position to scrutinise the specific staging of *Fallujah* – that is part art installation, part soundscape, part re-enacted actions and part verbatim testimony – and attempt to articulate how the aesthetic of post-realism flows and spreads over its verbatim components.

4.3.6 Staging

If the principal aim of a post-realist staging is to destabilise the accepted terms of documentary realism in verbatim theatre, *Fallujah* certainly more than qualifies for inclusion within this programme, as Holmes' statement in the introduction to the performance text clearly exemplifies:

> The play begins naturalistically, but this fragments as the piece progresses. There are screens on each wall. The piece should be accompanied by a sophisticated soundscape throughout. This should not be underscore, or mere naturalistic assistance, but an active player in the action. (148)

Fallujah therefore viscerally reflects the fractured experience of reality as well as the inescapable gap between "the real" and its representations, between the actual and the fictional. Arguably, there is already an inherent disjunction at work by placing the verbatim material in a site and engaging with the domain of site-specific and immersive performance. However, this state of affairs can either be minimised or maximised in practice by the creative team, depending on the kind of aesthetic experience they wish to generate. In the case of Jericho House's *Fallujah* – as is clear by now – I have begun with the premise that the theatre-makers have sought to effect a post-realist aesthetic despite the fact that they were also aiming at bringing about actual-world effects. Indeed, Holmes in the introduction to the text explains the kind of reaction he is hoping to provoke: "watching the play is intrinsically a political act" (144). Correspondingly, in the general introduction to the sourcebook *Fallujah: Eyewitness Testimony from Iraq's Besieged City* in which the performance text is published, Holmes explicitly states his aim in revisiting the emotionally loaded situation in Fallujah:

> There are two principal atrocities in Fallujah. The first is the criminal behaviour of American troops. The second is the failure of the West to condemn that behaviour. [...] There remains only the opportunity to publicize the disgrace, and to condemn it noisily. [...] For me, a collective act of witnessing is what the theatre necessarily involves, one that is inescapably ethical, as it requires us to take responsibility for our response to what we experience. (xiii-xiv)

Quite clearly, even if the words are verbatim and the piece endeavours to "give a clear and subliminal sense of what it is to be in Fallujah at this time" (148), the resulting performance of *Fallujah* never quite completes the documentary realist circuit. Ultimately, the conditions that made documentary realism possible are superseded by a series of post-realist strategies that open up the gaps between the layers of verbatim material. In addition, typical of the "post-realist" mode – as seen in the two preceding chapters – is the display of a changed perspective in the use of the verbatim material that abandons mimicry to the benefit of asymmetry, showcasing a response to the material itself. This shift is also more pronounced in the visible manipulation (and perhaps also deconstruction) of the verbatim words in performance. In the opening scene of *Fallujah* this interplay is already triggered as the recreation of the October 2001 interview of Condoleezza Rice by a journalist working for Al-Jazeera takes place live, but is also "*filmed and shown on a screen at the rear of the stage*" (155). In Act I, Scene Three, this time a more complex apparatus is in place as one of the actors/characters turns on a small television set so that the following scene "*is 'broadcast' on it, and on large screens on the wall*" (162). In Act III, Scene 15, another stratagem is used to distort the verbatim material on the occasion of Rice's speech in Stanford. The actress playing Rice is in the space whilst all the screens and the radio broadcast her speech at the same time and more disturbingly "*the first paragraph is spoken both live and pre-recorded*" (183). Again, in Act III Scene 22, another manipulation operates as the character of Sasha (the Canadian reporter) describes what she found in Fallujah whilst "[*m*]*uch of what she describes can be heard taking place throughout the scene*" (193) but "*the stage remains bare*" (193). This alone contradicts one of the founding principles of verbatim theatre according to which the creative team has the choice between two treatments of the verbatim material: a recreation of the exact context in which the words were originally uttered (documentary realism) or a dramatisation of what is said (new realism). Taken together, such manipulations are therefore perceived as parasitic to the normal operations of both verbatim theatre and realism. Admittedly, these strategies also serve to augment the performative potential of the verbatim process that documentary realism sough to annihilate entirely.

To come back to the aforementioned strategies of post-realism and concomitant with the elements of manipulation of the last paragraph, one finds within *Fallujah* an ongoing strategy of disruptive confusion that I see as more profound and damaging than the ones present in the new realist verbatim performances of Chapter 3. In Act II, Scene Eight, for example, "[*t*]*he scene begins as an apparent soliloquy; we realize only towards the end that she is filing a report*" (170). In the following scene, and in a more dramatic way, the audience is "*unclear to the source of this, it hovers 'above' the action as it were*" (173), "this" being an interview with a recently released French journalist. From Act II, Scene 13, our sense of confusion starts to grow as everything begins "*to appear subtly heightened*" (176). It follows that as in Act III, Scene 19, the scenes "*run speedily and energetically* [...] *the effect should be quite disorienting*" (190).

Another core feature of post-realism in verbatim theatre is a slow and internal current of destruction not only in theme – the "city of mosques" is being razed to the ground – but in form. Indeed in *Fallujah* we move from so called "straightforward scenes" to hybrids of performative ontologies created through a specific use of the soundscape, scenography, performers, elaborate digital technology and lighting. Significantly, this inheres in Act II, Scene 11:

> *The stage goes black, and heavy metal music is played very, very loudly. There is also the sound of artillery and small arms fire. This should last only for a minute or two.* (175)

Likewise, the staging of multi-synchronous episodes is a clear sign of its break from both documentary realism and new realism. For instance, Act II, Scene 17 that happens on screen only, is made to "*overlap with the previous scene, and act as a bridge into the more straightforwardly staged scene that comes next*" (188). Whilst Hare's *Stuff Happens* in Chapter 3.1 engaged with a certain duality by having two levels of narrative sometimes occurring at the same time, one of them arguably contained the other and the overall structure was fairly stable. In *Fallujah,* we reach a level of complexity that makes it almost impossible to discern which is which. In fact, one could argue that *Fallujah* is not only post-realist through its use of some specific strategies in performance, but through a structure that is more complex than it may at first seem. This is, in my view, very akin to a Russian-doll structure, except that one never has access to the first doll or perhaps – even worse – there may be no first doll at all!

To make these abstract considerations more concrete, let me refer to Scene 28 in Act IV. In this scene, Holmes purposely destabilises the relationship between "the real" and the fictional in *Fallujah* by introducing an unexpected twist that threatens the verbatim equilibrium and the idea of a stable and originary referent:

> At the end of this scene the voices should fade seamlessly on to track and then on to a tape recorder held by RANA, who switches it off and hands it to SASHA as they move into the next scene. The available interpretation is that all that has gone before has been the result of their reporting, and that we are now in the present. (210)

This post-realist gesture thus puts quotation marks around the realism of the verbatim narrative itself, making one even more aware of its frames and thereby participating in what Carol Martin has termed – in relation to other documentary performances – "an aesthetic and analytical discourse that represents the real in order to call it into question" (2013, 174). I would like to suggest that there are many frames in *Fallujah* that critically and subtly refer to the constant movement of citationality that is inherent to the verbatim practice. Here lies, it would seem, the crucial difference between documentary realism, new realism and post-realism. In the documentary realist performances of Chapter 2 this movement was more or less hidden from view and, if it was at times "hinted at," this was done in a manner that avoided drawing any attention to itself, the movement in question being constructed as straightforward. In the new realist pieces of Chapter 3, the movement was openly acknowledged but was placed within a broader construction so as not to truly destabilise the edifice of verbatim theatre. Now, in the verbatim works that engage with what I call “post-realism”, such a movement is never hidden from view and profoundly undermines the underpinning authority of the overall verbatim construction. Consistent with this picture, one finds different, intertwined manifestations of citationality at work in the space: those produced by the actors, those produced by the characters – when they are themselves involved in a movement of citationality as in Act III, Scene 17 in which the character of Sasha switches on the TV (188) – those produced by the act of translation from Arabic to English at various times in the piece, the manifold and competing mediating devices (speakers, microphones, screens, camera, computer, phone, sound system, tape recorder etc.), to some extent those produced by the audience as they react in real-time to what is happening around them and, finally, those produced by the author who is arguably made more visible here through, among many other things, a change of pace in the course of the performance.[135]

This opening to doubt and uncertainty typical of a post-realist practice may seem initially like a paradox in the context of a very political piece of verbatim performance. In respect to this position, I would like to propose instead a different reading of these strategies and assess whether they may be in fact introducing a kind of radicality hitherto not envisioned in verbatim theatre. Thus, in its site-specific format, verbatim theatre is not only about politics but embodies politics in action. On the most basic level, one could argue that, here, a radical change is coming directly at us.

4.3.7 On the Radical

For the most part, the usual means of arguing away site-specific theatre, still alluring to some critics and the British theatre industry more broadly, is to perceive it as forward-looking and entirely distinct from other forms of theatrical productions, as an ideal vessel for some radical reactions to and against the tropes of late global capitalism.[136]Catherine Love sums up this argument by stating in

---

135 In a way, form and content match in *Fallujah*. This pattern of endless mediation and remediation is a direct reference to the advent of digital technology at the time of the Iraq war – often called the YouTube War – that fostered news ways of seeing notably through soldier-generated video material and documentaries.

136 See Mangan (196) for an articulation of this view. David Lane expresses some reservations by stating that site-specificity only leads “in some cases to ‘radical’ popular theatre” (97) while

broad terms that "'non-text-based' work has frequently been seen as alternative, radical, progressive" ("Page") and Neil Bartlett explains that "[t]here was an unspoken assumption in the late 1980s and early 1990s that 'large-scale' radical work could only be produced on a site-specific basis" (Keidan and Mitchell 114). To some extent, this recalls Baz Kershaw's initial thesis in his *The Radical in Performance* that considered "'performance beyond theatre' as a more fruitful domain for radicalism than performance in theatre" (16). Indeed, these performances typically surrender control, demanding a more active engagement from its audience in opposition to the long-seated undisturbed and passive consumption that some quarters see in mainstream theatre-going habits.[137]Moreover, they do not leave any residue or tradable commodity after "the event," no publishable piece of writing that can ever hope to capture the experience of being there. An interrelated argument holds, as Dan Rebellato reminds us, that "the fact that these performances exist only in relation to the space they are in makes site-specific performance a form of theatre that is even harder to commodify than usual" (*Globalization* 54).

Having said that, one must also acknowledge that site-specific theatre is no longer political in the way it was in its early years in the late 1960s and early 1970s. As Marvin Carlson reminds us, site-specific performances' investment in particular locations was strongly associated with an engagement with working-class audiences (see Carlson 76). In other words, this was a type of theatre that played outside the mainstream stages and borders of official culture. Looking at today's site-specific performance works, one would surely argue, however, that one key aspect is the ease with which the mainstream marketing apparatus has recuperated its rhetoric. Simultaneously, its current popularity would seem to have exhausted its initial subversive potential as Bertie Ferdman argues:

> *Site-specific* has itself become usurped into mainstream marketing lingo: a promotional catchword and trendy signifier for innovation and pseudo-alternative experiences that promise to jolt audiences, to shake them up, so to speak, in 'new' or 'found' spaces. (5)

In effect, its modus operandi now shares some affinities with what is often referred to as "the growing experience industry" (Alston, "Audience" 128). One might have some difficulty of a different nature with this argument and it is necessary

---

Liz Tomlin unapologetically writes that "there no longer appears to be anything inherently radical about performing outside of theatre spaces" (*Companies* 115).

137 This belief is identified by Helen Freshwater as "one of the most cherished orthodoxies in theatre studies" (*Audience* 3) and one that misguidedly assumes "a connection between audience participation and political empowerment" (*Audience* 3).

to forestall any such criticism of what is proposed here by pointing out that the crux of the matter lies elsewhere, not least in the slippage between production and reception. My surmise here is that there is a third possibility, one that most reflects the experience of post-realism and contemporary site-specific verbatim theatre. These works create a kind of radicalism that is differently wired to the ones hitherto seen in this book. More specifically, this section is interested in the sort of relationship this kind of performance is establishing with its audience and its immediate contexts and how different types of spatial arrangements in verbatim can impact on our understanding of the radical. Finally, this section wants to interrogate some of the assumptions and prejudices that shape our attitudes towards this type of performance in order to explore how these produce our sense of what a radical verbatim practice is, has been, and should be.

But before going any further and attempting to untangle these various threads, it is important to restate how on one level, Jonathan Holmes' *Fallujah* engages with the notion of political performance. At first glance, this case study, one might argue, relates to Jenny Spencer's conception of post-9/11 political theatre according to which it can "help change and extend the public conversation, break down unhelpful social dichotomies, offer alternative viewpoints, expose political collusion, and provide more nuanced political analyses of complex situations" (13). Indeed, Holmes himself explains that he intended *Fallujah* to offer "the most multivocal representation possible of a largely unreported event in the war in Iraq" (Botham, "Deconstruction" 308).[138]The audience's experience of a multiplicity of voices and an endless variation of standpoints is ample testimony to this. Moreover, *Fallujah* challenges and questions the familiar words used by mainstream media as well as politicians. In the epilogue to her book *Don't Shoot the Clowns* where she recounts her experience in Iraq, Jo Wilding – who is turned into a "character" in Holmes' *Fallujah* – explains that in the US this manufacturing of language happened in a systematic way:

> The Pentagon's 'Office of Special Plans' produced 'talking points', lists of phrases which would appear in presidential speeches. The same sets of words would be repeated over and over in press briefings until they were sold to the journalists, who passed them on to the public by repetition in print and broadcast media. Ordinary people then began using them in their everyday conversation because they had insinuated themselves into public consciousness and from there they became ubiquitous. (268)

138 This quotation is taken from the programme of the symposium "Verbatim Practices in Contemporary Theatre" that was held at the Central School of Speech and Drama in London on 13 and 14 July 2006.

In other words, as Edward S. Herman and Noam Chomsky have claimed, “the ‘societal purpose’ of the media is to inculcate and defend the economic, social and political agenda of privileged groups that dominate the domestic society and the state” (298). And so, *Fallujah* serves as a counterpoint to this. However, as Holmes has himself asserted on a different occasion, the piece was equally committed “to encourage the audience to empathize with the predicament of those trapped in Fallujah” (144) by giving a strong sense of the personalities of the real people behind the characters and by attempting to convey the conviction with which the verbatim words were originally spoken. As is often the case with militant verbatim theatre, such as Philip Ralph’s *Deep Cut* that had remarkable “real world” effects, (“[t]he opening of a new inquest into Pte James’ death this year suggests that Ralph’s devastatingly presented evidence may yet become one of the few pieces of theatre to achieve political change” (Lawson “Westminster”)), *Fallujah* had a measurable impact besides the death threats its author received. More specifically, it revealed new evidence about what happened in *Fallujah* during the sieges and helped in bringing the use of chemical weapons further up the news agenda. The material in the source book – in which the performance text is published – also pointed to what could have been done in order to achieve a less tragic outcome with a focus on the human psychological effect of violence and it tried to shed some light on what the American aims were at the time, while also demanding that people be held to account for the large-scale violation of human rights.[139]Ultimately, *Fallujah* prompted a review of British army COIN training – as the piece was seen by army generals – and contributed to the successful defence of campaigners wrongly accused of incitement to terrorism by British solicitor and human rights activist Gareth Peirce.[140]As has been argued previously in this volume, the visceral aesthetic of post-realism is inseparable from this project and must therefore be framed within any consideration of the radical in site-specific verbatim performance. As discussed earlier, within this set up, the spectators may reach a state of unease when they are themselves put in question, or unsettled by the radical contingency of what appeared to be otherwise.

Now, one of the most recurrent assumptions is that these verbatim works are automatically radical by virtue of being site-specific. Writing in 2007 on his blog in his usual good-humoured style, British theatre critic Andrew Haydon de-

---

139 Essentially, the city of Fallujah was made an example of and the US employed divide and rule strategies. There is also an economic argument as the Americans built permanent military bases and gained access to the oil supply to China.

140 COIN designates counter-insurgency efforts and strategies that are aimed at both defeating and containing insurgency in the context of war and/or occupation.

scribes the contrast between the standard British verbatim play and these site-specific verbatim experiments as follows:

> It is easy enough to imagine the British verbatim play on the subject: conduct the interviews, cherry-pick the strongest moments, edit the whole into a satisfyingly play-like dramatic shape, complete with thesis and easily grasped message and deliver the finished script to Max Stafford-Clark. The approach here is somewhat more radical. Yes, they have conducted video interviews, and they have people relating stories, but they have arranged them over two sterile floors of a largely disused downtown office building, Punchdrunk-style.

In effect, since gaining unprecedented mainstream currency and success in the 2000s, the conception of British verbatim theatre as a practice of critical intervention has been increasingly under attack (Barnett "Poverty," Jenkins). In this context, site-specificity in verbatim performance can be read as a way for verbatim theatre to refresh itself and sidestep this critique. A similar line of thought would therefore conveniently align these performances with the post-dramatic's narrative of radicalism.

Perhaps most tellingly post-realist of all, is Act III, Scene 21 in which one of the night-time aerial attacks on Fallujah is simulated, complete with light show and sound. In this incredibly daring scene, the lights go off and the whole site is in complete darkness. The audience hears – perhaps even aurally visualises – faintly at first, gradually growing louder, the simulated noise of bombardment, rocket fire, intense gunfire and low-lying helicopters:

> *Excruciating noise. The soldiers leave, and SASHA and a few civilians cower in terror. Explosions repeat and are punctuated with white noise – so-called 'noise bombs' – which is unbearable. Gunshots, mortars and machine gun fire are heard and clearly unarmed civilians collapse. We run the aural gamut of bombardment, though we see very little. This goes on for several minutes before subsiding.* (193)

In this astonishing scene, accompanied by a claustrophobic soundscape – among the most disturbing ones in the performance – *Fallujah* shifts verbatim theatre irrevocably onto that level of radicalism which distinguishes it in this book's corpus. More specifically, this Baudrillardian simulation goes beyond a mere documentary realist representation of an experience as it "leaves open to supposition that, above and beyond its object, *law and order themselves might be nothing but simulation*" (Baudrillard, *Simulacra* 20). These disturbing and tangible situations firmly predicated on corporeality can also arguably be considered as triggering a stronger response from the audience. In this case, this may activate "an experiential immediacy in the performance moment where sentient and sensuous shareability enables an embodied knowledge of other(ed) identities and experiences" (Machon, *Visceral* 23). Crucially for Josephine Machon in her *(Syn)aesthetics: Redefining Visceral Performance*, "embodied knowledge can engage in a unique and actively political way with the marginal and transgressive" (23–24). This works on a premise that claims that "[t]he more the body is engaged on its own terms with the work, the more those attending are involved, implicated and affected by the experience, and so the further any ethical potential is developed" (Holmes, *Katrina*).[141]

Another assumption is that these works make it possible for spectators to inhabit sporadically, but for real, a life beyond the alienating global circuits of everyday existence under neoliberalism. In this respect, Jonathan Holmes' conception of theatre seems in alignment with this idea:

> theatre is really a kind of refuge, not a market; it is naturally hospitable, not mercantile [...] a re-imagining of the role of the audience. It ceases to be a band of consumers, paying for the familiar, and becomes instead a collective of witnesses, vital to the event's integrity. (Holmes, *Katrina*)

This brings one to Bourriaud's *Relational Aesthetics*, in the sense that site-specific verbatim theatre may be radical when it endeavours to move into the realm of sociability:

> What they produce are relational space-time elements, inter-human experiences trying to rid themselves of the straight jacket of the ideology of mass communications, in a way, of the places where alternative forms of sociability, critical models and moments of constructed conviviality are worked out. (44)

Concomitant with this picture, in academic and professional discourse alike, the radical potential of performances in traditional theatre buildings is generally seen as in thrall to the economic and cultural systems with which they operate. This is for instance the position taken by Baz Kershaw in his *The Politics of Performance* (1992) and by John McGrath in his *A Good Night Out* (1981). Liz Tomlin sums up this idea for the field of site-specific performance in the following manner: "it is intended to offer the spectator, who now becomes the participant, access to an experiential real of their own making, as a more radical alternative" (*Acts* 148).

Arguably, we are in a quotidian time of ongoing and bitter relational crisis within our contemporary conditions of individualism and cultural division. We

---

141 At this juncture, although I concede that these intentions are highly ethical, it seems to me that there is an underlying danger in all this.

no longer have a space socially and culturally in which to feel the fears and the sadness that a neoliberal context may cause and site-specific verbatim theatre can temporally be that space in which to experience these emotions. Jen Harvie writes in *Fair Play* that there is a need for contemporary performances to seek "strategies for preserving fairness, constructive social relations and individual agency while diminishing inequality and selfish individualism, despite the massive power and insidious spread of global neoliberal capitalism" (4). Similarly, Pierre Bourdieu perceives neoliberalism as questioning "*any and all collective structures* that could serve as an obstacle to the logic of the pure market" (his emphasis, "Utopia") whilst Nicolas Bourriaud argues that, within this context, "the social bond has turned into a standardised artefact" (9). In this way, *Fallujah* is particularly sensitive to the forces and social coordinates at work and creates theatrical situations, and attempts with the immediate social body of the audience, to momentarily reconfigure the established and predominant relational field in a given space, to reopen or imagine new ways of being together beyond the historical distribution of power currently exercising outside the performance space. This is what Mark Chou and Roland Bleiker have called (in relation to verbatim theatre) "a form of prefigurative politics" (248), to be understood as "microcosms of democratic experiments" (248) that are conceived as part of a restorative democratic project. If, traditionally, verbatim theatre is more or less "defined by a democracy of process" (Hughes 92) in the sense that actors and interviewees may be seen as co-authors, site-specific verbatim performance opens it to the members of the audience.[142]Perhaps, more than in any of the other strands of verbatim theatre analysed in this volume, the audience finds itself in what Lehmann has termed a "grey zone of overlapping experiences and perceptions" ("Tragedy" 100) which occurs, he continues, when there is:

> [a] necessity for the participants to make a decision about the nature of what they live through or witness. They find themselves in a double bind, calling for an aesthetic appreciation and at the same time for a reaction of responsibility which would be to some degree 'real'. (100)

Indeed, within the non conventional viewing arrangements of *Fallujah* this is exactly where one finds the most fertile seeds of a radical act at its fullest, an instance of theatre that may lead to a fundamental restructuring of the social and its already existing and overdetermined restrictive power structures. For instance in Act III, Scene 15 that recreates a speech by Rice in Stanford (183), Jonathan

142 For its detractors, this may be seen as a process of dispossession of both politics and aesthetics.

Holmes and his team set up situations in which the participant – if I may call him/her so – is not compelled into fulfilling the artists' preordained requirements and exigencies as far as interaction, relationship and interference are concerned. Thus understood, site-specific verbatim performances rehearse the democratic process by encouraging a new cultural landscape of thought and citizenship and deepening a sense of togetherness within the imbrication of the social and the aesthetic made possible by the work itself through interaction with the audience in attendance.

I would suggest here that this goes beyond what Nick Long calls "ethical imagination" understood as "reflect[ing] on what [the audience's] usual relations would be with the "real people" that lie behind the performance" (315). Arguably, such critical thoughts are produced when the audience is outside the work – even in the course of the performance – neglecting the affordances of visceral reaction, immediacy, turbulence, confusion, conscious or unconscious thoughts that site-specific theatre is yearning for. This recalls what Nicolas Bourriaud saw in certain contemporary art forms in his *Relational Aesthetics* when he claimed that they were "developing a political project" (17). In this way, such performances oppose and trouble existing and prevailing arrangements of power and thoughts – if only for a moment – with what Alain Badiou calls "*theater-ideas*" (his emphasis, 72) that are ideas that "cannot be produced in any other place or by any other means" (72).[143]In other words, they think the political anew as it "asks participants to imagine alternative configurations of performance, place and society" (McKinnie 24). In this spirit, one may argue that one is moving from the conventional conceptualisation of mimesis as a "world-reflecting model" (Halliwell 23) to its less often acknowledged understanding as "a 'world-simulating' or 'world-creating' conception of artistic representation" (Halliwell 23).

At the end of *Fallujah*, and after the last echo of prayer books rustling has completely faded, participants/witnesses are left with a sense of possibility, alternatives and hope that illuminates new spheres of action and routes out of a capitalist, neoliberal agenda or any finished vision of the real. For David Jubb, Joint Artistic Director of BAC, "theatrical situations where genuine human connection occurs holds the greatest potential for transformation" (Machon, *Immersive* 23) beyond the performance. Josephine Machon goes on to argue that there is what she calls a "lasting ephemerality" (*Immersive*, 44) in the sense that such

143 This is completely different from the idea of a play having a thesis as Badiou insists that the text alone cannot produce theatre-ideas. He further explains that: "[t]he idea arises in and by the performance, through the act of theatrical representation" (72) and that "[t]he public is part of what completes the idea" (74).

performances are "fleeting and only of the moment, never to be repeated in any form" (*Immersive*, 44) yet they often leave "a pleasurable and/or disquieting impression" (*Immersive*, 44).

However, even so construed, this is not a guaranteed effect at the point of reception. Indeed, there are evidently important objections to be made to this, perhaps romanticised, narrative of radicalism. As Kimberly Jannarone so adroitly notices, one must be wary of what she calls the "political fallacy of vanguard performance" to account for an eye-catching tendency "to associate experimental performance with progressive results and ideas" (2). Similarly, W. B. Worthen perceives one exemplary site-specific/immersive performance – Punchdrunk's *Sleep No More* (2003) – as "a rather depoliticized vision of Jacques Rancière's active, emancipated spectator" (82) and Jonas Tinius reminds us, via Claire Bishop, that "participatory art forms can also produce very powerful forms of social semantic oppression [and] faux social cohesion." Finally, Adam Alston goes even further by suggesting that "immersive theatre shares particular values with neoliberalism, such as entrepreneurialism as well as the valorization of risk, agency and responsibility" ("Audience" 128). Indeed, this shaky analogy needs to be critically questioned in the context of a global capitalist society that extols the virtues of the "unique experience" – the flagship of the service economy – as an end in itself. In the light of this danger, I am in no way suggesting that this is always the case but, instead, that one should examine these post-realist structures critically as they have equally the potential to invite a more palatable form of the status quo. Under these circumstances, Liz Tomlin rightly observes that: "Rather than seeking a deconstruction or destruction of the existing power structures, they offer a palliative model of how we might better survive them" (*Acts* 190).

To sum up, fundamentally, the capacity of site-specific verbatim performance to wholly surround its audience can be said to facilitate a comprehension of others through a focus on the interior structures of experience, a significant departure from documentary realism which tended to focus on the exterior. In a way, this can be seen as an intensification of the headphone-verbatim technique of Chapter 3.2 that took audiences and actors closer to the constellation of words and patterns of thinking of others. This immersion here is proposed in four dimensions. Of course such a reconstruction is an impossible project, as post-realism tells us by being ontologically perched above it. Thus framed, verbatim performance in this configuration clearly moves away from a conception of political theatre associated with a belief that showing alternative narratives from the past is in itself enough to trigger change in the post-performance world. One is therefore witnessing a radical change of gear insofar as the premium is put on the new context in which these verbatim words and stories will finally be heard. One final illustration of the complex interaction between radicalism and site-specific verbatim performance is putting the final touch to the "programme." In this, another model of site-specific verbatim theatre is proposed and I would like to suggest that one can find, here, a stronger locus for these post-realist concerns. This performance made an even more shocking attack on convention and seems to me to exemplify one of the directions taken by contemporary verbatim theatre-makers that stretch these works to their furthest extent whilst restoring the importance given to the arbitrariness of speech.

Jonathan Holmes' *De Gabay* was a one-day National Theatre Wales production that was entirely free of charge, albeit ticketed,[144] on Sunday 3 March 2013 and was also a durational piece more ambitious in scale than *Fallujah*.[145] The piece invited an audience into the private living rooms of the Somali community of Butetown, Cardiff, especially that of its young poets and the generation that came before them. Their words had not been previously recorded and/or transcribed and edited into a performance text and learnt. They were simply their words to another person that happened to be an audience member. Still, at this point, one might well argue that these situations were not spontaneous enough and rightly belonged to the domain of fantasy. To borrow a formulation by Judith Butler from a completely different context, fantasy here "is not equated with what is not real, but rather with what is not *yet* real, what is possible or futural, or what belongs to a different version of the real" ("Fantasy" 105). The creative team's intervention can hardly be more intensely felt than in the radical gesture of gathering an audience to witness the contrapuntal spoken words of minority voices in the same space. For Holmes, this is a profound difference going beyond the mere portrayal of change and towards the enactment of it (2009). Put another way, these works can trigger effects that endure long after the performance/event has been witnessed, should the audience allow reality to conform to yet another fiction.

---

144 "De Gabay" is a Somali term for "the poem".

145 A durational piece typically lasts a lot longer than a conventional theatre performance. It can be a matter of hours or even "days, weeks or longer" (Kalb, "Solo" 154). The longest verbatim performance to date in Britain may be the performance "Iraq Out and Loud" that started at the Fringe in Edinburgh from 6pm on Monday 8 August 2016 in a garden shed and read all 2.6 million words (twelve volumes) of the Chilcot Report in a 24-hour and 7 days a week non-stop recital. By comparison, Neil Bartlett's *24 hours of peace* (2019), that started on Remembrance Sunday at 11 o'clock at the Royal Exhange Theatre in Manchester and ended twenty-four hours later was "only" 237000 words long.

4.3.8 Conclusion

In this part, despite the fact that the term “site-specific” “has become hackneyed and meaningless through use and abuse” (Buren 79), I nonetheless studied the impact of its discourses on contemporary British verbatim performance so as to help formulate new questions about what a post-realist verbatim practice can be. First, we discovered that what I have called “site-specific verbatim performance” happens in extremely varied venues that are often non-theatrical, non-social and unusual but real rural or urban spaces (for example abandoned and derelict industrial sites or country houses), being recoded by the performance and its temporary occupation by an audience. In short, these works draw verbatim theatre out of the documentary realist and new realist “empty space” and back into life, finally permitting it to be fully itself, that is to say an ongoing moment-to-moment experience: something that *happens* to us. Second, I observed that, due to the impossibility of recording these theatrical events which only admit a few spectators at a time for a very limited run, the intersection of verbatim and site-specific performance has remained relatively ignored. For the researcher, therefore, missing the live event virtually equates to acknowledging the failure of any possible critical engagement with these sporadic, fleeting and ephemeral pieces on an otherwise text-based horizon.[146] Third, I posited that the development of site-specificy in verbatim theatre significantly renegotiates its aesthetic boundaries. The emphasis on an experiential real, that is to say, where all senses are engaged and a greater or lesser degree of unpredictability is key to its aesthetic effects, as opposed to an almost exclusive reliance on mediated representations of the real, and this indicated a clear departure from both documentary realism and new realism, as they were defined in Chapters 2 and 3 respectively. This difference may be so striking for an audience that Josephine Machon speaks of a “lasting ephemerality” in her *Immersive Theatres* to account for the impact of these works: “[t]he experience bleeds into the real world, you are aware of attending to detail, sensation being heightened as you wend your way home” (55). By and large, verbatim theatre’s quintessential engagement with the real, on the other hand, is a more paradoxical one: it typically operates in its very betrayal and absence – absence of the original speakers, events and locations, etc. – a real that is no longer there.

Fourth, site-specific verbatim theatre is likewise emblematic of the postdramatic turn in the sense that its “dramaturgical elements [...] obey no hierarchy of

146 If verbatim scripts tend to be less publishable than their new writing counterparts, this phenomenon is even more intensified in its site-specific format.

text, performer, stage, props and viewing audiences" (Pearson and Shanks 101) and also in relation to its "aesthetics of undecidability" (Lehmann, *Postdramatic* 100). In other words, the ground rules have shifted and the verbatim words are decentred as the defining element in performance.

Fifth, as part of this investigation, I took issue with some of the criticism this type of work seems to be attracting and I explored more specifically the focus on how it supposedly shifts the performance from theatrical illusion to "real" event. I then argued that the combination of site-specificity with verbatim material does not necessarily make "a more truthful art" (Bradfield 127) or become "a further means of creating authenticity" (Shaughnessy 89) – understood as documentary realism – as these authors have claimed in two of the only publications addressing the subject. In fact, this is because their terms are fundamentally different on an ontological level and therefore ultimately short-circuit the possibility of a heightened documentary realism, whatever this may be in practice. Instead, this overlap in performance appears to exceed verbatim theatre's claim to the real that crudely promises that what you are experiencing is a repeat or a condensed version of it. Here, the materiality of the site is real and has bearing on how a given audience experiences the reconstitution of a supposedly authentic past and may arguably involve considerable deviation. In short, we discovered that verbatim theatre in this new format was closer to the notion of site-specific performance advocated by Mike Pearson, "where the laws and bye-laws, the decorum and learnt contracts of theatre can be suspended" ("Chin" 40). It pointed towards a radical expansion as well as a problematisation in performance of the perceived aesthetic limitations of verbatim theatre" (Lane 76). In light of this, my surmise in this chapter has been that site-specific verbatim theatre, as a special kind of social practice, triggers new kinds of interaction with the real, one may even suggest a whole new matrix of interactions that subverts old, prescribed and received monopolies of verbatim representation by operating with a different representational space and placing the audience within the experience. In a recent article, Caroline Wake calls this "a theatre of multiple or intersectional realities, which is to say performance that combines one or more of 'real events,' 'real people,' 'real sites,' 'real tasks,' and occasionally 'real pain.'" ("Review" 120).

Sixth, it would appear that the spatial quality of these works – far from being a mere backdrop – does not merely function as a site of semiotic intersections or postmodern collages but, instead, as a potential site of intervention that provides fertile grounds for re-engineering our social relationships. These site-specific verbatim works can and do, quite often, provide an experimental framework to engage with different structures of experience by realising an aesthetic which gives shape to unprecedented ways of being together communally as well as culturally. Such performances, then, remove the last spectatorial safety net of observing a distanced slice of life since here, as in *Fallujah*, “a spectator is forced to decide whether she engages directly with the actor speaking to her, or removes herself further from the action” (144).[147]The tension achieved by these post-realist strategies makes the verbatim performance less pre-ordained, regimented or even sanitised and the retreat from engagement with the visceral political reality on display is virtually impossible. In *Fallujah*, more specifically, this reflects Holmes’ stated conviction of the necessity of collective commitment in relation to atrocities: “it seems to me an evasion of responsibility – even perhaps unethical – not to take a stand on the morality of the event” (144–145). Audiences in site-specific verbatim works are therefore invited to physically take a stand on a comfortable-uncomfortable continuum as they experience different narratives, voices, perspectives and information in direct opposition to the familiar articulacy of the hegemonic opinion-forming global corporations:

> auditors and spectators are free to choose how closely they engage with the material and how involved they become in the unfolding events [...] we are part of that event [...] immersed in its sensual space and sharing responsibility for its occurrence. (Holmes, *Katrina*)

Seventh, the relational proposal of these pieces – that I perceive as a key component of post-realism – is more embedded in a creative process willing to accommodate the additional and external stimuli that inevitably permeate each encounter with a live audience. The intersection of verbatim carriers with signs of liveness as reality (or its simulation) seeps in reorients the audience’s active engagement and enriches the verbatim theatre of participation they are hoping to provoke. Verbatim theatre invokes, then, a different performance apparatus that I have called “post-realism”, becoming a fundamentally unstable construct, a field of experimentation where perceptual habits are tested. In *Fallujah*, this is clear from the first stage direction that is explicitly more open than verbatim theatre tends to allow for: “*the space can be in the process of being lit, and technicians can be getting things sorted even as the audience enters* [...] [*t*]*he audience is encouraged to applaud as the interviewee enters*” (155). To say it again with Jonathan Holmes: “if there is leakage, so be it” (148). Admittedly, this reveals something important about the radical potential of these practices: we

147 Here, verbatim theatre stops resembling conventional theatre and its familiar rituals in the way George and Portia Kernodle have defined them: “this strange actor-god is close, often very close, and he is safe to watch only because the audience knows he will stay in his conventional playing area, start and stop at conventional signals, and speak not directly from the gods or even his own person as an actor, but will repeat imaginary lines of a created character” (5).

need to see the radical as a matter of actively playing with space, that is to say a form of situated practice, being both critical and spatial. This different mapping of radicalism is exactly what is arresting and thought-provoking about these works. In *Fallujah*, this manifests precisely when discussions concerning what took place in Iraq's besieged city extend to investigate the sites of relation between two diametrically opposed subjects. In effect, the audience physically experiences the often-changing positions one occupies: "the briefings and lecture [...] occur on a different side of the space, so that the audience perspective is constantly shifting" (148). Such a contingency – both real and perceived – that is more or less suggestive in verbatim theatre, was especially brought to the fore during a performance of Michael Wynne's *Who Cares*, though billed as a promenade piece, that I attended at the Royal Court Theatre in April 2015.[148] In Scene Two – that takes place between three actors recreating a coffee break with NHS nurses – the actor playing Carl fell over by accident, to which one of the other actors in character instantly responded by saying "mind the rug." Clearly, through their ambulatory journey in the "real world", the performers and members of the audience in these works are more vulnerable to such mishaps and at times they require an immediate "real-world" reaction, beyond the routine demands of the performing and spectatorial activity and causing an adjustment of response.

To conclude, site-specific verbatim theatre characterises an evanescent, sensory and communal mode of performance and spatial distribution in which the audience is at its most mobile in order to encounter previously uttered verbatim words as well as unexpected viewpoints (sometimes at the edges of one's vision). Importantly, these site-specific verbatim works are conceived of as collaborative public acts, event experiences and perhaps even as a sort of elevating civic exercise fostering encounters outside our habitual frames. The aim of *Fallujah* is to make one think about what one is really doing when one thinks one has understood a situation, forcing one to consider the possibility of other narratives one has (perhaps unconsciously) occluded in the process, and thereby resisting the satisfactory stasis privileged by the status quo. The relational quality of these performances pertains to a temporary discontinuity and rupture with its wider surrounding social landscape and the meta-text of exclusionary power which is designed to enlarge our experience. As I have noted, this is a very poorly documented area of verbatim practice, but one – I contend – richly deserving of research, as a key to the latest aesthetic developments of contemporary verbatim theatre.

148 Wynn's *Who Cares* (2015) unfolded all over the Royal Court Theatre's building.

4.4 Conclusion: Post-Realism as the By-product of a Meeting of Worlds

Within the current "simulacral smoke-screen of a post-real society" (Gonzalez xvi), postdramatic discourses and the absorption of increasingly sophisticated performance practices (prompted by ongoing developments in new technologies) have seeded a new mode of contemporary verbatim theatre as well as a new way of conceiving verbatim theatre beyond its documentary role, original meaning and predicates, eschewing many of its familiar conventions and espousing instead fragmentation, hybridity, instability of meaning and a tendency to undo itself. These post-realist pieces can be seen as methodical and conceptual verbatim projects whose core status is that of a language game whereby every syllable can potentially function as a sign. In other words, the verbatim material changes dynamically throughout the performance and the only "norm" that can be asserted is the verbatim material that underlays it. All the case studies in this part pushed the limits of what had previously been considered verbatim theatre, making new syntheses of traditions and forms that had hitherto been considered as separate. Each of them elaborated a very different understanding of it, requiring a new and remarkable set of skills each time. Yet, they all seemed to partake of a common project that took the verbatim material seriously, whilst pushing it to its limits from within until it began to contradict itself. In other words, they exploited verbatim theatre's internal antagonism through an emphasis on the performative reality over mimetic accuracy. My aim here has thus been to reflect the plurality of a kind of verbatim theatre – as a group or entity held together by no formal ties or particular commitments but shrouded in unease – that goes beyond both documentary realism and new realism (with its mere ruses and tricks), as I have defined them in Chapters 2 and 3. At the same time, I have attempted to define the through-lines of principles to be found in what I have called "post-realism."

Significantly, some scholars have identified within contemporary verbatim theatre the growing visibility of these post-realist works. In the literature – even though no theorisation on this scale has yet been produced – it appears that such a phenomenon has been framed initially within the postdramatic, if understood as aiming "to produce a shock treatment to destabilise the spectators, immersing them in a world where they need to learn afresh how to listen, to see, to feel" (Pavis, *Routledge* 5). Timothy Youker calls any such verbatim work "a dissensual documentary", that is to say a performance that, "while not denying the existence of facts, treats documents primarily as collections of symbols that have unstable or contestable correspondences with facts" (220). Put in another way, post-realist verbatim pieces seek for a form of verbatim performance beyond direct reference and coherence but do not fully abandon the will to be verbatim.

Now, to come back to my earlier claim in the introduction to this part – according to which the post-realist paradigm was strikingly "an outright aggression within the frame" – one must recall that the term "verbatim" itself is a constant reminder of the importance given "to speaking as opposed to showing, to the replacement of visual recreation with verbal recollection" (Soncini, *Conflict* 79). In other words, verbatim theatre's power is typically seen to entirely reside in the immediacy of the words being spoken on a stage, hence the usual rejection of experimentation with language and aesthetic strategies of presentation. Contrary to what was seen in the two previous parts of this book, theatre-makers' willingness to think differently about the potentiality of verbatim performances and to offer alternatives to conventional verbatim practices have explicitly drawn on new physical and visual vocabularies. Indeed, the verbatim material is no longer used solely to represent the world but to propose different ways of living within it as we clearly saw in the final chapter on site-specific verbatim theatre. Their gesture – sometimes at the very limits of intelligibility – is one of verbatim representation defeated that ceases to be recognisable, thus giving us a shock of vanishing materiality. These performances participate in verbatim theatre and yet seem to reject it, casting a questionable light on its prevailing foundations.

As we have just seen, post-realism is an open field of practice and a web of interdependence that poses the biggest challenge to an attempt at theorising the latest alleged aesthetic direction taken by British verbatim theatre in the 21st century – not least because it insists all the more tenaciously on being ontologically illuminated only as the concept of "verbatim" and the exhilarating verbal show it yearns for almost disappear – and it might well be the case that the work and inquiry in this part could only hypothesise a sort of unvoiced practice, adumbrated orientation, a threshold of entry into the enabling assumption of a speculative post-realist paradigm. Suffice it to recall that, theoretically, a true post-realist verbatim practice would confidently and disquietingly (perhaps even masochistically) go so far as to seek to escape relating to realism by way of adaptation or transgression: this would produce a verbatim practice capable of erasing itself as it goes along. Perhaps, then, and only then, verbatim theatre would no longer need to announce itself as verbatim, unburdened by any system of reference or gruelling narrative. To some extent, this is not a substantially attainable goal.

Crucially though, post-realism has radically widened the parameters of verbatim theatre to its furthest limit, from the fetishisation of the verbatim and its desired and cherished documentary effect that was once the main point, to almost any kind of uninvigilated freewheeling formal experimentation that asserts its contingency within the unsettling expanded field of performance and the postdramatic de-definition of verbatim. Put another way, post-realist verbatim theatre takes a parallax view, it is a transgressive gesture that moves the strand into a postdramatic open practice that seemingly abstracts itself from the aesthetic of realism through a panoply of new "disciplinary" encounters (be they dance, the visual arts, experimental film, the musical or site-specific/immersive art to name only those represented in this book's case studies) and that act as complicating factors within its presentational apparatus.[149] In other words, these encounters place pressure on the verbatim text and interrogate the manifold representational strategies that can be used to make sense of competing narratives within our late capitalist environment.

Before going further, it should be noted that this theorisation of the phenomenon – as the attentive reader will have noticed – had almost all the advantages that a comfortable and safe position provides. In other words, I did not need to consider transgressive examples this time as they would certainly appear redundant within an aesthetic paradigm that prides itself on being transgressive *par excellence*. And if we return again to the example of DV8's physical verbatim theatre, decidedly emblematic of the anomalous situation of the immiscible concept of verbatim theatre within performance that is made to work alongside various stage elements that stretch it beyond endurance, it appears that my conception of post-realism was not able to fully comprehend its aesthetic complexity. Similarly, *London Road* was in many ways post-realist and yet, at times, it resisted this reading. In this light, one is forced to confront an unsettling truth and a seeming impasse: a looming residual realism is also ghosting the forces and flows of the post-realist strand as a kind of never-ceasing implacable matrix and subtext that is still very much the sticking point in most, if not all, verbatim performances. In *Fallujah*, this ambiguous position is encapsulated by the introduction to the play itself. In it, Jonathan Holmes admits that "documentary realism is not possible in this case, and [his] feeling is that it would be disingenuous to attempt it" (142). Yet, six pages later, he names documentary realism as one of "the two dominant dramatic modes of the piece" (148).

While there is a case to be made for the capacity of post-realism to erode with remarkable speed one's sense of documentary realism in a given verbatim performance, one must also pursue a counter-argument: even a hard-core post-realist verbatim theatre practice can never fully abandon realism, as it must deeply engage with its tropes in order to attempt to supersede it. The crime was almost perfect, all trace of it had nearly disappeared. But as Peter Boenisch seems to suggest, this was in fact to be expected as "residues of conventional

149 Lehmann characterises the postdramatic as "a meeting point of the arts" (*Postdramatic* 31).

dramatic logic [are to be found] even in contemporary experimental modes of performance" (163). It follows that post-realism only made a temporary dent in documentary realism. In DV8's *To Be Straight with You* that adopts a mode of spectacular expression very far from conventional understandings of theatrical realism, one finds particular passages that move into realism. Indeed, at times the real body of the performers are employed in the service of realism and, at others, the performers seemingly retain their actual status of performers. In *London Road*, the verbatim words are used at intervals to recreate with exactitude the London Road sitting rooms in moments that could arguably pass as poster scenes for my concept of documentary realism. In *Fallujah*, the site of the performance highlights this tension even more. The audience may well experience the performance happening in the site and, in this case, the site serves realism because it represents something else (that is to say, a clinic in Fallujah, a press conference in Baghdad or a checkpoint). However, the site can never be fully written over by realism, and in this case, the audience experiences the performance happening *with* the site.[150] In short, one always has a sense of the aesthetics gently laid over the real site and bodies. As we have seen, for Lehmann there is a noticeable pattern that takes one from realism to the real and that markedly breaks with its crucial unity, that is to say "the close connection between the text of an action, report or process and the theatrical representation oriented towards it" (*Postdramatic* 56), or, put even more simply by Lehmann, "a representation of that which it [the audience] takes for recognizable reality" (*Postdramatic* 58). This facetiously brings one back to the earlier affinities with the notion of realism, as seen in the documentary realist performances of Chapter 2 and the new realist ones of Chapter 3. Now, what I maintain is what I have already implied: while verbatim theatre has clearly shifted into a creature that ill-fits its parentage, the issue of realism is still capital to its operations. And one may wonder whether realism in these postdramatic orchestrations of the verbatim material in performance has a place that contradicts the ideas put forward by Lehmann. In some ways, this may be how post-realism in contemporary verbatim performance mostly affirms and sharpens its departure from the postdramatic.

In conclusion, the analysis of the three case studies mapped out a panorama of post-realist tendencies in verbatim theatre and provided ample evidence of the radical visual styles of its enhanced performance matrix. These selected examples from recent British productions have demonstrated that the salient feature

---

150 Mike Pearson and Cliff McLucas usefully invoke the terms of "host" and "ghost" to describe this phenomenon: "The host site is haunted for a time by a ghost that the theatre-makers create. Like all ghosts it is transparent and the host can be seen through the ghost" (McLucas 128).

of post-realist verbatim theatre has been its relational nature as something in motion, in a constant process of negotiation rather than a state of being. The central feature of post-realism is not in itself whether there is physical, musical or site-specific theatre with which to construct a verbatim performance, but rather the vulnerability with which the creative team works on the verbatim material. These generous acts of manipulation and extraneous elements do not necessarily "go" with the verbatim material, but they arguably open the space for a radical practice that seems to have been lost since verbatim theatre turned into a methodology during its contemporary renaissance.

5 Conclusion

It is time, now, to make an assessment despite the fact that countless other verbatim productions in Britain – and these come around more and more frequently – could have further enriched the main argument of this book and perhaps even challenged my claims. In 2016, for instance, the National Theatre produced a new kind of show using both the verbatim methodology and virtual reality headsets. Its first project, *HOME: AAMIR*, was "an immersive, 360-degree, verbatim documentary" (Snow 2016) that promised to unsettle anew the aesthetics of verbatim theatre, thereby creating the need for a fourth aesthetic category that is yet to be imagined, as VR continues to develop and increasingly engages with theatre practices.

5.1 Summary of Findings and Contribution

For better or worse, verbatim theatre's unprecedented expansion has meant that it occupies today an increasingly conspicuous presence in the realm of theatre and performance, and dare I say in mainstream culture writ large. What it means today, and has meant for several years or decades, is fairly specific. In the 2000s, verbatim theatre became reframed as a predominantly realist art form, even though its origins are also to be found outside the realist tradition. As was signalled in the Introduction, and as will have become clear throughout the entirety of the book, realism is a key concept for contemporary British verbatim theatre. This study of seven British verbatim performances has shown that the concept of realism continues to provide tools for understanding the complexity of new aesthetic forms and communities of words, challenging us to rethink "its perceived formal and ideological conservatism" (Aston, "Room" 17) as expressed in much critical writing.

The main characteristics and aesthetic developments of contemporary British verbatim theatre since the mid-1990s can be summarised as follows: its entire orientation within the mainstream theatre ecology has decisively moved away from documentary realism but, at the same time, this book does not assert a caesura, as the strand is still very much alive and the plays studied in Chapter 2.2 are continually being considered as "the hallmark of the new spate of documentary work on the British stage" (Soncini, "War" 393).[1] Although the documentary

1 As contemporary verbatim theatre-makers have been mostly detached from the earlier documentary and verbatim traditions in Britain, they created a new "documentary realist" tradition

https://doi.org/10.1515/9783110715767-007

realist strand started to slow down in the 2000s, its enduring and resonant presence arguably troubles a purely linear temporality that would end at its vanishing point. The performances studied here have demonstrated rather effortlessly – with every surreptitious incursion of strangeness – that the use of verbatim material in performance was no barrier to aesthetic innovation, calling into question the standard opposition that has pitched verbatim theatre against new writing.

More specifically, as the case studies have also shown, compared to the immense possibilities of British verbatim theatre in the past, contemporary verbatim performances appeared primitively inflexible within the usual means of a stringent documentary realist formulation (its most comprehensive paradigm), presenting an unconsciously ingrained embarrassment and diffidence at being theatre. Engorged as they are with documentary realism, the plays studied in Chapter 2 may wrongly seem at times as if they are just there, without any aesthetic programme at all, and it is hoped that the present work has complicated and refined that picture. At the same time, it was surely to be expected that verbatim theatre would become absorbed very rapidly into the pluralistic contemporary British theatre scene and its complex theoretical issues. If the second part of this book analysed its poignant resistance in performance to the devastating prospects of its own critique, interrogating the ontological claim itself, the third contained the possibility of its suspension, plainly demonstrating that contemporary British verbatim theatre is a great deal more varied and uncertain than is often thought, its energy being a direct challenge to its easy confinement into effete aesthetic sterility. In fact, the whole book was deliberately designed as an attempt to look afresh at a field of research, practice and debate in which familiar lines of inquiry have already been established – in the words of Timothy Youker there is a "frustratingly circumscribed critical discourse that currently surrounds documentary dramaturgy" (222) – thus challenging my own problematic, unpurged assumptions about verbatim theatre and those of others with, it is hoped, a healthy dose of scepticism.

In the light of these observations, the present study attempted to provide an adequate critical framework so as to reconsider the aesthetic qualities of the verbatim theatre of this period that I perceived as constantly evolving. In the course of this study, it has also become clear that a decline in documentary realism does not need to be accepted fatalistically. The initial hypothetical assertion

---

which originated in Nicolas Kent's tribunal plays at the Tricycle. Every practitioner I spoke to in the course of this study acknowledged Kent's huge influence and that everything for them more or less started with his work.

that recent verbatim performances solely attempted to "go beyond" documentary realism was met by the equally valid realisation that they were also re-inventing documentary realist positions. To this extent, post-realist verbatim performances can also be seen as realist because something survives of the tradition they seek to efface. With the analysis of a site-specific verbatim performance, this journey through contemporary British verbatim theatre and its oscillation between the aesthetic conventions of realism and avant-garde experimentation ends.

The seven performances examined in this volume represent the major aesthetic directions contemporary verbatim theatre has followed in Britain. Together, they revealed a very wide spectrum of questions and demands that go into verbatim theatre-making. There are hopeful signs that the moment of verbatim theatre has not passed and it is very likely that subsequent research in the field may drive the frontier back still further and suggest a new facet and/or another angle of approach in attempting to navigate its bizarre labyrinth.[2] Crucially, this volume has refuted the over-simplistic divisions between verbatim theatre and aesthetic experimentation, whilst casting light on some previously under-researched practices, such as headphone, physical and site-specific verbatim performances.

It is hoped that this research has provided the reader with an accurate, if incomplete, impression of what contemporary verbatim theatre in Britain is like today. Furthermore, I hope that it has also given a sense of what verbatim theatre will be like tomorrow, when some of the theatre-makers discussed here take up their places at the very top of the profession whilst new ones start to emerge. As the writing of this book was being concluded and the first and second decade of the 21st century are receding in memory, verbatim theatre has continued to fluctuate, transform and exceed its familiar parameters through bold acts yielding unforeseen possibilities, trajectories and courses, thus urging us to rethink its general aesthetic coordinates beyond the project of documentary realism. Within this last remark lies – if I must venture a hypothesis – the possibility that verbatim theatre, as we know it today, may prove to be a turning point and thus undergo a profound aesthetic reformulation in the future. For now, it might be less contentious to focus again retrospectively on its current reformulation and it remains to ask the question of whether or not verbatim theatre is edging towards "post-postmodernism".

2 For some critics, the resurgence of verbatim theatre is an anomaly or parenthesis in the history of contemporary British theatre but it seems to me that its continuous endurance and pressing relevance (despite the number of its detractors) proves that it is in fact symptomatic of our present time.

5.2 Towards Post-postmodernism?

5.2.1 Preliminary Remarks

In this last section, I would like to posit that post-postmodernism is the conceptual link between the three aesthetic positions of verbatim theatre and their paradoxical demands – documentary realism, new realism and post-realism – I have been discussing so far. My own suggestion is that post-postmodernism may be what makes contemporary British verbatim theatre both fascinating and distinctive.

If, for Elisabeth Angel-Perez and others, verbatim theatre constitutes "a new aesthetics" ("Nation" 65), it may be the case that it is challenging the critical orthodoxies of contemporary British theatre and reimagining its discourses. It is also, most certainly, creating a timely theoretical malaise by drawing on a radical lexicon and seeming to become everything that Postmodernism said contemporary art practices could never be. These observations therefore lead one to the last troubling question of this book: what challenges to postmodernism and to theatrical representation arise from contemporary British verbatim theatre as presented here?

It should be clear by now that this book has been trying to argue for a more complicated understanding of how contemporary British verbatim theatre appears to be wedded to a realist project, how its materials are organised around such aesthetic and ideological considerations. To get the full measure of this argument, one must at once retrieve verbatim theatre from the spurious limbo of naivety to which some critics have so eagerly consigned it. Instead, the figure of the verbatim theatre-maker should be perceived as one of richly ambiguous mastery, at once questionably detached and in control of the performance text. As seen throughout this book, verbatim theatre – even within what I have called "post-realism" – never fully endorses a postmodern stance, shifting one's theoretical speculation in other directions. In other words, there has not really been a search for a "viable" postmodern verbatim theatre or any postmodern radicalisation of verbatim theatre, even though it has not completely avoided the postmodern judgement on its practices and its misplaced penchant for a precarious equilibrium soon to be disturbed. If good points are made in some of the antagonistic criticisms offered since the mid-1990s, there are also perhaps some points that can be taken aboard without giving up the endeavour of verbatim theatre.

What I am clumsily trying to get at here is that verbatim theatre appears more endangered in theory than in practice. With this assertion, verbatim works in this period clothe themselves in the language and rhetoric of postmodernism and postdramatic discourse while at the same time seeking a new aesthetic frame, as a movement still in the process of emerging. From this perspective, verbatim theatre can be understood as a theatre of postmodern means (as opposed to ends), taking place within the ruins of earlier versions of postmodernism, repurposing them for its own use.[3]

5.2.2 Defining "a Beyond Postmodernism" with Verbatim Theatre

Postmodernism has, it would seem, passed its historic moment and no longer serves as a catalyst for new models of theatrical performances.[4]Some critics have asserted that the cultural formation known as "postmodernism" is undoubtedly past its sell-by date or "already dead" (Brînzeu 39) and, on close inspection, it seems that it is no longer perceived as "a major breakthrough to new kinds of knowledge" (Kershaw, *Radical* 6) or "the current version of the avant-garde" (Auslander, "Review" 197).[5]Indeed, one reading of so-called postmodern performances identifies a downward spiral leading to "the harnessing of aesthetics to a superficial postmodern ideology" (Gonzalez xvi), one which contents itself with a constant acknowledgment of its implication in the rhetorical game to ostensibly fulfil a late-capitalist transaction.

Writing in 2015, Liz Tomlin notices on the British stage an "increasing turn away from the postmodern" (*Companies* 119) and its decline would seem to reveal that a new space is thus opening out for "a reassessment of radicalism" ("Radicalism" 7) – as Dave Beech has argued – and perhaps even a new radical practice. Indeed, it seems difficult today to adhere to Patrice Pavis's past contention that "[o]ur time is more ready for postmodern relativism, for all that can be experimented with without a conflict to resolve, without any hope of changing the world" ("Gestus" 188). The fact that publications such as Chris Goode's

---

3 Similarly, Andy Lavender perceives that we have arrived "beyond postmodernism at a changed cultural paradigm, albeit one attuned to continuing postmodern tactics and techniques" (4).

4 The first collection of essays on twenty-first century drama by Siân Adiseshiah and Louise LePage (2016) uncompromisingly opens its first part with a consideration of a beyond postmodernism.

5 The 2000s have seen a mainstream upsurge of attempts to describe something beyond postmodernism. See for instance Josh Toth's *The Passing of Postmodernism* (2010), or Nicoline Timmer's *Do You Feel It Too?: The Post-Postmodern Syndrome in American Fiction at the Turn of the Millennium* (2010), or even the epilogue in the second edition of Linda Hutcheon's *The Politics of Postmodernism* (2002). In the realm of contemporary theatre, Daniel Schulze makes a similar claim in *Authenticity in Contemporary Theatre and Performance: Make it Real* (2017).

The Forest and the Field: Changing Theatre in a Changing World (2015) dedicate a whole chapter to "changing the world", undoubtedly marks a certain reconsideration – if not reformation – of agency. This striking difference – along with others I cannot explore here – indicate(s) that we are in the midst of yet another narrative that one can choose to call post-postmodernism.

In assessing the aesthetic contribution of contemporary verbatim theatre, I contend that the strand is at the cusp of much broader theoretical issues. Clearly, there is something particularly striking about an art form in which two (almost?) incommensurable aesthetic and theoretical positions can coexist and overlap as I have written elsewhere: "[i]n our opinion, these performances retain realism and the aim of political efficacy within a postmodern aesthetic practice that seeks strategies to evade or directly challenge the systematic contradictions of late capitalism" (Garson and Gonzalez 2015). If, indeed, contemporary verbatim theatre unleashes a complexity that overcomes the shortcomings of postmodern relativity, then it is incumbent upon us to consider the possibility of post-postmodernism which I perceive as coinciding with the rehabilitation of "old concepts", a rehabilitation that most versions of postmodernism sought to overthrow and that might reap rich rewards. By rehabilitation, I do not mean a reactionary restoration of some sort nor that these performances are relapsing into a nostalgic past before postmodernism, as some critics have been quick to declare.[6]Among them, one finds principally: returns to "the real", realism, truth, authenticity, engagement with the social process, precisely those concepts that the narrative(s) of postmodernism have repeatedly dismissed.

It follows, perhaps, that the only way to be transgressive in our hollow "postmodern" context is to return to realism, an aesthetic framing that has become particularly conversant with verbatim theatre and withstands the shock. This is exactly what Bernd Stegeman advocates in his *Lob de Realismus* (2015) when he claims that "[i]n order to find a way out of the neoliberal 'misery' and its postmodern 'mist of relativism', theatre needs to embrace a realism that recognizes the contradictions of today's world" (Tinius 87).[7]This full frontal assault on postmodernism, however, does not fail to recognise that there have been some changes in the audience's expectations and perceptions. More specifically, this is a kind of realism that adapts somehow to the postmodern critiques, or rather suggests that postmodernism is intrinsic to its operations. In other words, verbatim theatre does not seek to undo the postmodern critique, or shy

6 I mean that they refuse the definition of these terms as conceptualised by postmodernism.
7 To some extent, this seems to be close to the concept of "traumatic realism" advocated by Michael Rothberg (albeit in a very different context) that he defines as capable of exceeding "both classical realism and the poststructuralist critique of representation" (100).

away from it, but attempts to recover what is arguably overlooked in these critiques. One may even go further by suggesting that these competing paradigms are in fact a foregrounding of the repressed in postmodernism. In sum, verbatim theatre works here from an understanding of what it cannot do.

Beyond being an exasperating act of academic wishful thinking on my part or a brief and weak blip on the radar of contemporary British theatre, the term "post-postmodernism"[8] has already been used in relation to verbatim theatre and its capacity to find a way forward, an alternative to a postmodern frame of interpretation.[9] To give just a few examples (since this book only intends here to discuss how this concept has direct bearing on its project and is not suddenly turning into a study of the post-postmodern impulse – or worse "agenda" – in contemporary British verbatim theatre), Ursula Canton in her *Biographical Theatre* points out that "a Postmodernist framework cannot suffice for the discussion of [verbatim] plays" (38) and highlights the need for a new approach that "would have to acknowledge that Truth is no longer an uncontested concept without abandoning it altogether" (38). By stating that it is possible that an approach "overcomes the dead end of Postmodernist Relativism without returning to the false certainties of a 'Modern' understanding of the world" (42), Canton revives the notion of radical verbatim performance.[10] More recently, and with even more assurance, Siân Adiseshiah and Louise LePage wrote in their introduction that some performance works "orientate us beyond a postmodernist moment bound to the later decades of the last century" (4) and that "[c]ertainly the huge abundance, popularity and success of verbatim theatre in the twenty-first century speak to a contemporary desire for 'real life' to permeate performance, for performance to be situated in the worlds audiences recognize" (4). Most tellingly, following the interim TaPRA (Theatre and Performance Research Association) event of the Performer Training Working Group hosted by the University of Northumbria on 11 May 2016, Maria Kapsali asked the following powerful ques-

---

8 Mark Fisher uses another term "capitalist realism" that he distinguishes from postmodernism in the following manner: "Capitalist realism no longer stages [a] kind of confrontation with modernism. On the contrary, it takes the vanquishing of modernism for granted: modernism is now something that can periodically return, but only as a frozen aesthetic style, never as an ideal for living" (8).

9 What is being suggested here is that the phenomenon of post-postmodernism is not a response to what constitutes original postmodern thought for some, but rather takes issue with how these ideas have been internalised and what they have become today in a context where they seem to have reached the status of cultural norms, a social convention that has lost its much-needed critical function.

10 Like Canton, others are keen to make this connection (see Botham 2008, Kershaw and Ley 2008, Lipovetsky and Beumers 2008).

tions online: "[d]oes verbatim theatre complete a full circle to naturalism (in theatre) and hyperrealist painting (in visual arts)? Is it a post-postmodern response to the conceptual abstraction that has dominated 20th c[entury] art?."[11] To date, her tempting invitation is still awaiting comments. In some respects, this conclusion seems a propitious moment to write a first response, now that a lengthly discussion concerning the complex aesthetics of contemporary British verbatim theatre has taken place all through this book.

First of all, it can be pointed out that, back in 2007, Michał Lachman already answered the second question in this manner: "Well, perhaps yes because it [verbatim theatre] innocently believes that the objective truth of facts can be reconstructed from the chaotic pool of rioting, subjective voices" (321). Some context is needed, however, as Lachman says practically in the same breath that "[v]erbatim drama is post-dramatic as much as it is post-literary, post-fiction and post-journalism" (321), so one can at least expect a certain dose of irony in this last remark. More seriously, I tentatively argued in 2015 that "verbatim theatre, as a fascinating theoretical anomaly, might represent at least a preliminary step towards a post-postmodern theatre" (Garson and Gonzalez 2015), concluding that:

> No doubt this is the kernel of a new aesthetic attitude that deploys postmodern strategies within a more nuanced and sophisticated version of realism in order to produce a theatre that reconnects and reconstructs, this re-engaging with social issues. (Garson and Gonzalez 2015)

Second, of course being verbatim theatre does not intrinsically equate to being post-postmodern or edging towards post-postmodernism – whatever it may mean in practice – simply because verbatim theatre or a version of verbatim theatre has always been there in Britain (at least since the 1960s, depending on one's definition). The only difference is that since the mid-1990s and the beginning of the 2000s verbatim theatre has reached the mainstream and experienced an unprecedented popularity. Generally speaking[12], one can draw some lines between its mainstream emergence and its position today and suggest that there

---

11 These questions were left on a blog dedicated to welcoming comments during and after the TaPRA event: http://theatredanceperformancetraining.org/2016/05/training-to-give-evidence-space-to-leave-comments/.

12 This narrative is of course flawed and oversimplifies a more complex history. As studied in Chapter 2 of this book, pre-1990s verbatim theatre was in fact very different from anything resembling realism.

are at least two waves of contemporary British verbatim theatre.[13] The first wave certainly engaged with postmodernism but in a more tacit way, that was often associated with an alarming naivety.[14]The second wave of works that started in the 2000s was noticeably different, as more and more theatre-makers explored verbatim's aesthetic possibilities. Yet, these works also incorporated the criticism that the first wave had received and therefore became more playful and experimental.[15] Crucially, they used postmodern techniques to invoke the real.

Third, the term post-postmodernism is arguably more familiar in the domain of contemporary fiction as was the case with postmodernism in the 1980s.[16]On this subject, Robert Eaglestone writes of "post-postmodernism" the following:

> if the word 'post-postmodernism' wasn't too silly in itself, it wouldn't be accurate precisely because, while these writers have clearly learned a great deal from the experimentalism of postmodernism and its forebears, they have integrated it, domesticated it, and returned some way to the more traditional forms of the novel. (15)

Whilst it would be perilous to base an understanding of post-postmodernism on its articulation in a completely different art form, I would like to contend that verbatim theatre has also absorbed the aesthetics of postmodernism – or at least a version of it – in order to do something different with it. In this view, post-postmodernism as I conceive it does not really mark a break with postmodernism either, in the same way as what I called "post-realism" did not entirely with realism.

Fourth, as I see it, Post-postmodernism is not a *fait accompli* but I intend to make as explicit as possible – within the limited space of the remainder of this conclusion – some of the aesthetic aspects of contemporary verbatim theatre (as distilled from a close analysis of this book's case studies) that can be considered ingredients of such a setting.

---

13 Other critics have observed this phenomenon. For instance, Markus Wessendorf divides verbatim theatre into two categories: a non-postmodern one and "a postmodern version of documentary drama" (340).

14 Roughly speaking, this is the documentary realism I analysed in Chapter 2.

15 Following my argument in this study, one can discern two movements within the second wave. The new realist one and the post-realist one (when verbatim theatre undergoes so much change that it becomes barely recognisable as verbatim).

16 In the literature, the term "post-postmodernism" is in competition with others such as "renewalism" that has been used to designate a recent abandonment of the aesthetic characteristics of postmodernism.

Fifth, it may well be the case that the term "post-postmodernism" is merely a lumping together of a loose set of positions that aspire to go beyond postmodernism, or rather against a version of postmodernism, advocated by self-proclaimed postmodernists, that has subsequently gone wrong. Following this argument, it may be useful here to invoke the term "constructivist postmodernism" used by Carol Martin to designate a *specific* articulation of "a post-postmodern theoretical perspective" (*Dramaturgy* 3). Martin employs the term to convey "the recognition that although postmodern techniques are largely shared by many cosmopolitan places in the world, these techniques can be and are used for very different ends" (*Dramaturgy* 3). More specifically, she claims that "[s]ceptism and irony are still present but no longer center stage" (*Dramaturgy* 4), a point with which I find myself in entire agreement as far as verbatim theatre is concerned. Indeed, a cluster of clichés have become the paramount activities of current versions of postmodernism, clichés such as: irony, radical scepticism, indifference to society and dizzying relativism, a merely playful and immaterial nature as well as a propensity to revel in verbal gymnastics. These have become so aggravated and chronic that, in the theatre world, "postmodernism" has come to represent an art form bereft of purpose as no adequate theory of agency seems possible.

I would like to propose that contemporary British verbatim theatre has a proclivity to make new connections across the fragments, re-constructing the social, and this is where the performances discussed in this study diverge from postmodernism most significantly and clearly. They manage to overcome the defeatist limitations and impasse of some postmodernist positions, they are what Derek Paget calls "Seekers After Truth" ("Documentarism" 140). But these verbatim works even take a further step. Verbatim theatre brings its own sophistication as a source of discontinuous insight into post-postmodernism via its repeated and sustained recourse to realism. The particular realism of verbatim theatre is not a return to a kind of traditional realism. It is a realism that takes into account the interventions of postmodernism whilst being critical of it. Conceived of in this way, realism in verbatim theatre becomes another aesthetic game and one may thus deduct that we are still in postmodernism, the only difference being that some of its prominent features have now receded whilst some previously obscured characteristics have now come to the fore. At this stage, it is still arduous to determine whether or not verbatim theatre is typically post-postmodern, but one can safely say, however, that it re-appropriates or adapts the techniques of both postmodernism and realism to the specific exigencies of the present and may even speak back to post-postmodernism.

5.3 Five Main Theses

In the remainder of this volume and for practical purposes, I will compress here the main findings I obtained through the close study of seven representative British verbatim performances into five summative arguments or theses.

Although I employed a number of salient examples – some of which had never been discussed in the critical literature –, one of my main concerns has been to draw out questions in relation to recent theories of verbatim theatre's distinctiveness as a representational form. In particular, I suggest that poststructuralist discussions often develop a fairly reductive conception of verbatim theatre and have not always engaged with verbatim theatre's potential to explore the complexities of its own positions, that is to say that its distinctiveness as an art form may be affirmed in terms of its ability to play out the possibilities and problematics of the real before an audience. My main goal in this monograph has thus been to flesh out the scope and the complexity of the topic, and prepare the ground for my own suggestions as to how the aesthetics of verbatim theatre might be theorised in a context that saw its problematic incorporation into the mainstream. More specifically, I have endeavoured to create a typology of contemporary British verbatim theatre and settled for three main aesthetic channels or tropes that carry its markers of authenticity in a completely different way (albeit with the realisation that the borders of these aesthetic channels are necessarily porous).

The first one, "documentary realism" marked my search for an adequate vocabulary to grasp the intricacies of the presentation and reception of most verbatim works in Britain. The tribunal plays studied in Section 2.2 can largely be said to have brought verbatim theatre to public recognition and they constituted, in my view, a powerful landmark in the codification of this aesthetic trend. In addition, I also argued that documentary realism should not be seen as the essence of contemporary British verbatim theatre, but as the subject of a constantly shifting interplay between realism, the real and theatricality. In other words, documentary realism is a potential on which verbatim theatre can capitalise. However, a given verbatim performance may organise these elements somewhat differently meanwhile and this is surely the case with my second category, what I have called "new realism" which, on the contrary, acknowledged in performance the inevitable moments of disjuncture when attempting to mediate these verbatim testimonies. And finally, "post-realism", that was essentially pushing conventional understandings of verbatim theatre to their limits, illustrating a determination to use the means of the stage to continually unsettle its audience by shifting tone and style, aggressively juxtaposing different types of verbatim material and performativity in an attempt to shake up the audience's ordinary perspective on the world.

These three parts (or three chapters: 2, 3 and 4) were designed to draw broad distinctions between different verbatim methodologies, aiming to bring a sense of order to an otherwise, dare I write, chaotic horizon. However, as I claimed on various occasions in this volume, and if I may overstate the case for the purposes of clarity, these parts were not intended to draw fixed distinctions between practitioners as well styles of performance. Importantly, I strove to keep a healthy distance from my own pronouncements, manifesting a self-reflexive awareness of the limitations of this approach, but also, it is hoped its many strengths.

The title of this book expresses my primary hypothesis according to which verbatim theatre may have moved away from documentary realism. And it is surely right to posit that this move may be seen in retrospect as one of its most important aesthetic developments. At the same time, I tried to make clear throughout that this was in no way a straightforward move and that documentary realism had not completely disappeared from the British stage and instead, we had witnessed a more subtle renewal of its positions.

What is most remarkable, perhaps, is the idea that verbatim theatre's engagement with realism may be seen as an indicator of its entry into the realm of postpostmodernism. Put another way, verbatim theatre's realism takes into account the interventions of postmodernism whilst being critical of it.

Last, but not least, verbatim theatre is never completely epic, dramatic or postdramatic and this is why it may be seen as important to address its specificity through a more tailored analysis, which I believe I have provided in this work.

To put it more succinctly, this study has demonstrated that in order to better understand the continued relevance of verbatim theatre, we need to reconcile a set of concerns surrounding notions such as realism, postmodernism and the possibility of making political performances today. My central argument is therefore that verbatim theatre is far more complex than is frequently suggested in the critical literature and that the systematic study of its complex intertwinement with realism produces valuable new thinking.

Arguably, an alternative snapshot might reveal something very different about the still evolving and multifaceted contemporary verbatim theatre, but it is my contention that this volume in the CDE series has made an original contribution to the knowledge in my field. I do not claim to have offered an exhaustive survey or explanation of realism in verbatim theatre – potentially every play or performance throws up fresh questions with regard to this notion – but it does aim to outline and develop perspectives on some key performances relating to the topic.

5.4 Epilogue: Contemporary British Verbatim Theatre: What Next?

Four years have now elapsed since I wrote the bulk of this manuscript in the context of my PhD dissertation, and in that time, the picture of verbatim theatre and contemporary British theatre has changed again quite appreciably, and not only in response to one of the worst political crises in recent British history: Brexit. There seems to also be on the British stage much less clear patterning and aesthetic grouping than ever before, as verbatim theatre seemingly "lost steam" (Pankratz 314).

As I have argued throughout, the points made here only apply to the contemporary British stage and cannot expect to necessarily translate into other geographical areas currently experiencing a verbatim boom such as Canada, Australia, the US, Russia, Germany and New Zealand, although I hope that there may be some fascinating parallels to be drawn from a comparative analysis that may or may not lead to the discovery of some global patterns.[17]

In addition, this study also revealed some gaps in British verbatim theatre history and another result of this project may be to simply reignite the urge to look into the past in order to understand the present moment. Another one may be to look at physical verbatim theatre, site-specific verbatim performances and the verbatim musical from a different perspective so as to assess in more detail how the verbatim material affects the category of practice known as "physical theatre", "site-specific performance" and "the musical." In addition, I could not consider another version of site-specific verbatim performance within the scope of this book, that is to say what happens when a verbatim text is made specifically for a non-physical site such as the internet or other virtual realms.

As the twenty-first century moves into its third decade, it is very mildly controversial to claim that new writing is spattered with elements derived from verbatim theatre, that verbatim theatre has irrevocably affected the power structures and aesthetics of British theatre-making as well as some of the most fondly respected playwrights. In fact, my surmise, is that despite the vitality, the changes and innovations I have been describing in this volume, verbatim theatre has also solidified into a playwriting style in the 21st century. Most visibly, I would argue

17 Perhaps, this may indeed be a new trend in verbatim theatre scholarship as productions, scholars and publications travel internationally. For instance, in a recent article, even though Emma Cox focused on an Australian corpus of verbatim pieces, she interrupted the flow of her critical narrative to contrast these with British verbatim theatre (2016). Amanda Stuart Fisher has also just published a monograph (*Performing the Testimonial*, 2020) on verbatim theatre that encompasses the UK, Australia, America and South Africa.

that one can notice this in the way some playwrights script, or rather reproduce, speech in plays. To make matters more concrete, if the aforementioned style is used throughout a “new writing” play, it is then called mock or faux verbatim. It is therefore hoped that some of the points I have made about contemporary verbatim theatre can also illuminate the ever-changing field of new writing as well as provoke further research and discussion.

Work Cited

Primary Literature

Allen, Jim. *Perdition.* Ithaca Press, 1987.

Anderson, Davey. *Blackout.* Samuel French, 2009.

Armitage, Simon. *Black Roses: The Killing of Sophie Lancaster.* Pomona Books, 2012.

Bartlett, Alice. *Not in my Name.* Oberon Books, 2009.

Bell, Liam Murray. *The Busker.* Myriad Editions, 2014.

Bentley, Eric. *Are You Now or Have You Ever Been. Rallying Cries: Three Plays by Eric Bentley.* Northwestern University Press, 1997, pp. 1–76.

Bissett, Cora and Yusra Warsama. *Rites.* Oberon Books, 2015.

Blair, Jon, and Norman Fenton. *The Biko Inquest.* Rex Collings, 1978.

Blythe, Alecky, and Adam Cork. *London Road.* Nick Hern Books, 2011.

Blythe, Alecky. *Come Out Eli. The Methuen Drama Anthology of Testimonial Plays*, edited by Alison Forsyth, Methuen Drama, 2014, pp. 127–171.

Blythe, Alecky. *Cruising.* Nick Hern Books, 2006.

Blythe, Alecky. *Little Revolution.* Nick Hern Books, 2014.

Blythe, Alecky. *The Girlfriend Experience.* Nick Hern Books, 2008.

Blythe, Alecky. *Voices from the Mosque. Decade: Twenty New Plays.* Nick Hern Books, 2011.

Brittain, Victoria, and Gillian Slovo. *Guantanamo: "Honor Bound to Defend Freedom."* Oberon Books, 2004.

Brittain, Victoria. *The Meaning of Waiting.* Oberon Books, 2010.

Bhuchar, Sudha. *My Name Is ….* Methuen Drama, 2014.

Bhuchar, Sudha, Kristine Landon-Smith and Louise Wallinger. *The Trouble with Asian Men.* Aurora Metro Books, 2012.

Bourne, Bette, and Mark Ravenhill. *A Life in Three Acts.* Methuen Drama, 2009.

Buffini, Moira *et al. Greenland.* Faber and Faber, 2011.

Burke, Gregory. *Black Watch.* Faber and Faber, 2010.

Cheeseman, Peter. *The Knotty: A Musical Documentary.* Methuen, 1970.

Dunbar, Andrea. *Rita, Sue and Bob too. Rita, Sue and Bob too & A State Affair.* Methuen, 2000.

Dunn, Nell. *Cancer Tales.* Amber Lane Press, 2002.

Edgar, David. *Playing with Fire.* Nick Hern Books, 2005.

Emmott, Stephen. *Ten Billion.* Vintage Books, 2013.

Fritz, James. *Lines.* Unpublished manuscript, 2010.

Gay Sweatshop. *Care and Control. Strike While the Iron is Hot: Three Plays on Sexual Politics*, edited by Michelene Wandor, The Journeyman Press, 1980, pp. 63–113.

Gilroy, Steve. *Motherland.* Oberon Books, 2009.

Goode, Chris. *Monkey Bars.* Oberon Books, 2012.

Grosvenor, Catherine. *Cherry Blossom.* Nick Hern Books, 2008.

Gupta, Tanika. *Gladiator Games.* Oberon Books, 2005.

Hammond, Lisa, Rachael Spence and Lee Simpson. *Still No Idea.* Oberon Books, 2018.

Hare, David. *Behind the Beautiful Forevers.* Faber and Faber, 2014.

Hare, David. *Berlin/Wall.* Faber and Faber, 2009.

Hare, David. *The Asian Plays: Fanshen, Saigon, A Map of the World.* Faber and Faber, 1986.

https://doi.org/10.1515/9783110715767-008

Hare, David. *Stuff Happens*. Faber and Faber, 2004.
Hare, David. *The Permanent Way or la Voie Anglaise*. Faber and Faber, 2003.
Hare, David. *The Power of Yes*. Faber and Faber, 2009.
Hare, David. *Via Dolorosa*. Faber and Faber, 1998.
Holmes, Jonathan. *Fallujah: Eyewitness Testimony from Iraq's Besieged City*. Constable and Robinson, 2007.
Holmes, Jonathan. *Katrina: A Play of New Orleans*. Methuen Drama, 2009.
Kelly, Dennis. *Taking Care of Baby*. Oberon Books, 2007.
Kent, Nicolas. *Srebrenica*. Oberon Books, 2005.
Lafferty, Zoe. *The Fear of Breathing*. Oberon Books, 2012.
Levy, Simon. *What I heard about Iraq*. Unpublished manuscript, 2005.
Marlow, Bethan. *Sgint*. Sherman Cymru, 2012.
McCafferty, Owen. *Titanic: Scenes from the British Wreck Commissioner's Inquiry, 1912*. Faber and Faber, 2012.
McLean, Duncan. *Aalst*. Methuen Drama, 2008.
McPherson, Neil. *I Wish to Die Singing: Voices from the Armenian Genocide*. Oberon Books, 2015.
Morgan, Peter. *Frost/Nixon*. Faber and Faber, 2006.
Nevitt, Roy. *All Change: A Musical Documentary. The Story of Early Wolverton*. The People's Press of Milton Keynes, 1977.
Norton-Taylor, Richard, and John McGrath. *Half the Picture. Truth is a Difficult Concept: Inside the Scott Inquiry*, edited by Richard Norton-Taylor and Mark Lloyd. Fourth Estate, 1995, pp. 211–274.
Norton-Taylor, Richard, and Matt Woodhead. *Chilcot*. Oberon Books, 2016.
Norton-Taylor, Richard, and Nicolas Kent. *Called to Account: The Indictment of Anthony Charles Lynton Blair for the Crime of Aggression against Iraq – a Hearing. The Tricycle: Collected Tribunal Plays 1994–2012*. Oberon Books, 2014, pp. 663–770.
Norton-Taylor, Richard. *Justifying War: Scenes from the Hutton Inquiry. The Tricycle: Collected Tribunal Plays 1994–2012*. Oberon Books, 2014, pp. 417–510.
Norton-Taylor, Richard. *Nuremberg: The War Crimes Trial: Transcript*. Oberon Books, 1997.
Norton-Taylor, Richard. *The Colour of Justice: Based on the Transcripts of the Stephen Lawrence Inquiry. The Tricycle: Collected Tribunal Plays 1994–2012*. Oberon Books, 2014, pp. 291–415.
Owen, Gary. *Big Hopes and Other Plays*. Gary Owen, 2009, pp. 199–220.
Price, Tim. *The Radicalisation of Bradley Manning*. Methuen Drama, 2012.
Ralph, Philip. *Deep Cut*. Oberon Books, 2008.
Rapley, Chris and Duncan Macmillan. *2071: The World We'll Leave our Grandchildren*. John Murray, 2015.
Ravenhill, Mark. *Shoot/Get/Treasure/Repeat*. Methuen Drama, 2008.
Rickman, Alan and Katherine Viner. *My Name is Rachel Corrie*. Nick Hern Books, 2005.
Ridley, Philip. *Radiant Vermin*. Methuen Drama, 2015.
Rowbotham, Sheila. *Friends of Alice Wheelon*. Pluto Press, 1986.
Sheers, Owen. *The Two Worlds of Charlie F.* Faber and Faber, 2012.
Silva, Hannah. "Opposition: Extracts from a Political Play on Words." *Forms of Protest*. Penned in the Margins, 2013, pp. 60–68.
Slovo, Gillian. *The Riots: From Spoken Evidence*. Oberon Books, 2011.

Smith, Andy. *The Preston Bill.* Oberon Books, 2015.
Smith, Anna Deavere. *Fires in the Mirror.* Anchor Books, 1993.
Smith, Anna Deavere. *House Arrest: A Search for American Character In and Around the White House, Past and Present.* Anchor Books, 2004.
Smith, Anna Deavere. *Let Me Down Easy.* Dramatists Play Service Inc, 2016.
Soans, Robin. *A State Affair.* Methuen Drama, 2000.
Soans, Robin. *Crouch, Touch, Pause, Engage.* Oberon Books, 2015.
Soans, Robin. *Deep Heat: Encounters with the Famous, the Infamous and the Unknown.* Oberon Books, 2011.
Soans, Robin. *Life after Scandal.* Oberon Books, 2007.
Soans, Robin. *Mixed Up North.* Oberon Books, 2009.
Soans, Robin. *Talking to Terrorists.* Oberon Books, 2005.
Soans, Robin. *The Arab-Israeli Cookbook.* Aurora Metro Books, 2004.
Stephens, Simon. *Plays 1.* Methuen Drama, 2005.
Summerskill, Claire. *Hearing Voices.* Tollington Press, 2010.
Summerskill, Claire. *Rights of Passage.* Tollington Press, 2016.
Theatre Workshop. *Oh, What a Lovely War.* Methuen Drama, 1965.
Unwin, Paul and Sarah Beck. *This Much is True.* Unpublished manuscript, 2009.
Walker, Neil. *Do We Do the Right Thing?.* Unpublished manuscript, 2015.
Wilson, Esther *et al. Unprotected.* Josef Weinberger, 2006.
Woolf, Dennis. *Beyond Belief: Scenes from the Shipman Inquiry*
Wynne, Michael. *Who Cares.* Faber and Faber, 2015.

Secondary Literature

Adiseshiah, Siân, and Louise LePage. *Twenty-First Century Drama: What Happens Now.* Palgrave Macmillan, 2016.
Adiseshiah, Siân. "Political Returns on the Twenty-first Century Stage: Caryl Churchill's *Far Away, Drunk Enough to Say I Love You?* and *Seven Jewish Children.*" *C21 Literature: Journal of 21st-century Writings*, vol. 1, no. 1, 2012, pp. 103–121.
Albright, Ann Cooper. "Split Intimacies: Corporeality in Contemporary Theater and Dance.". *The Oxford Handbook of Dance and Theater*, edited by Nadine George-Graves. Oxford UP, 2015, pp. 19–34.
Alcoff, Linda. "The Problem of Speaking for Others." *Cultural Critique*, vol. 20, 1991–1992, pp. 5–32.
Allain, Paul, and Jen Harvie. *The Routledge Companion to Theatre and Perf*ormance. Routledge, 2006.
Alston, Adam. "Audience Participation and Neoliberal Value: Risk, Agency and Responsibility in Immersive Theatre." *Performance Research*, vol. 18, no. 2, 2013, pp. 128–138.
Alston, Adam. *Beyond Immersive Theatre.* Palgrave Macmillan, 2016.
Alston, Adam. "Funding, Product Placement and Drunkenness in Punchdrunk's The Black Diamond." *Studies in Theatre and Performance*, vol. 32, no. 2, 2012, pp. 193–208.
Alterman, Eric. *When Presidents Lie: A History of Official Deception and its Consequences.* Viking, 2004.

Anderson, Michael, and Linden Wilkinson. "A Resurgence of Verbatim Theatre: Authenticity, Empathy and Transformation." *Australian Drama Studies*, vol. 50, 2007, pp. 153–169.
Angel-Perez, Elisabeth. "Deconstructing the nation ? British Theatre in the Age of Postmodernism." *Théâtre et Nation*, edited by Jeffrey Hopes and Hélène Lecossois. PU Rennes, 2011, pp. 61–76.
Angel-Perez, Elisabeth. "La Contrainte comme Artifice sur la Scène Anglaise Contemporaine: Tom Stoppard, Martin Crimp, et Caryl Churchill." *Sillages Critiques*, vol. 10, 2009.
Angel-Perez, Elisabeth. "Language Games and Literary Constraints: Playing with Tragedy in the Theatre of Caryl Churchill and Martin Crimp." *Contemporary British Theatre: Breaking New Ground*, edited by Vicky Angelaki. Palgrave Macmillan, 2013, pp. 79–95.
Aragay, Mireia and Enric Montforte. "Racial Violence, Witnessing and Emancipated Spectatorship in *The Colour of Justice, Fallout* and *Random*." *Contemporary British Theatre: Breaking New Ground*, edited by Vicky Angelaki. Palgrave Macmillan, 2013, pp. 96–120.
Aston, Elaine, and Janelle Reinelt. "Building Bridges: Life on Dunbar's Arbor, Past and Present." *Theatre Research International*, vol. 26, no. 3, 2001, pp. 285–293.
Aston, Elaine. "Room for Realism?" *Twenty-First Century Drama: What Happens Now*, edited by Siân Adiseshiah and Louise LePage. Palgrave Macmillan, 2016, pp. 17–35.
Audi, Robert. "The Place of Testimony in the Fabric of Knowledge and Justification." *American Philosophy Quarterly*, vol. 34, no. 4, 1997, pp. 405–422.
Auslander, Philip. "'Just Be your Self': Logocentrism and Difference in Performance Theory." *Acting (Re)Considered: A Theoretical and Practical Guide (2nd Edition)*. Routledge, 2002 [1995], pp. 53–61.
Auslander, Philip. "Review." *The Drama Review: TDR*, vol. 37, no. 3, 1993, pp.196–201.
Auslander, Philip. "Toward a Concept of the Political in Postmodern Theatre." *Theater Journal*, vol. 39, no. 1, 1987, pp. 20–34.
Austin, John Langshaw. *How to Do Things with Words*. Clarendon Press, 1962.
Badiou, Alain. *Handbook of Inaesthetics*, translated by Alberto Toscano. Stanford University Press, 2005.
Bahl, Ankur and Uri Omi. Can We Talk About This?: Education Pack. Royal National Theatre, 2012, https://www.dv8.co.uk/media/files/CWTAT_DV8_Workpack.pdf.
Bahl, Ankur. "A Performer's Experience." To Be Straight With You: Education Workpack. Royal National Theatre, 2008, https://www.dv8.co.uk/media/files/To%20Be%20Straight%20With%20You%20FI NAL.pdf.
Balme, Christopher B. *The Cambridge Introduction to Theatre Studies*. CUP, 2008.
Baraniecka, Elżbieta. *Sublime Drama: British Theatre of the 1990s*. CDE Studies Vol. 23. De Gruyter, 2013.
Barker, Howard. "The Ethics of Relevance and the Triumph of the Literal." *Contemporary Drama in English*, vol. 9, 2002, pp. 85–90.
Barker, Howard. "The Spoken Body and the Utopian Regard." *Gramma: Journal of Theory and Criticism*, vol. 17, 2009, pp. 15–16.
Barnett, David. *A History of the Berliner Ensemble*. CUP, 2015.
Barnett, David. "Performing Dialectics in an Age of Uncertainty, or: Why Post-Brechtian ≠ Postdramatic." *Postdramatic Theatre and the Political: International Perspectives on Contemporary Performance*, edited by Karen Jürs-Munby, Jerome Carroll and Steve Giles. Methuen Drama, 2013, pp. 47–66.

Barnett, David. "The Poverty of Verbatim Theatre." *Irish Theatre Magazine*, vol. 25, 2005, pp. 16–19.

Barnett, David. "Toward a Definition of Post-Brechtian Performance: The Example of In the Jungle of the Cities at the Berliner Ensemble, 1971." *Modern Drama*, vol. 54, no. 3, 2011, pp. 333–356.

Barnett, David. "When is a Play not a Drama? Two Examples of Postdramatic Theatre Texts." *New Theatre Quarterly*, vol. 24, no. 1, 2008, pp. 14–23.

Barthes, Roland. *The Rustle of Language*, translated by Richard Howard. University of California Press, 1989.

Basset, Kate. "Betty Blue Eyes, Novello, London
 London Road, NT Cottesloe, London
 Moonlight, Donmar, London." *Independent*, 16 April 2011, http://www.independent.co.uk/arts-entertainment/theatre-dance/reviews/betty-blue-eyes-novello- londonlondon-road-nt-cottesloe- londonmoonlight-donmar-london-2268834.html.

Baudrillard, Jean. *Simulacra and Simulation*, translated by Sheila Faria Glaser. University of Michigan Press, 1994.

Baudrillard, Jean. *Symbolic Exchange and Death*, translated by Iain Hamilton Grant. Sage Publications, 1993.

Beech, Dave. "Include Me Out." *Art Monthly*, no. 315, 2008, pp. 1–4.

Beech, Dave. "Recovering Radicalism: Dave Beech on Critical Art after Postmodernism." *Art Monthly*, no. 323, 2009, pp. 7–10.

Bennett, David. *Brecht in Practice: Theatre, Theory and Performance.* Methuen Drama, 2015.

Bentley, Eric. *The Theory of the Modern Stage.* Penguin Books, 1968.

Bentley, Eric. "Writing for a Political Theatre." *Performing Arts Journal*, vol. 9, no. 2/3, 1985, pp. 45–58.

Ben-Zvi, Linda. "Staging the Other Israel: The Documentary Theatre of Nola Chilton." *The Drama Review: TDR*, vol. 50, no. 3, 2006, pp. 42–55.

Bernbaum, Joel. *What They Said: Verbatim Theatre's Relationship to Journalism.* Carleton University, 2010. PhD thesis manuscript.

Berninger, Mark, Christoph Henke and Bernhard Reitz. "Introduction." *Contemporary Drama in English*, vol. 19, 2012, pp. 9–16.

Beswick, Katie. "Bola Agbaje's *Off the Endz.* Authentic Voices, Representing the Council Estate: Politics, Authorship, and the Ethics of Representation." *Journal of Contemporary Drama in English*, vol. 2, no. 1, 2014, pp. 97–112.

Bignell, Jonathan, Derek Paget, Heather Sutherland and Lib Taylor. "Narrativising the Facts: Acting in Screen and Stage Docudrama." *Contemporary Drama in English*, vol. 18, 2011, pp. 21–52.

Billington, Michael. "*Bloody Sunday: Scenes from the Saville Inquiry.*" *Guardian,* 12 avril 2005, *Theatre Records,* 9–22 April 2005, 470.

Billington, Michael. "Introduction." *The Tricycle: Collected Tribunal Plays 1994–2012.* Oberon Books, 2014, pp. 1–3.

Billington, Michael. *State of the Nation: British Theatre since 1945.* Faber and Faber, 2007.

Billington, Michael. "The Girlfriend Experience; Guardian 25.9.08." *Theatre Record*, vol. 22 September-5 October 2008, p. 1087.

Billington, Michael. "To Be Straight With You." *Guardian,* 3 November 2008, p. 36.

Billington, Michael. “V is for Verbatim Theatre.” *Guardian*, 8 May 2012, http://www.theguardian.com/stage/2012/may/08/michael-billington-verbatim-theatre.
Blesser, Barry and Linda-Ruth Salter. *Spaces Speak, Are You Listening?: Experiencing Aural Architecture*. MIT Press, 2007.
Blix, Hans. “Reviewing Myself.” *Guardian*, 14 September 2004, http://www.theguardian.com/stage/2004/sep/14/theatre.politicaltheatre.
Bluhm, Annika. *The Theatre Arts Audition Book for Men*. Routledge, 2003.
Bluhm, Annika. *The Theatre Arts Audition Book for Women*. Routledge, 2003.
Blythe, Alecky. “Come Out Eli Programme.” Tristan Bates Theatre, 2003.
Blythe, Alecky. “Interview with Alecky Blythe by Lib Taylor on 16th September 2011.” https://www.reading.ac.uk/web/FILES/ftt/Alecky_Blythe_16th_September_2011.pdf.
Blythe, Alecky. “How to Create Verbatim Theatre.” *Ideastap*, 2 August 2013, http://www.ideastap.com/ideasmag/theknowledge/alecky-blythe- london-road-on-verbatim-theatre.
Blythe, Alecky. “Singing for Real.” National Theatre Programme for London Road. NT, 2011, pp. 6–7.
Blythe, Alecky. “Where Have I Been All my Life (Programme).” New Vic, 2012.
Boenisch, Peter M. “Towards a Theatre of Encounter and Experience: Reflexive Dramaturgies and Classic Texts.” *Contemporary Theatre Review*, vol. 20, no. 2, 2010, pp. 162–172.
Böger, Astrid. *People’s Lives, Public Images: The New Deal Documentary Aesthetic*. Gunter Narr, 2001.
Boll, Julia. *The New War Plays: From Kane to Harris*. Palgrave Macmillan, 2013.
Boon, Richard. *About Hare: The Playwright and the Work*. Faber and Faber, 2003.
Bosworth, Lori. “Canadian Stage’s London Road Paves the Way for Healing.” Toronto City, 3 February 2014, http://www.torontonicity.com/2014/02/03/canadian-stages-london-road-paves-the-way-for-healing/.
Botham, Paola. “From Deconstruction to Reconstruction: A Habermasian Framework for Contemporary Political Theatre.” *Contemporary Theatre Review*, vol. 18, no. 3, 2008, pp. 307–317.
Botham, Paola. “Witnesses in the Public Sphere: Bloody Sunday and the Redefinition of Political Theatre.” *Political Performances: Theory and Practice*, edited by Susan C. Haedicke, Deirdre Heddon, Avraham Oz and E.J. Westlake. Rodopi, 2009, pp. 35–54.
Bottoms, Stephen. “Putting the Document into Documentary: An Unwelcome Corrective?” *The Drama Review*, vol. 50, no. 3, 2006, pp. 56–68.
Bottoms, Stephen. “Staging the Democratic Deficit: Or, Three Uses of the Interview.” *Contemporary Theatre Review*, vol. 25, no. 2, 2015, pp. 190–204.
Bourdieu, Pierre. “The Essence of Neoliberalism: Utopia of Endless Exploitation.” Translated by Jeremy J. Shapiro. *Le Monde Diplomatique*, December 1998, http://www.homme-moderne.org/societe/socio/bourdieu/varia/essneoUK.html.
Bourdieu, Pierre. *Outline of a Theory of Practice*, translated by Richard Nice. CUP, (1977) 2013.
Bourriaud, Nicolas. *Relational Aesthetics. Les Presses du Réel*, [1998] 2002.
Bradfield, Jon. “Keeping it Real.” *Mixed Up North*, edited by Robin Soans. Oberon Books, 2009, pp. 125–127.
Bradley, Laura. *Brecht and Political Theatre: The Mother on Stage*. Clarendon Press, 2006.

Bradshaw, Peter. "*The Arbor*– review." *Guardian*, 21 October 2010, http://www.theguardian.com/film/2010/oct/21/the-arbor- film-review.
Bradshaw, Peter. "*London Road* Review – Gripping, Macabre Verbatim Musical About Real-life Serial Murders." *Guardian*, 11 June 2015, http://www.theguardian.com/film/2015/jun/11/london-road-film-review-gripping-macabre-verbatim- musical.
Brazier, Tori. "*London Road* – Interview with Anita Dobson." *One Room with a View*, 9 June 2015, https://oneroomwithaview.com/2015/06/09/london-road-interview-anita-dobson/.
Brecht, Bertolt. "Against George Lukács." *Aesthetics and Politics*. Verso Books, 1977, pp. 68 –85.
Brecht, Bertolt. *The Messingkauf Dialogues*, edited and translated by John Willett. Methuen, 1965.
Brînzeu, Pia. "Post-postmodernism: An Ugly Wor(l)d?." *The European English Messenger*, vol. 24, no. 2, 2015, pp. 39–42.
British Broadcasting Corporation. "Suffolk Muders Play Angers Tania Nicol's Mother." *BBC*, 16 February 2011, http://www.bbc.co.uk/news/uk-england-suffolk-12480330.
Brook, Peter. *The Empty Space*. Touchstone, [1968] 1996.
Brown, Georgina. "*London Road*." *Theatre Record* (9 – 22 April 2011), p. 411.
Brown, Ismene. "The Choreographer that does things to tango couples that Relate would not recommend." *Spectator*, 20 June 2015, http://www.spectator.co.uk/2015/06/the-choreographer-that-does-things-to-tango-couples-that-relate-would-not-recommend/.
Brown, Kate. Beyond The Show Do We Do the Right Thing?: Information Pack. BeFrank, 2015.
Brown, Ross. *Sound: A Reader in Theatre Practice*. Palgrave Macmillan, 2010.
Buckland, Fiona. "Towards a Language of the Stage: the Work of DV8 Physical Theatre." *New Theatre Quarterly*, vol. 11, no. 44, 1995, pp. 371–380.
Buren, Daniel. "Like a Palimpsest; or, The Metamorphosis of an Image." *Contemporary Sculpture Projects in Münster*, edited by Klaus Bussmann, Kasper König and Florian Matzner Verlag Gerd Hatje 1997, p. 79.
Bush, Sophie. *The Theatre of Timberlake Wertenbaker*. Methuen Drama, 2013.
Butler, Judith. *Frames of War: When is Life Grievable?* Verso Books, 2009.
Butler, Judith. "Performative Acts and Gender Constitution: An Essay in Phenomenology and Feminist Theory." *Theatre Journal*, vol. 40, no. 4, 1988, pp. 519–531.
Butler, Judith. "The Force of Fantasy: Feminism, Mapplethorpe, and Discursive Excess." *Differences: A Journal of Feminist Cultural Studies*, vol. 2, no. 2, 1990, pp. 105–25.
Butler, Judith. *Undoing Gender*. Routledge, 2004.
Butler, Robert. "Trial by Tricycle Theatre." *Independent*, 17 January 1999, p. 4.
Butler, Tom. "Alecky Blythe, Writer of London Road, Tells us about the Show being at the National Theatre." *London Calling*, 25 July 2012, http://londoncalling.com/show/alecky-blythe-writer-of- london-road-tells- us-about-the-show-being-at-the- na/.
Calkin, Jessamy. "London Road: How the Ipswich Murders Inspired an Acclaimed Musical." *The Telegraph*, 1 June 2015, http://www.telegraph.co.uk/culture/film/11635588/London-Road-how-the-Ipswich-murders-inspired-an-acclaimed-musical.html.
Callery, Dymphna. *Through the Body: A Practical Guide to Physical Theatre*. Nick Hern Books, 2001.
Callow, Simon. "Previews Must Go!." *Guardian*, 8 September 2004, p. 17.
Cameron, Gail. "Unprotected (Review)." *Oral History* , vol. 34, no. 2, 2006, pp. 107–108.

Campbell, Peter A. “My Naïve American Postdramatic.” *Contemporary Theatre Review*, vol. 25, no. 3, *Backpages*, 2015, pp. 446–464.
Canton, Ursula. *Biographical Theatre: Re-Presenting Real People?*. Palgrave Macmillan, 2011.
Canton, Ursula. “We May not Know Reality, but it Still Matters–A Functional Analysis of ‘Factual Elements’ in the Theatre.” *Contemporary Theatre Review*, vol. 18, no. 3, 2008, pp. 318–327.
Cantrell, Tom. *Acting in Documentary Theatre*. Palgrave Macmillan, 2013.
Cantrell, Tom. *Acting Processes in Contemporary British Verbatim Theatre*. University of York, 2010. PhD thesis manuscript.
Cantrell, Tom and Mary Luckhurst. *Playing for Real*. Palgrave Macmillan, 2010.
Cantrell, Tom. “Playing for Real in Max Stafford-Clark’s *Talking to Terrorists*.” *Studies in Theatre & Performance*, vol. 31, no. 2, 2011, pp. 167–180.
Carlson, Marvin. *Shattering Hamlet’s Mirror: Theatre and Reality*. University of Michigan Press, 2016.
Carney, Sean. *The Politics and Poetics of Contemporary English Tragedy*. University of Toronto Press, 2013.
Carroll, Jerome, Karen Jürs-Munby and Steve Giles. “Introduction: Postdramatic Theatre and the Political.” *Postdramatic Theatre and the Political: International Perspectives on Contemporary Performance*, edited by Karen Jürs-Munby, Jerome Carroll and Steve Giles. Methuen Drama, 2013, pp. 1–30.
Carroll, Noël. “Non- fiction Film and Postmodernist Skepticism.” *Post-theory: Reconstructing Film Studies*, edited by David Bordwell and Noël Carroll. UWP, 1996, pp. 283–306.
Cavendish, Dominic. “DV8’s Can We Talk About This: The Riskier Show of the Year?” *The Telegraph*, 28 February 2012, http://www.telegraph.co.uk/culture/theatre/theatre-features/9101513/DV8s-Can-We-Talk-Abut-This-the-riskiest-show-of-the-year.html.
Chamberlain, Franc. “Gesturing towards Post-physical Performance.” *Physical Theatres: A Critical Reader*, edited by John Keefe and Simon Murray. Routledge, 2007, pp. 117–22.
Chambers, Colin (ed.). *The Continuum Companion to Twentieth Century Theatre*. Continuum, 2002.
Cheeseman, Peter. “on documentary theatre.” *Talking to Terrorists*, edited by Robin Soans. Oberon Books, 2005, pp. 104–107.
Cheeseman, Peter. “Peter Cheeseman.” *On Directing: Interviews with Directors*, edited by Gabriella Giannachi and Mary Luckhurst. St. Martin’s Griffin, 1999, pp. 13–18.
Cheeseman, Peter. “Production Casebook No. 1: The Staffordshire Rebels.” *Theatre Quarterly*, vol. 1, no. 1, 1971, pp. 86–102.
Chinna, Stephen. *Performance: Recasting the Political in Theatre and Beyond*. Peter Lang, 2003.
Chisholm, Alex. “Story-Truth and Factual- Truth.” *Contemporary Theatre Review. Backpages*, vol. 17, no. 4, 2007, p. 601–602.
Chou, Mark and Roland Bleiker. “Betrayed by Democracy: Verbatim Theater as Prefigurative Politics.” *Doing Democracy: Activist Art and Cultural Politics*, edited by Nancy S. Love and Mark Mattern. State University of New York Press, 2013, pp. 231–255.
Chou, Mark, Jean-Paul Gagnon and Lesley Pruitt. “Putting Participation on Stage: Examining Participatory Theatre as Alternative Site for Political Participation.” *Policy Studies*, vol. 36, no. 6, 2015, pp. 607–622.

Cieślak, Magdalena and Agnieszka Rasmus. "Introduction." *Against and Beyond Transgression in Mass Media, Popular Culture and Performance*, edited by Magdalena Cieślak and Agnieszka Rasmus. Cambridge Scholars, 2012, pp. 1–4.
Clapp, Susannah. "Bloody Sunday: Scenes from the Saville Inquiry. Observer 17.4.05." *Theatre Record*, 9–22 April 2005, p. 469.
Clapp, Susannah. "Double Indemnity." *Guardian*, 29 April 2007, p. 19.
Clapp, Susannah. "In Cold Blood." *The Observer*, 17 April 2005, p. 13.
Claycomb, Ryan. "Voices of the Other: Documentary and Oral History Performance in Post-9/11 British Theatre." *Political and Protest Theatre after 9/11: Patriotic Dissent*, edited by Jenny Spencer. Routledge, 2012, pp. 93–107.
Clements, Rachel. "*Cherry Blossom* – Report on the Performance and Post-show Discussion." 21 October 2008, http://www.gla.ac.uk/schools/socialpolitical/crcees/newsevents/news/newsarchive2 008/headline_208846_en.html.
Cohen-Cruz, Jan. "Motion of the Ocean: The Shifting Face of U.S. Theater for Social Change since the 1960s." *Theater*, vol. 31, no. 3, 2001, pp. 95–107.
Cohn, Ruby. *Retreats from Realism in Recent English Drama*. CUP, 1991.
Coles, Esther. "Esther Coles – interviewed by Lib Taylor – 12 August 2009." https://www.reading.ac.uk/web/FILES/ftt/Esther_Coles_12th_August_2009.pdf.
Colleran, Jeanne. *Theatre and War: Theatrical Responses since 1991*. Palgrave Macmillan, 2012.
Connor, Steven. "Vocus Pocus." *This Is a Voice*, edited by Jeremy Fisher and Gillyanne Kayes. Wellcome Collection, 2016, pp. 6–16.
Cook, Robin et al. "For or Against." *Guardian*, 3 September 2004, p. 2.
Cooke, Lez. British Television Drama: A History. British Film Institute, 2015.
Cooper, Charlie. "Critical Pedagogy in Higher Education." *Socially Just, Radical Alternatives for Education and Youth Work Practices: Re-imagining Waysof Working with Young People*, edited by Charlie Cooper, Sinéad Gormally and Gill Hughes. Palgrave Macmillan, 2015, pp. 39–64.
Cork, Adam. "Saying in Tune." *London Road*, Alecky Blythe and Adam Cork. Nick Hern Books, 2011, pp. viii–x.
Costa, Maddy. "London Road: Murder, They Sang." *The Guardian*, 11 April 2011, p. 20.
Coulthard, Malcolm. "Author Identification, Idiolect, and Linguistic Uniqueness." *Applied Linguistics*, vol. 25, no. 4, 2004, pp. 431–447.
Cox, Emma. *Performing Noncitizenship: Asylum Seekers in Australian Theatre, Film and Activism*. Anthem Press, 2015.
Cox, Emma. *Theatre & Migration*. Palgrave Macmillan, 2014.
Cox, Emma. "The Politics of Innocence in Contemporary Theatre about Refugees." *Twenty-First Century Drama: What Happens Now*, edited by Siân Adiseshiah and Louise LePage. Palgrave Macmillan, 2016, pp. 213–236.
Croall, Jonathan. *Buzz Buzz! Playwrights, Actors and Directors at the National Theatre*. Methuen Drama, 2008.
Crompton, Sarah. "The Arts Column: False Start." *The Telegraph*, 15 September 2004, http://www.telegraph.co.uk/culture/theatre/drama/3623958/The-arts-column- false-start.html.
Crompton, Sarah. "What is the Appeal of Islamic State to Young People?" *The Guardian*, 8 April 2016, http://www.theguardian.com/stage/2016/apr/08/nicolas-kent-gillian-slovo-national-theatre-production-another-country- islamic-state-uk- young-people.

CSSD. “Verbatim practices in contemporary theatre at the CSSD.” Promotional email sent by Sophie Nield to the SCUDD List on Friday 26 May 2006.
Daily Mail Reporter. “Theatre Show about the Suffolk Strangler’s Killing Spree to be Staged.” *Daily Mail*, 27 January 2011, http://www.dailymail.co.uk/news/article-1350943/London-Road-musical-Suffolk-Stranglers-killing-spree-staged-National-Theatre.html.
Davis, Rib. “Rib Davis.” *Documentary Arts Report*, vol. 2, 1986, p. 10.
Davis, Rib. “London Road.” *Oral History*, vol. 41, no. 1, 2013, pp. 125–127.
Davis, Rib. *Writing Dialogue for Scripts (4th Edition)*. Bloomsbury, 2016.
Dawson, Gary Fisher. *Documentary Theatre in the United States: An Historical Survey and Analysis of its Content, Form and Stagecraft*. Greenwood Press, 1999.
Deeney, John. “David Hare and Political Playwriting: Between the Third Way and the Permanent Way.” *A Companion to Modern British and Irish Drama*, edited by Mary Luckhurst. Blackwell, 2006, pp. 429–440.
De Jongh, Nicholas. “Bloody Sunday: Scenes from the Saville Inquiry ; *Evening Standard* 12.4.05.” *Theatre Record*, 9 – 22 April 2005, pp. 472–473.
De Jongh, Nicholas. “On the Game in Girlfriend Experience.” *Evening Standard*, 24 September 2008, p. 46.
Delgado-García, Cristina. “Making Time: The Prefigurative Politics of Quarantine’s Entitled.” *Performances of Capitalism, Crises and Resistance : Inside/Outside Europe*, edited by Marilena Zaroulia and Philip Hager. Palgrave Macmillan, 2015, pp. 94–112.
Delgado-García, Cristina. *Rethinking Character in Contemporary British Theatre: Aesthetics, Politics, Subjectivity*. De Gruyter, 2015.
De Martino, Alessandra. “Cultural Translation and Stage Language: A Case: Eduardo De Filippo.” *Differences on Stage*, edited by Alessandra De Martino, Paolo Puppa and Paola Toninato. Cambridge Scholars Press, 2013, pp. 170–190.
Demastes, William W. *The Cambridge Introduction to Tom Stoppard*. CUP, 2012.
Derbyshire, Harry and Loveday Hodson. “Performing Injustice: Human Rights and Verbatim Theatre.” *Law and Humanities*, vol. 2, no.2, 2008, pp. 191–211.
D’Eristal, Louise. “La Décadence Littéraire.” *La Querelle de la Statue de Baudelaire*, edited by André Guyaux. Presses de l’Université Paris-Sorbonne, 2007, pp. 409–10.
Derrida, Jacques and Anne Dufourmantelle. *Of Hospitality*, translated by Rachel Bowlby. SUP, 2000.
Derrida, Jacques. *Archive Fever: A Freudian Impression*, translated by Eric Prenowitz. The University of Chicago Press, 1996.
Derrida, Jacques. “The Law of Genre, translated by Avital Ronell.” *Glyth*, vol. 7, 1980, pp. 202 –232.
Derrida, Jacques. “The Theater of Cruelty, translated by Alan Bass.” *Theater*, vol. 9, no.3, 1977, pp. 6–19.
de Waal, Ariane. *Theatre on Terror: Subject Positions in British Drama*. De Gruyter, 2017.
Diamond, Elin. “Identity Politics Then and Now.” *Theatre Research International*, vol. 37, no. 1, 2012, pp. 64–67.
Diamond, Elin. *Unmaking Mimesis: Essays on Feminism and Theater*. Routledge, 1997.
Dolan, Jill. “‘Finding our Feet in the Shoes of (One An) Other’: Multiple Character Solo Performers and Utopian Performatives.” *Modern Drama*, vol. 45, no. 4, 2002, pp. 495 –518.
Dolan, Jill. *Utopia in Performance: Finding Hope at the Theater*. University of Michigan, 2005.

Dorney, Kate and Frances Gray. *Played in Britain: Modern Theatre in 100 Plays*. Methuen Drama, 2012.

Dorney, Kate. *The Changing Language of Modern English Drama 1945–2005*. Palgrave Macmillan, 2009.

Drees, Danielle. "The Sleeping Spectator: A Sleep Cultures Critique of Punchdrunk's Sleep No More." *Performance Research*, vol. 21, no. 1, 2016, pp. 101–105.

Dromgoole, Dominic. "Reality Check." *Guardian* 23 October 2004, p. 19.

Dugdale, Sasha. "Preface." *Performing Violence: Literary and Theatrical Experiments of New Russian Drama*, edited by Birgit Beumers and Mark Lipovetsky. Intellect Books, 2009, pp. 13–25.

Duggan, Patrick. "Others, Spectatorship and the Ethics of Verbatim Performance." *New Theatre Quarterly*, vol. 29, no. 2, 2013, pp. 146–158.

Eaglestone, Robert. *Contemporary Fiction: A Very Short Introduction*. OUP, 2013.

Economist. "But is it Theatre?." *Economist* (1 July 2004), http://www.economist.com/node/2876778.

Eddershaw, Margaret. "Actors on Brecht." *The Cambridge Companion to Brecht*, Second Edition, edited by Peter Thomson and Glendyr Sacks. CUP, 2006, pp. 278–296.

Edgar, David and Amanda Whittington. *The Working Playwright: Engaging with Theatres*. The Writers' Guild of Great Britain, 2015.

Edgar, David. "Doc and Dram." *Guardian*, 27 September 2008, http://www.theguardian.com/stage/2008/sep/27/theatre.davidedgar.

Edgar, David. "Secretary of the Times." *Irish Pages*, 2007, pp. 105–116.

Edgar, David. *The Second Times as Farce: Reflections of the Drama of Mean Times*. Lawrence and Wishart, 1988.

Edgar, David. "Verbatim Theatre (Verbatim)." *The Tricycle: Collected Tribunal Plays 1994–2012*. Oberon Books, 2014, pp. 4–40.

Edwardes, Jane. "Keeping it Real; Theatre: Preview." *Time Out* (3 March 2004). LexisNexis Database.

Eldridge, David. "Massive Attack." *Guardian* (27 June 2005).

Elvgren Jr., Gilette A. "Documentary Theatre at Stoke-on-Trent." Educational Theatre Journal, vol. 26, no. 1, 1974, pp. 86–98.

Erickson, Jon. "Defining Political Performance with Foucault and Habermas: Strategic and Communicative Action." *Theatricality*, edited by Tracy C Davis and Thomas Postlewait. CUP, 2003, pp. 156–185.

Esch-Van Kan, Anneka. "The Documentary Turn in Contemporary British Drama and the Return of the Political: David Hare's Stuff Happens and Richard Norton-Taylor's Called to Account." *A History of British Drama: Genres – Developments – Model Interpretations*, edited by Sibylle Baumbach, Birgit Neumann and Ansgar Nünning. Wissenschaftlicher Verlag Trier, 2011, pp. 413–28.

Eyre, Richard. *Talking Theatre: Interviews with Theatre People*. Nick Hern Books, 2009.

Favorini, Attilio and Gillette Elvgren. "I Sing of Cities: The Musical Documentary." *Performing Democracy: International Perspectives on Urban Community-Based Performance*, edited by Susan C. Haedicke and Tobin Nellhaus. The University of Michigan Press, 2001, pp. 181–96.

Favorini, Attilio. "Introduction. After the Facts: Theater and the Documentary Impulse." *Voicings: Ten Plays from the Documentary Theatre*, edited by Attilio Favorini. Ecco Press, 1995, pp. xi–xxxix.

Favorini, Attilio. "Collective Creation in Documentary Theatre." *A History of Collective Creation*, edited by Kathryn Mederos Syssoyeva and Scott Proudfit. Palgrave Macmillan, 2013, pp. 97–112.

Féral, Josette. "From Event to Extreme Reality: The Aesthetic of Shock." Translated by Leslie Wickes. *The Drama Review: TDR*, vol. 55, no. 4, 2011, pp. 51–63.

Ferdman, Bertie. "A New Journey Through Other Spaces: Contemporary Performance beyond 'Site-Specific'." *Theater*, vol. 43, no. 2, 2013, pp. 5–25.

Festjens, Thijs. "Aural Energies in Rimini Protokoll's Call Cutta: Sound, Documentary, Performance and Narratological Aspects of 'The World's First Mobile Phone Theatre'." *Audionarratology: Interfaces of Sound and Narrative*, edited by Jarmila Mildorf and Till Kinzel. De Gruyter, 2016, pp. 165–182.

Filewod, Alan. *Collective Encounters: Documentary Theatre in English Canada*. TUP, 1987.

Filmer, Andrew. "Book Review: Performing Site-Specific Theatre: Politics, Place, Practice, edited by Anna Birch and Joanne Tompkins. Palgrave Macmillan, 2012)." *Performance Paradigm*, vol.11, 2015, pp. 128–130

Finburgh, Clare. "Theatre's Other: Event and Testimony in British Verbatim Plays." *Sillages Critiques*, vol. 18, 2014.

Finney, Simone. "Common Distractions." Carmen Disruption: Official Programme. Almeida Theatre, 2015, pp. 8–9.

Fisher, Amanda Stuart. *Performing the Testimonial: Rethinking Verbatim Dramaturgies*. Manchester University Press, 2020.

Fisher, Mark. *Capitalist Realism: Is There no Alternative?*. Zero Books, 2009.

Fisher, Mark. *How to Write about Theatre: A Manual for Critics, Students and Bloggers*. Methuen Drama, 2015.

Fiske, John. *Television Culture*. Routledge, 1987.

Forsyth, Alison and Chris Megson. "Introduction." *Get Real: Documentary Theatre Past and Present*. Palgrave Macmillan, 2009, pp. 1–5.

Forsyth, Alison and Chris Megson. "Preface to the Paperback Edition." *Get Real: Documentary Theatre Past and Present*. Palgrave Macmillan, 2011, pp. ix–xv.

Fortier, Mark. *Theory/Theatre: An Introduction*. Routledge, 2016.

Fosha, Diana. *The Transformative Power of Affect: A Model for Accelerated Change*. BasicBooks, 2000.

Foster, Hal. "For a Concept of the Political in Contemporary Art." *Recodings: Art, Spectacle, Cultural Politics*, edited by Hal Foster. Bay Press, 1985, pp. 139–155.

Foster, Susan Leigh. *Choreographing Empathy. Kinaesthesia in Performance*. Routledge, 2011.

Foucault, Michel. "A Preface to Transgression." *Language, Counter-memory, Practice: Selected Essays and Interviews by Michel Foucault*, edited and translated by Donald F. Bouchard. Cornell University Press, 1977, pp. 29–52.

Freshwater, Helen. "Physical Theatre: Complicite and the Question of Authority." *A Concise Companion to Contemporary British and Irish Drama*, edited by Nadine Holdsworth and Mary Luckhurst. Blackwell Publishing, 2008, pp. 171–199.

Freshwater, Helen. *Theatre Censorship in Britain: Silencing, Censure and Suppression*. Palgrave Macmillan, 2009.

Freshwater, Helen. *Theatre & Audience*. Palgrave Macmillan, 2009.
Fricker, Miranda. *Epistemic Injustice: Power and the Ethics of Knowing*. OUP, 2007.
Friedman, Jeff. "Fractious Action: Oral History-Based Performance." *Handbook of Oral History*, edited by Thomas L. Charlton, Lois E. Myers and Rebecca Sharpless. Altamira Press, 2006, pp. 465–509.
Gaines, Jane. "The Production of Outrage: The Iraq War and the Radical Documentary Tradition." *Framework: The Journal of Cinema and Media*, vol. 48, no. 2, 2007, pp. 36–55.
Gale, Maggie B. and John F. Deeney. "Introduction: Early Political Theatres." *The Routledge Drama Anthology and Sourcebook: From Modernism to Contemporary Performance*, edited by Maggie B. Gale and John F. Deeney. Routledge, 2010, pp. 290–309.
Garde, Ulrike and Meg Mumford. "Postdramatic Reality Theatre and Productive Insecurity: Destabilising Encounters with the Unfamiliar in Theatre from Sidney and Berlin." *Postdramatic Theatre and the Political: International Perspectives on Contemporary Performance*, edited by Karen Jürs-Munby, Jerome Carroll and Steve Giles. Methuen Drama, 2013, pp. 147–164.
Garde, Ulrike and Meg Mumford. *Theatre of Real People: Diverse Encounters at Berlin's Hebbel am Ufer and Beyond*. Methuen Drama, 2016.
Gardner, Lyn. "Does Verbatim Theatre Still Talk the Talk?" *Guardian*, 7 May 2007, http://www.theguardian.com/stage/theatreblog/2007/may/07/foreditorsdoesverbati mthea.
Gardner, Lyn. "To Be Straight with You." *Guardian*, 15 April 2008, p. 34.
Garner, Jr., Stanton B. *Bodied Spaces: Phenomenology and Performance in Contemporary Drama*. Cornell University Press, 1994.
Garson Cyrielle and Madelena Gonzalez. "'What a Carve up !': The Eclectic Aesthetics of Postmodernism and the Politics of Diversity in some Examples of Contemporary British Verbatim Theatre." *Etudes Britanniques Contemporaines*, vol. 49, 2015.
Garson, Cyrielle. "Continuité et Discontinuité sur la Scène Britannique Contemporaine: le Cas du Théâtre 'Verbatim'." *Sphères*, vol. 2, 2014, pp. 36–45.
Garson, Cyrielle. "La scène britannique au XXIe siècle : crise de l'auteur dramatique ou changement de paradigme ?" *Coup de Théâtre*, vol. 29, 2015, pp. 193–221.
Garson, Cyrielle. "Remixing Politics: The Case of Headphone-Verbatim Theatre in Britain." *Journal of Contemporary Drama in English*, vol. 2, no. 1, 2014, pp. 50–62.
Garson, Cyrielle. "The Changing Aesthetics of Verbatim Theatre: From Engaged Theatre to Globalisation." *Aesthetics and Ideology in Contemporary Literature and Drama*, edited by Madelena Gonzalez and René Agostini. Cambridge Scholars, 2015, pp. 175–192.
Garson, Cyrielle. "Verbatim Theatre and New Writing in Britain: A State of 'Kindred Strangers'?." *Etudes Britanniques Contemporaines*, vol. 48, 2015.
Garson, Cyrielle. "Verbatim Theatre in Translation: A Triple Metamorphosis." *Sphères*, vol. 1, 2013, pp. 13–22.
Gaskill, William. *A Sense of Direction: Life at the Royal Court*. Faber and Faber, 1988.
Gelderman, Carol. "Hyperrealism in Contemporary Drama: Retrogressive or Avant-Garde." *Modern Drama*, vol. 26, no. 3, 1983, pp. 357–367.
Genette, Gérard. *Palimpsests: Literature in the Second Degree*, translated by Channa Newman and Claude Doubinsky. University of Nebraska Press, [1982] 1997.
Ginman, John. "Out of Joint: Max Stafford-Clark and 'the Temper of our Time'." *Contemporary Theatre Review*, vol. 13, no. 3, 2003, pp. 16–23.

Glinga, Werner. *Legacy of Empire: A Journey Through British Society.* MUP, 1986.
Gob Squad. *Gob Squad and the Impossible Attempt to Make Sense of it all.* Gob Squad, 2010.
Goode, Chris. *The Forest and the Field: Changing Theatre in a Changing World.* Oberon Books, 2015.
Goodman, Amy and Denis Moynihan. "Foreword." *Beyond the Green Zone: Dispatches from an Unembedded Journalist in Occupied Iraq*, edited by Dahr Jamail. Haymarket Books, 2007, pp. xi–xvi.
Goodman, Amy. "Dr. Jay Lifton: American Psychological Association Should 'Prohibit any Involvement' of Psychologists in Interrogations." *Democracy Now*, 12 June 2006, http://www.democracynow.org/2006/6/12/dr_robert_jay_lifton_american_psychological.
Gonzalez, Madelena. "Introduction: The Return of the Real in Art and Politics – Putting Ideals into Practice." *Aesthetic and Ideology in Contemporary Literature and Drama*, edited by Madelena Gonzalez and René Agostini. Cambridge Scholars, 2015, pp. xi–xxviii.
Goss, Lindsay. "Postdramatic Theatre and the Political: International Perspectives on Contemporary Performance (Review)." *Theatre Survey*, vol. 57, 2016, pp. 160–162.
Gowers, Andrew. "Stuff that Happens in the Build-up to War." *Financial Times*, 10 September 2004, p. 5.
Greig, David. "A Tyrant for all Time." *Guardian*, 28 April 2003, p. 14.
Grellet, Françoise. *A Handbook of Literary Terms: Introduction au Vocabulaire Littéraire Anglais.* Hachette, 2013.
Grossman, Elwira M. "Transnational or Bi-cultural? Challenges in Reading post-1989 Drama 'Written outside the Nation'." *Polish Literature in Transformation*, edited by Ursula Philips. Lit, 2013, pp. 241–250.
Guynn, William. "The Non-fiction Film and its Spectators." *The Semiotic Bridge: Trends from California*, edited by Irmengard Rauch and Gerald F. Carr. Mouton de Gruyter, 1989, pp. 33–48.
Hainaux, René. "Le Théâtre Documentaire." *World Theatre/Théâtre dans le Monde*, vol. 17, no. 5–6, 1968, pp. 375.
Halba, Hilary, and Stuart Young. "Making Verbatim Theatre: Processes of Mediation in Hush: A Verbatim Play about Family Violence." *Le Théâtre Neo-documentaire: Résurgence ou Réinvention?*, edited by Lucie Kempf, and Tania Moguilevskaia. PUN – Editions Universitaire de Loraine, 2013, pp. 171–80.
Hall, Joan Wylie. "'Everybody's Talking': Anna Deavere Smith's Documentary Theatre." *Contemporary African American Women Playwrights: A Casebook*, edited by Philip C. Kolin. Routledge, 2007, pp. 150–166.
Halliwell, Stephen. *The Aesthetics of Mimesis: Ancient Texts and Modern Problems.* Princeton University Press, 2002.
Hamidi-Kim, Bérénice. *Les Cités du Théâtre Politique en France depuis 1989.* L'Entretemps, 2013.
Hamidi-Kim, Bérénice. "Post-Political Theatre versus the Theatre of Political Struggle." *New Theatre Quarterly*, vol. 24, no. 1, 2008, pp. 41–50.
Hammond, Philip. "To Be Straight with You, I Left the Theatre Angry." *Belfast Telegraph*, 22 October 2009, http://www.belfasttelegraph.co.uk/entertainment/belfast-festival/reviews/to-be-straight-with-you-i-left-the-theatre-angry-14538205.html.

Hammond, Will and Dan Steward.*Verbatim, Verbatim: Contemporary Documentary Theatre*. Oberon Books, 2008.

Hannan, Jason. *Truth in the Public Sphere*. Lexington Books, 2016.

Harbinger, Richard. "Trial by drama." *Judicature*, vol. 55, 1971, pp. 122–128.

Hare, David. *Acting Up: A Diary*. Faber and Faber, 1999.

Hare, David. *Asking Around: Background to the David Hare Trilogy*, edited by Lyn Haill. Faber and Faber, 1993.

Hare, David. "David Hare Live Webchat – Post Your Questions Here." *The Guardian*, 21 March 2012, http://www.theguardian.com/culture/theatreblog/2012/mar/21/david- hare-live-webchat-questions.

Hare, David. "Don't Look for a Reason." *The Guardian*, 12 April 2003, p. 21.

Hare, David. "Mere Fact, Mere Fiction." *South Downs*. Faber and Faber, 2011, pp. 57–78.

Hare, David. "Obedience, Struggle & Revolt." *Obedience, Struggle & Revolt: Lectures on Theatre*. Faber and Faber, 2005, pp. 9–32.

Hare, David. "on factual theatre." *Talking to Terrorists*, edited by Robin Soans. Oberon Books, 2005, pp. 111–113.

Hare, David. Writing Left-Handed: Collected Essays. Faber and Faber, 1991.

Harvie, Jen. "Introduction: Contemporary Theatre in the Making." *Making Contemporary Theatre: International Rehearsal Processes*, edited by Jen Harvie and Andy Lavender. Manchester University Press, 2010, pp. 1–16.

Harvie, Jen. "DV8's Can We Afford This: The Cost of Devising on Site for Global Markets." *Theatre Research International*, vol. 27, no. 1, 2002, pp. 68–77.

Harvie, Jen. *Fair Play – Art, Performance and Neoliberalism*. Palgrave Macmillan, 2013.

Hauser, Arnold. "The Origin of Domestic Drama." *The Theory of the Modern Stage*, edited by Eric Bentley. Penguin Books, 1968, pp. 403–419.

Hauthal, Janine. "Realisms in British Drama since the 1990s: Anthony Neilson's Realism and Gregory Burke's Black Watch." *Realisms in Contemporary Culture: Theories, Politics, and Medial Configurations*, edited by Dorothee Birke and Stella Butter. De Gruyter, 2013, pp. 146–177.

Haydon, Andrew. "Site-specific Verbatim Theatre without Actors." Postcards from the Gods, 23 November 2007, http://postcardsgods.blogspot.fr/2007/11/munich-part-one.html.

Haydon, Andrew. "Talking to Terrorists." *Culturewars*, 6 July 2005, http://www.culturewars.org.uk/2005 – 01/terrorists.htm.

Haydon, Andrew. "Theatre in the 2000s." *Modern British Playwriting 2000 – 2009*, edited by Dan Rebellato. Methuen Drama, 2013, pp. 40–98.

Heddon, Deidre. *Autobiography and Performance*. Palgrave Macmillan, 2008.

Heddon, Deidre. "To Absent Friends: Ethics in the Field of Auto/Biography." *Political Performances: Theory and Practice*, edited by Susan C. Haedicke, Deirdre Heddon, Avraham Oz and E.J. Westlake. Rodopi, 2009, pp. 111–133.

Herman, Edward S. and Noam Chomsky. *Manufacturing Consent: The Political Economy and the Mass Media*. Pantheon Books, [1988] 2002.

Hickling, Alfred. "Beyond Belief." *Guardian*, 28 October 2004, p. 28.

Hingorani, Dominic. "Creating Theatre Work for a Diverse Teenage Audience." *Theatre for Young Audiences: A Critical Handbook*, edited by Tom Maguire and Karian Schuitema. Trenthan Book, 2012, pp. 59–68.

Hinton, William. *Fanshen: A Documentary of Revolution in a Chinese Village.* Monthly Review Press, 1966.

Hitchings, Henry. "London Road is an Astonishing Tale." *Evening Standard*, 15 April 2011, http://www.standard.co.uk/goingout/theatre/london-road-is-an-astonishing-tale-7424497.html.

Hoggard, Liz. "Out of Crises, A Drama." *The Observer*, 27 March 2005, p. 8.

Holmes, Jonathan. "Introduction to the Play." *Fallujah: Eyewitness Testimony from Iraq's Besieged City*, edited by Jonathan Holmes, Constable and Robinson, 2007, pp. 113–137.

Holmes, Jonathan. "Theatre and Experience." *Katrina: A Play of New Orleans.* Kindle ed.: Methuen Drama, 2009.

Homden, Carol. *The Plays of David Hare.* CUP, 1995.

Hopton, Tricia. "Adapting Verbatim Theater: David Hare's *Stuff Happens.*" *Pockets of Change: Adaptation and Cultural Transition*, edited by Tricia Hopton, Adam Atkinson, Jane Stadler, Peta Mitchell. Lexington Books, 2011, pp. 15–28.

Hoyland, Bill. "Bill Hoyland interviewed by Derek Paget." University of Reading: Unpublished interview, 2008, http://www.reading.ac.uk/web/files/ftt/bill_hoyland_14th_july_2008.pdf.

Hoyt, Clark. "Was there Napalm in Fallujah?." *New York Times*, 18 July 2007, http://publiceditor.blogs.nytimes.com/2007/07/18/was-there-napalm- in-fallujah/?_r=0.

Hubble, Denise and Amanda Stuart Fisher. "What Happened Next?." *From the Mouths of Mothers*, edited by Amanda Stuart Fisher. Aurora Metro Books, 2013.

Hughes, Jenny. Performance in a Time of Terror: Critical Mimesis and the Age of Uncertainty. MUP, 2011.

Hur, Soon-ja. "The Aesthetic Characteristics of Verbatim Theatre." *Journal of Korean Theatre Education*, vol. 23, 2013, pp. 201–236.

Hutcheon, Linda. *The Politics of Postmodernism*, Second Edition. Routledge, 2002.

Hutchison, Yvette. "Verbatim Theatre in South Africa: 'Living History in a Person's Performance'." *Get Real: Documentary Theatre Past and Present*, edited by Alison Forsyth and Chris Megson. Palgrave Macmillan, 2009, pp. 209–223.

Inchley, Maggie. *Voice and New Writing, 1997–2007: Articulating the Demos.* Palgrave Macmillan, 2015.

Inman, Nick. Politipedia: *A Compendium of Useful and Curious Facts about British Politics.* Harriman House, 2007.

Innes, Christopher. *A Sourcebook on Naturalist Theatre.* Routledge, 2000.

Innes, Christopher. *Modern British Drama: The Twentieth Century.* CUP, 2002.

Innes, Christopher. *Modern German Drama: A Study in Form.* CUP, 1979.

Innes, Christopher. "Towards a Post-millennial Mainstream? Documents of the Times." *Modern Drama*, vol. 50, no. 3, 2007, pp. 435–452.

Irmer, Thomas. "A Search for New Realities: Documentary Theatre in Germany." *The Drama Review: TDR*, vol. 50, no. 3, 2006, pp. 16–28.

Iser, Wolfgang. *The Fictive and the Imaginary: Charting Literary Anthropology.* Johns Hopkins University Press, 1993.

Jameson, Frederic. *Brecht and Method.* Verso, [1998] 1999.

Jameson, Frederic. *Signatures of the Visible.* Routledge, 1992.

Jameson, Frederic. *The Political Unconscious.* Routledge, [1981] 1989

Jannarone, Kimberly. "Introduction: The Political Fallacy of Vanguard Performance." *Vanguard Performance Beyond Left and Right*, edited by Kimberly Jannarone. University of Michigan Press, 2015, pp. 16–31.

Jauss, Hans Robert. *Toward An Aesthetic of Reception*, translated by Timothy Bahti. University of Minnesota Press, 1982.

Jeffers, Alison. "Interview with Chris Honer." Unpublished interview conducted on 11 March 2005.

Jeffers, Alison. "Looking for Esrafil: Witnessing 'Refugitive' Bodies in I've got something to show you." *Get Real: Documentary Theatre Past and Present*, edited by Alison Forsyth and Chris Megson. Palgrave Macmillan, 2009, pp. 91–106.

Jeffers, Alison. "Refugee Perspectives: The Practice and Ethics of Verbatim Theatre and Refugee Stories." *Platform*, vol. 1, no. 1, 2006, pp. 1–17.

Jeffers, Alison. *Refugees, Theatre and Crisis: Performing Global Identities*. Palgrave Macmillan, 2011.

Jenkins, Tiffany. "Political Theatre's Final Curtain." *Independent*, 28 December 2011, p. 34.

Jha, Prem Shankar. *The Twilight of the Nation State: Globalisation, Chaos and War*. Pluto Press, 2006.

Johnson, Beth. "Art Cinema and The Arbor: Tape-recorded Testimony, Film-Art and Feminism."*Journal of British Cinema and Television*, vol. 13, 2016, pp. 278–291.

Jones, Alice. "A Faulty Light Inspires Louise Wallinger's New Noisy Neighbour Show; The Diary." *Arts and Books Review*. (22 June 2013), pp. 10.

Jones, Owen. "On Sky News Last Night, I Realised How Far Some Will Go to Ignore Homophobia." *Guardian*, 13 June 2016, https://www.theguardian.com/commentisfree/2016/jun/13/sky- news-homophobia-orlando-sexuality.

Julien, Martin. "From There to Here: Canadian's Stage Production of London Road." *TDR: The Drama Review*, vol. 59, no. 3, 2015, pp. 161–168.

Jürs-Munby, Karen. "Introduction." *Postdramatic Theatre*, Hans-Thies Lehmann, translated by Karen Jürs-Munby. Routledge, 2006, pp. 1–15.

Kalb, Jonathan. "Documentary Solo Performance: the Politics of the Mirrored Self." *Theater*, vol. 31, no. 3, 2001, pp. 13–29.

Kalb, Jonathan. *Great Lengths: Seven Works of Marathon Theater*. The University of Michigan Press, 2011.

Kaneko Lucas, Valerie. "'There's No Justice – Just us': Black Britons, British Asians, and the Criminal Justice System in Verbatim Drama." *"Black" British Aesthetics Today*, edited by R. Victoria Arana. Cambridge Scholars, 2007, pp. 262–282.

Kaplan, E. Ann. "Theories and Strategies of the Feminist Documentary." *New Challenges for Documentary*, edited by Alan Rosenthal. University of California Press, 1988, pp. 78 –102.

Katsouraki, Eve. "Affirmative Antinomies in Naturalistic and Realist Theatre." *Performance Research: A Journal of the Performative Arts*, vol. 19, no. 2, 2014, pp. 63–73.

Keir, Elam. *The Semiotics of Theatre and Drama*. Routledge, 1980.

Kelleher, Joe. *Theatre & Politics*. Palgrave Macmillan, 2009.

Keidan, Lois and CJ Mitchell (eds.). *Programme Notes: Case Studies for Locating Experimental Theatre*. Oberon Books, 2013.

Kellaway, Kate. "Theatre of War." *Guardian*, 29 August 2004, p. 5.

Kempf, Lucie and Tania Moguilevskaia. *Le Théâtre Néo-documentaire: Résurgence ou Réinvention?.* Presses Universitaires de Nancy, 2013.
Kent, Nicolas. "Tricycle: The Power of Radical Empathy." *Peacenews.info*, June 2011, http://peacenews.info/node/6203/tricycle-power-radical-empathy.
Kent, Nicolas. "Verbatim Theatre (Verbatim)." *The Tricycle: Collected Tribunal Plays 1994–2012.* Oberon Books, 2014, pp. 4–40.
Kernodle, George and Portia Kernodle. *Invitation to the Theatre (brief second edition).* Harcourt Brace Javanovich, [1971] 1978.
Kershaw, Baz. "Alternative Theatres, 1946–2000." *The Cambridge History of British Theatre: Volume 3: Since 1895*, edited by Baz Kershaw. CUP, 2004, pp. 291–325.
Kershaw, Baz. *The Politics of Performance.* Routledge, 1992.
Kershaw, Baz. *The Radical in Performance: Between Brecht and Baudrillard.* Routledge, 1999.
Kershaw Baz and Graham Ley. "Editorial." *Contemporary Theatre Review*, vol. 18, no. 3, 2008, pp. 289–292.
Kershaw, Liz. "The Estates of the Nation." *Independent*, 28 June 2001, p. 12.
King, Janet. "Shocking Play Still as Electric." *Western Morning News* (Plymouth, 30 November 2000), p. 2.
Kinghorn, Shane. "Crossing the Line: Reconstruction/Reconciliation." *Symbolon*, vol. 22, 2012, pp.14–21.
Kipphardt, Heinar. "Heinar Kipphardt Pays Tribute to Erwin Piscator." *Le Théâtre dans le Monde*, vol. 17, 1968, pp. 302–311.
Kirby, Michael. "On Political Theatre." *The Drama Review: TDR*, vol. 19, no. 2, 1975, pp. 129–135.
Klein, Bernhard. "'*Stuff Happens*': David Hare and Verbatim Theatre." *Cases of Intervention: The Great Variety of British Cultural Studies*, edited by Marie Hologa, Christian Lenz, Cyprian Piskurek and Stefan Schlensag. Cambridge Scholars, 2013, pp. 207–222.
Kondo, Dorinne. "(Re)Visions of Race: Contemporary Race Theory and the Cultural Politics of Racial Crossover in Documentary Theatre." *Theatre Journal*, vol. 52, no. 1, 2000, pp. 81–107.
Kondo, Dorinne. "Shades of Twilight: Anna Deavere Smith and Twilight: Los Angeles 1992." *Connected: Engagements with Media*, edited by George E. Marcus. The University of Chicago Press, 1996, pp. 313–346.
Kritzer, Amelia Howe. *Political Theatre in Post-Thatcher Britain. New Writing: 1995–2005.* Palgrave Macmillan, 2008.
Kushnir, Andrew. "If You Mingle: Thought on How Theatre Humanizes the Audience." *Defence of Theatre: Aesthetic Practices and Social Interventions*, edited by Kathleen Gallagher and Barry Freeman. UTP, 2016, pp. 83–98.
Kustow, Michael. *Theatre@risk.* Methuen Drama, 2000.
Lacey, Stephen. *British Realist Theatre: The New Wave in its Context 1956–1965.* Routledge, 1995.
Lacey, Stephen. "Staging the Contemporary: Politics and Practice in Post-War Social Realist Theatre." *British Social Realism in the Arts since 1940*, edited by David Tucker. Palgrave Macmillan, 2011, pp. 57–80.
Lachman, Michał. "History or Journalism: Two Narrative Paradigms in Bloody Sunday. Scenes from the Saville Inquiry by Richard Norton-Taylor." *Studia Anglica Posnaniensia*, vol. 43, 2007, pp. 305–314.

Lachman, Michał. "The Colours of History or Scenes from the Inquiry into Verbatim Drama." *Contemporary Drama in English*, vol. 14, 2007, pp. 311–324.
Landon-Smith, Kristine. "Headphone Verbatim Masterclass." 27 March 2012, https://www.youtube.com/watch?v=ULHkrMjtrZM.
Landy, Robert. J. and David T. Montgomery. *Theatre for Change: Education, Social Action and Therapy.* Palgrave Macmillan, 2012.
Lane, David. *Contemporary British Drama.* Edinburgh UP, 2010.
Lansdale, Janet. "Ancestral and Authorial Voices in Lloyd Newson and DV8's 'Strange Fish'." *New Theatre Quarterly*, vol. 20, no. 2, 2004, pp. 117–126.
Lavender, Andy. *Performance in the Twenty-First Century: Theatres of Engagement.* Routledge, 2016.
Lawson, Mark. "From the West End to Westminster: which Playwrights should run Britain?." *Guardian,* 6 November 2015, http://www.theguardian.com/stage/2015/nov/06/west-end-westminster-which-playwrights-run-britain-waste#comment-62850607.
Lawson, Mark. "Musicals We Love: London Road." *Guardian*, 10 February 2014, https://www.theguardian.com/stage/2014/feb/10/musicals-we- love- london-road-mark-lawson.
Leask, Josephine. "Lloyd Newton." *Fifty Contemporary Choreographers*, edited by Martha Bremser. Routledge, 1999, pp. 228–231.
Lehmann, Hans-Thies. "A Future for Tragedy? Remarks on the Political and the Postdramatic." *Postdramatic Theatre and the Political: International Perspectives on Contemporary Performance*, edited by Karen Jürs-Munby, Jerome Carroll and Steve Giles. Methuen Drama, 2013, pp. 87–109.
Lehmann, Hans-Thies. *Postdramatic Theatre.* Translated by Karen Jürs-Munby, Routledge, 2006.
Lehmann, Hans-Thies. "Word and Stage in Postdramatic Theatre." *Contemporary Drama in English*, vol. 14, 2007, pp. 37–54.
Lemoine, Xavier. "Embodying Hybridity: Anna Deveare Smith's Identity Cross-Overs." *Understanding Blackness through Performance*, edited by Anne Crémieux, Xavier Lemoine and Jean-Paul Rocchi. Palgrave Macmillan, 2013, pp. 237–261.
Letts, Quentin. "Bloody Sunday: Scenes from the Saville Inquiry; Daily Mail 12.4.05." *Theatre Record*, 9–22 April 2005, p. 470.
Leung, Wayne. "MSN [MEN SEEKING MEN] (Lemontree creations) 2013 Toronto Fringe Review." 5 July 2013, http://www.mooneyontheatre.com/2013/07/05/msm- men-seeking-men-lemontree-creations-2013-toronto- fringe-review/.
Lipovetsky, Mark and Birgit Beumers. "Reality Performance: Documentary Trends in Post-Soviet Russian Theatre." *Contemporary Theatre Review*, vol. 18, no. 3, 2008, pp. 293–306.
Little, Suzanne. "In and Out of Tune with Reality: Opposed Strategies of Docume ntary Theatre." *Double Dialogue*, vol. 14, 2011.
Little, Suzanne. "Repeating Repetition." *Performance Research*, vol. 20, no. 5, 2015, pp. 44–50.
Lombardi, Chris. "How 'London Road' Became 2016's Most Unique Movie Musical." *Film Independent*, 9 September 2016, http://www.filmindependent.org/blog/how-london-road-became-2016s-most- unique- movie- musical/.
Long, Nicholas J. "For a Verbatim Ethnography." *Anthropology, Theatre, and Development*, edited by Alex Flynn and Jonas Tinius. Palgrave Macmillan, 2015, pp. 305–333.

Love, Catherine. “Are We on the Same Page.” *Exeunt*, 17 September 2015, http://exeuntmagazine.com/features/are-we-on-the-same-page/.
Love, Catherine. “Hippo World Guest Book, Caryl Churchill Theatre.” http://catherinelove.co.uk/2014/01/23/hippo-world-guest-book-caryl-churchill-theatre/
Luckhurst, Mary. “Introduction: Research, Acting Strategies, and Performer and Audience.” *Playing for Real: Actors on Playing Real People*, edited by Tom Cantrell and Mary Luckhurst. Palgrave Macmillan 2010, pp. 1–22.
Luckhurst, Mary. “Contemporary English Theatre: Why Realism.” *Contemporary Drama in English*, vol. 9, 2002, pp. 73–84.
Luckhurst, Mary. “Ethical Stress and Performing Real People.” *Performing Ethos*, vol. 1, no. 2, 2010, pp. 135–52.
Luckhurst, Mary. “Verbatim Theatre, Media Relations and Ethics.” *A Concise Companion to Contemporary British and Irish Drama*, edited by Nadine Holdsworth and Mary Luckhurst, Wiley-Blackwell, 2008, pp. 200–222.
Luckhurst, Mary and Chloe Veltman. *On Acting: Interviews with Actors*. Faber and Faber, 2001.
Macaulay, Alastair. “Bloody Sunday: Scenes from the Saville Inquiry ; *Financial Times* 14.4.05.” *Theatre Record*, 9–22 April 2005, p. 469.
Machon, Josephine. *Immersive Theatres: Intimacy and Immediacy in Contemporary Performance*. Palgrave Macmillan, 2013.
Machon, Josephine. “Punchdrunk.” *British Theatre Companies 1995–2014*, edited by Liz Tomlin. Methuen Drama, 2015, pp. 255–282.
Machon, Josephine. *(Syn)aesthetics: Redefining Visceral Performance*. Palgrave Macmillan, 2009.
Machon, Josephine. “Watching, Attending, Sence-making: Spectatorship in Immersive Theatres” *Journal of Contemporary Drama in English*, vol. 4, no. 1, 2016, pp. 34–48.
Madison, D. Soyini. “The Mike Daisey Affair: Labor and Performance.” *Communication and Critical/Cultural Studies*, vol. 9, no. 2, 2012, pp. 234–240.
Maedza, Pedzisai. *Performing Asylum: Theatre of Testimony in South Africa*. African Studies Centre Leiden, 2017.
Magelssen, Scott. *Simming: Participatory Performance and the Making of Meaning*. University of Michigan Press, 2014.
Magris, Erica, and Béatrice Picon-Vallin. *Les théâtres documentaires*. Editions Deuxième Époques, 2019.
Maguire, Tom. *Performing Story on the Contemporary Stage*. Palgrave Macmillan, 2015.
Mahut, Jérémy. “La Remise en Question des Modèles Épique et Dramatique par le Théâtre-témoignage: Résumé de la Communication.” Université de Reims, 2009, pp. 1–3, http://www.univ-reims.fr/site/laboratoire labellis/crimel/gallery_files/site/1/1697/3184/10102/20987/20990.pdf.
Malzacher, Florian. “There is a World for People Like You: Audience. The Spectator as Bad Witness and Bad Voyeur.” *“Not Even a Game Anymore”: The Theatre of Forced Entertainment*, edited by Judith Helmer and Florian Malzacher. Alexander Verlag, 2004, pp. 121–138.
Mangan, Michael. *The Drama, Theatre and Performance Companion*. Palgrave Macmillan, 2013.
Marlowe, Sam. “Come Out Eli.” *The Times*, 8 September 2004, http://www.recordeddelivery.net/ComeOutEliInfoPack.pdf.

Martin Carol and Anna Deavere Smith. "Anna Deavere Smith: The Word Becomes You. An Interview." *The Drama Review: TDR*, vol. 37, no. 4, 1993, pp. 45–62.
Martin, Carol. "Bearing Witness: Anna Deavere Smith from Community to Theatre to Mass Media." *A Sourcebook of Feminist Theatre and Performance: On and Beyond the Stage*, edited by Carol Martin. Routledge, 1996, pp. 81–93.
Martin, Carol. "Bodies of Evidence." *The Drama Review: TDR*, vol. 50, no. 3, 2006, pp. 8–15.
Martin, Carol. "Introduction: Dramaturgy of the Real." *Dramaturgy of the Real on the World Stage*, edited by Carol Martin. Palgrave Macmillan, 2010, pp. 1–14.
Martin, Carol. *Theatre of the Real*. Palgrave Macmillan, 2013.
Mason, Gregory. "Documentary Drama from the Revue to the Tribunal." *Modern Drama*, vol. 20, no. 3, 1977, pp. 263–278.
McCormack, Jess. *Choreography and Verbatim Theatre: Dancing Words*. Palgrave Pivot, 2018.
McCormick, Sheila. "Tribunal Theatre and the Voice of Dissent as seen in Bloody Sunday: Scenes from the Saville Inquiry." *Voicing Dissent: Perspectives in Irish Criticism*, edited by Sandrine Brisset and Noreen Doody. Irish Academic Press, 2012, pp. 61–68.
McCutcheon, Jade Rosina. "The Performance Mirror: Self, Consciousness and Verbatim Theatre." *Embodied Consciousness: Performance Technologies*, edited by Jade Rosina McCutcheon and Barbara Sellers-Young. Palgrave Macmillan, 2013, pp. 145–158.
McDermott, Phelim. "Physical Theatre and Text." *Physical Theatres: A Critical Reader*, edited by John Keefe and Simon Murray. Routledge, 2007, pp. 201–208.
McGrath, John. *A Good Night Out: Popular Theatre: Audience, Class and Form*. Eyre Methuen, 1981.
McKenzie, Jon. *Perform or Else: From Discipline to Performance*. Routledge, 2001
McKinnie, Michael. "Rethinking Site-Specificity: Monopoly, Urban Space, and the Cultural Economics of Site-Specific Performance." *Performing Site-Specific Theatre: Politics, Place, Practice*, edited by Anna Birch and Joanne Tompkins. Palgrave Macmillan, 2012, pp. 21–36.
McLean, Duncan. "An Introduction to Aalst." NationalTheatreScotland, February 2007, http://www.nationaltheatrescotland.com/content/default.asp?page=s278_1.
McLucas, Cliff. "Ten Feet and Three-quarters of an Inch of Theatre." *Site-specific Art*, edited by Nick Kaye. Routledge, 2000, pp. 125–137.
McMillan, Joyce. "Language Lessons." *The Scotsman*, 2 October 2008, p. 9.
McMillan, Joyce. "Fiction Can Tell Truths too." *The Scotsman*, 13 June 2015, http://www.scotsman.com/what-s-on/theatre-comedy-dance/joyce- mcmillan-fiction-can-tell-truths-too-1-3801540.
Megson, Chris and Janelle Reinelt. "Conversations with David Edgar." *Contemporary Theatre Review*, vol. 22, no. 3, 2012, pp. 372–91.
Megson, Chris. "'Acting with Facts': Actors Performing the Real in British Theatre and Television since 1990, Department of Film, Theatre and Television, University of Reading, 8 May 2009.' *Research in Drama Education: The Journal of Applied Theatre and Performance*, vol. 15, no. 1, 2010, pp. 133–35.
Megson, Chris. "Chris Megson in Conversation with Jonathan Holmes, Cora Bissett and Jem Wall: Performance and Ethics – A Debate." *Performing Ethos*, vol. 3, no. 1, 2012, pp. 59–75.

Megson, Chris. "Half the Picture: 'A Certain Frisson' at the Tricycle Theatre." *Get Real: Documentary Theatre Past and Present*, edited by Alison Forsyth and Chris Megson. Palgrave Macmillan, 2009, pp. 195–208.

Megson, Chris. "Hare, David." *The Oxford Encyclopedia of British Literature*, edited by David Scott Kastan. OUP, 2006, pp. 510–514.

Megson, Chris. "Introduction." *The Methuen Drama Book of Naturalistic Plays*, edited by Chris Megson. Methuen Drama, 2010.

Megson, Chris. "'The State We're in': Tribunal Theatre and British Politics in the 1990s." *Theatre of Thought: Theatre, Performance and Philosophy*, edited by Daniel Watt and Daniel Meyer-Dinkgräfe. Cambridge Scholars, 2007, pp. 110–127.

Megson, Chris. "The Theatres Act (1968), Documentary Theatre and the Actor's 'Overwhelming Reality'." *Studies in Theatre and Performance*, vol. 31, no. 2, 2011, pp. 137–151.

Megson, Chris. "This Is All Theatre: Iraq Centre Stage." *Contemporary Theatre Review*, vol. 15, no. 3, 2005, pp. 369–371.

Megson, Chris. "Tribunal Testimony and Political Process." Central School of Speech and Drama: Unpublished panel discussion with Richard Norton-Taylor and Thomas Wheatley, 2006, http://crco.cssd.ac.uk/24/1/CEN_V_tribunal_edit.pdf.

Megson, Chris. "Verbatim Practices in Contemporary Theatre: Symposium Report." *Contemporary Theatre Review*, vol. 16, no. 4, 2006, pp. 529–532.

Merahi, Danielle. *Théâtres du reel*. L'Entretemps, 2016.

Merlin, Bella. "Acting Hare: The Permanent Way." *The Cambridge Companion to David Hare*, edited by Richard Boon. CUP, 2007, pp. 123–37.

Merlin, Bella. "Bella Merlin interviewed by Heather Sutherland." University of Reading: Unpublished interview, 2007, http://www.reading.ac.uk/web/FILES/ftt/Bella_Merlin_11th_December_2007.pdf.

Merlin, Bella. *The Complete Stanislavski Toolkit*. Nick Hern Books, 2014.

Merlin, Bella. "The Permanent Way: NT Workpack." Out of Joint (2003), http://www.outofjoint.co.uk/wp-content/uploads/2010/09/PW_pac.pdf.

Middeke, Martin, Peter Paul Schnierer and Aleks Sierz (eds.). *The Methuen Drama Guide to Contemporary British Playwrights*. Methuen Drama, 2011.

Mienczakowski, Jim. "Ethnodrama: Performed Research – Limitations and Potential." *Emergent Methods in Social Research*, edited by Sharlene Nagy Hesse-Biber and Patricia Lavy. Sage Publications, 2006, pp. 235–252.

Mitchell, Katie. *The Director's Craft: A Handbook for the Theatre*. Routledge, 2009.

Monks, Aoife. "'This Painful Chapter': Performing the Law in Bloody Sunday: Scenes from the Saville Inquiry." *Contemporary Theatre Review*, vol. 23, no. 3, 2013, pp. 345–56.

Morgan, Kenneth O. "Epilogue (2000–2010)." *The Oxford History of Britain, Revised Edition*, edited by Kenneth O. Morgan. OUP, 2010, pp. 677–710.

Morgan, Richard. *Brith Gof: Y Llyfr Glas: 1988–1995*. Brith Gof, 1995.

Morris, Pam. *Realism*. [2003] 2009.

Mumford, Meg. *Bertolt Brecht*. Routledge, 2009.

Murray, Douglas. "Bloody Sunday, or the Theatre of Moral Corruption." *Open Democracy*, 12 May 2005, https://www.opendemocracy.net/node/2500.

Murray, Simon and John Keefe. *Physical Theatres: A Critical Introduction*. Routledge, 2007.

Murray, Simon and John Keefe. *Physical Theatres: A Critical Reader*. Routledge, 2007.

Murray, Timothy. "Introduction: The Mise-en-Scène of the Cultural.'*Mimesis, Masochism, and Mime: The Politics of Theatricality in Contemporary French Thought.* The University of Michigan Press, 1997, pp. 1–26.
Nancy, Jean-Luc. *Listening,* translated by Charlotte Mandell. Fordham University Press, [2002] 2007.
Nealon, Jeffrey. *Post-Postmodernism: Or, The Cultural Logic of Just-in-Time Capitalism.* SUP, 2012.
Nevitt, Roy. "Peter Cheeseman." *Documentary Arts Reports*, vol. 2, 1986, pp. 7–10.
Nevitt, Roy. "The Living Archive Project." *Englisches Theater der Gegenwart: Geschichte(n) und Strukturen*, edited by Klaus Peter Müller. Narr, 1993, pp. 145–154.
Nevitt, Roy. "The Theatre of Fact." *Theatre Papers*, vol. 5, no. 9, 1985, pp. 1–16.
Newson, Lloyd. "DV8 Physical Theatre. Company Promo." 10 March 2015, https://www.youtube.com/watch?v=AsKtaMrxguE.
Newson, Lloyd. "DV8's To Be Straight with You: Dancing Against Prejudice." *Guardian*, 28 October 2008, https://www.theguardian.com/stage/2008/oct/27/dv8-straight-with-you.
Newson, Lloyd. "Foreword." In *To Be Straight with You Workpack*, Ankur Bahl and Uri Omi. London: Royal National Theatre, 2008, 3, https://www.dv8.co.uk/media/files/To%20Be%20Straight%20With%20You%20FINAL.pdf.
Newson, Lloyd. "Interview: To Be Straight with You. Q&A," November 2007, https://www.dv8.co.uk/pages/interview-to-be-straight-with- you-qa.
Nichols, Bill. *Representing Reality: Issues and Concept in Documentary.* Indiana University Press, 1991.
Norton-Taylor, Richard. "Spirit of Inquiry." *NewStatesman*, 7 June 2004, http://www.newstatesman.com/node/148148.
Norton-Taylor, Richard. "Tribunal Testimony and Political Process." Central School of Speech and Drama: Unpublished panel discussion with Chris Megson and Thomas Wheatley, 2006, http://crco.cssd.ac.uk/24/1/CEN_V_tribunal_edit.pdf.
Norton-Taylor, Richard. "Verbatim Theatre Lets the Truth Speak for Itself." *Guardian,* 31 May 2011, http://www.theguardian.com/stage/2004/mar/17/theatre.pressandpublishing.
Nunn, Gary. "Verbatim Theatre is like good Reality TV on Stage." *Guardian*, 26 June 2015, http://www.theguardian.com/media/mind- your-language/2015/jun/26/verbatim-theatre-is- like-good-reality-tv-on-stage.
Nzewi, Meki. *A Contemporary Study of Musical Arts Informed by African Indigenous Knowledge Systems*, Vol. 3. Pretoria: Centre for Indigenous Instrumental African Music and Dance (Ciimda), 2007.
Oades, Roslyn. "Creating Headphone-Verbatim Performance." *Verbatim: Staging Memory and Community,* edited by Paul Brown. Strawberry Hills: Currency Press, 2010, pp. 84–87.
O'Connor, Jacqueline. *Documentary Trial Plays in Contemporary American Theater.* Southern Illinois University Press, 2013.
Oddey, Alison. *Devising Theatre: A Practical and Theoretical Handbook.* Routledge, 1994.
Olsson, Jesper. "The Audiographic Impulse: Doing Literature with the Tape Recorder." *Audiobooks, Literature and Sound Studies*, edited by Matthew Rubery. Routledge, 2011, pp. 61–75.
O'Malley, Tommy. "Verbatim Playwright Alecky Blythe Takes 'London Road' from the Street to the Stage to the Screen." *The Civilians*, 9 September 2016, http://

extendedplay.thecivilians.org/verbatim-playwright-alecky-blythe-takes-london-road-from-the-streets-to-the-stage-to-the-screen-90916/.
O' Reilly, Sally. "Fallujah: Old Truman Brewery London May 1 to June 2." *Arts Monthly*, vol. 308, 2007, pp. 41–42.
Ostrowska, Joanna. "Site-specific Theatre as a Form of Documentary Theatre." *Le Théâtre Néo-documentaire: Résurgence ou Réinvention?*, edited by Lucie Kempf and Tania Moguilevskaia. Presses Universitaires de Nancy, 2013, pp. 239–246.
Pankratz, Anette. "Political Drama : Edward Bond's *Saved* and David Edgar's *Maydays*." *A History of British Drama: Genres – Developments – Model Interpretations*, edited by Sibylle Baumbach, Birgit Neumann, Ansgar Nünning. Wissenschaftlicher Verlag Trier 2011, pp. 305–318.
Paget, Derek. "Acting with Facts: Actors Performing the Real in British Theatre and Television since 1990. A Preliminary Report on a New Research Project." *Studies in Documentary Film*, vol. 1, no. 2, 2007, pp. 165–76.
Paget, Derek. "Acts of Commitment: Activist Arts, the Rehearsed Reading, and Documentary Theatre." *New Theatre Quarterly*, vol. 26, no. 2, 2010, pp. 173–193.
Paget, Derek. "A Homage to Charles Chilton." *Documentary Arts Reports*, vol. 3, 1989, pp. 1 –4.
Paget, Derek. "Editorial." *Studies in Theatre & Performance*, vol. 31, no. 2, 2011, pp. 133 –135.
Paget, Derek. *"Oh What a Lovely War" and the Broken Tradition of Documentary Theatre: An Investigation into the Origins, Manifestations and Influence of Documentary Theatre in the UK*. University of Manchester: Unpublished PhD thesis, 1988.
Paget, Derek. "New Documentarism on Stage: Documentary Theatre in New Times." *Zeitschrift für Anglistik und Amerikanistik*, vol. 56, no. 2, 2008, pp. 129–141.
Paget, Derek. "The 'Broken Tradition' of Documentary Theatre and its Continued Powers of Endurance." *Get Real: Documentary Theatre Past and Present*, edited by Alison Forsyth and Chris Megson. Palgrave Macmillan, 2009, 224–238.
Paget, Derek. *True Stories? Documentary Drama on Radio, Screen and Stage*. MUP, 1990.
Paget, Derek. "'Verbatim Theatre': Oral History and Documentary Techniques." *New Theatre Quarterly*, vol. 3, no.12, 1987, pp. 317–336.
Palmer, Richard H. *The Contemporary British History Play*. Greenwood Press, 1998.
Patsalidis, Savas and Elizabeth Sakellaridou. "Introduction. The Text Srikes Back: The Dynamics of Performativity." *Gramma: Journal of Theory and Criticism*, vol. 17 2009, pp. 7–11.
Patterson, Michael. *Strategies of Political Theatre: Post-War British Playwrights*. CUP, 2003.
Pavis, Patrice. "Brechtian Gestus and its Avatars in Contemporary Theatre." *Brecht Yearbook*, vol. 24, 1999, pp. 177–190.
Pavis, Patrice. *Contemporary Mise en Scène: Staging Theatre Today*, translated by Joel Anderson. Routledge, 2012.
Pavis, Patrice. *Dictionary of the Theatre: Terms, Concepts, and Analysis*, translated by Christine Shantz. University of Toronto Press, 1998.
Pavis, Patrice. "The Classical Heritage of Modern Drama: The Case of Postmodern Theatre (trans. Loren Kruger)." *Modern Drama*, vol. 29, no. 1, 1986, pp. 1–22.
Pavis, Patrice. *The Routledge Dictionary of Performance and Contemporary Theatre*, translated by Andrew Brown. Routledge, 2016.

Peacock, Keith. *Thatcher's Theatre: British Theatre and Drama in the Eighties.* Greenwood Press, 1999.
Pearson, Mike and Michael Shanks. *Theatre/Archaeology.* Routledge, 2001.
Pearson, Mike. "My Balls/Your Chin." *Performance Research*, vol. 3, no. 2, 1998, pp. 35–41.
Pearson, Mike. *Site-specific Performance.* Palgrave Macmillan, 2010.
Perlez, Jane. "An Assault in Iraq, a Stage Hit in London." *New York Times*, 29 May 2007, p. 1.
Phelan, Peggy. *Unmarked: The Politics of Performance.* Routledge, 1993.
Pickering, Kenneth and Jayne Thompson. *Naturalism in Theatre: Its Development and Legacy.* Palgrave Macmillan, 2013.
Piscator, Erwin. "A Letter to the Weltbühne." *World Theatre,* vol. 17, no. 5–6, 1968, pp. 325 –329.
Piscator, Erwin. "Post 'Investigation'." *World Theatre*, vol. 17, no. 5–6, 1968, pp. 345–349.
Piscator, Erwin. *The Political Theatre*, translated by Hugh Rorrison. Avon, 1978.
Platinga, Carl. "Moving Pictures and the Rhetoric of Nonfiction Film: Two Approaches." *Post-Theory: Reconstructing Film studies*, edited by David Bordwell and Noël Carroll. University of Wisconsin Press, 1996, pp. 307–324.
Pollock, Della. "Introduction: Remembering." *Remembering: Oral History Performance*, edited by Della Pollock. Palgrave Macmillan, 2005, pp. 1–17.
Pressley, Nelson. *American Playwriting and the Antipolitical Prejudice.* Palgrave Macmillan, 2014.
Preston, Peter. "Extra! Extra!." *Guardian*, 17 March 2004, p. 12.
Priseman, Peter (ed.). *Documentary Realism: Painting in the Digital Age.* Seabrook Press, 2015.
Quinzaños, Isabel. Crouch, Touch, Pause, Engage Education Work Pack. Out of Joint, 2015, http://www.outofjoint.co.uk/wp-content/uploads/2010/07/Crouch-Workpack-19-Feb-2015.pdf.
Radin, Robert. "The Recuperation of The Theory-Death of the Avant-Garde." *Diacritics,* vol. 28, no. 2, 1998, pp. 41–51.
Radosavljević, Duška. *Theatre-Making: Interplay between Text and Performance in the 21st Century.* Palgrave Macmillan, 2013.
Rae, Paul. *Theatre & Human Rights.* Palgrave Macmillan, 2009.
Rancière, Jacques. "The Emancipated Spectator." *Artforum,* 2007, pp. 271–280.
Rebellato, Dan. "Dennis Kelly." *Contemporary Theatre Review. Backpages,* vol. 17, no. 4, 2007, pp. 603–608.
Rebellato, Dan. "Does the Mega- musical Boom Mean Theatre's Bust." *Guardian*, 18 January 2011, http://www.theguardian.com/stage/theatreblog/2011/jan/18/mega- musicals-theatre-west-end.
Rebellato, Dan. "Exit the Author." *Contemporary British Theatre: Breaking New Ground*, edited by Vicky Angelaki. Palgrave Macmillan, 2013, 9–31.
Rebellato, Dan. "From the State of the Nation to Globalization: Shifting Political Agendas in Contemporary British Playwriting." *A Concise Companion to Contemporary British and Irish Drama*, edited by Nadine Holdsworth and Mary Luckhurst. Blackwell, 2008, pp. 245 –62.
Rebellato, Dan. "Inaugural Lecture: Theatre, Sex & Zombies: The Strange Case of the Naturalistic Stage." Unpublished lecture delivered on 4 February 2014 at the Caryl Churchill Theatre, Royal Holloway University of London, http://static1.squarespace.com/

static/513c543ce4b0abff73bc0a82/t/52fd01b9e4b0649c39e12411/1392312761464/Inaugural+lecture.pdf.
Rebellato, Dan. "Introduction." *Citizenship, Mark Ravenhill.* Methuen Drama, 2015, pp. 3–17.
Rebellato, Dan. "Introduction: Naturalism and Symbolism: Early Modernist Practice." *The Routledge Drama Anthology and Sourcebook: From Modernism to Contemporary Performance*, edited by Maggie B. Gale and John F. Deeney. Routledge, 2010, pp. 6–24.
Rebellato, Dan. "Review: State of the Nation by Michael Billington." *Contemporary Theatre Review*, vol. 18, no. 4, 2008, pp. 528–530.
Rebellato, Dan. "Ten Years with Nothing in Common. " 2015 Bruntwood Prize Award Ceremony Programme, p. 5, http://www.danrebellato.co.uk/news/2015/12/1/bruntwood-prize.
Rebellato, Dan. *Theatre & Globalization.* Basingstoke: Palgrave Macmillan, 2009.
Rebellato, Dan. "Writing Writing: British Playwriting in the Twenty-First Century." *A Leaping Battle: Reading New Writing*, edited by Pat To Yan. International Association of Theatre Critics & On and On Theatre Workshop, 2015, pp. 128–78.
Reid, Trish. "Return, Revisit, Re-enact: Re-staging the Confessions for a Contemporary Audience." *Paul Bright's Confessions of a Justified Sinner*, edited by Untitled Projects and Pamela Carter. Oberon Books, 2015, 136–137.
Reinelt, Janelle and Gerald Hewitt. "Principles and Pragmatism in Political Theatre: David Edgar's Continental Divide." *Theatre Forum*, vol. 25, 2004, pp. 3–14.
Reinelt, Janelle and Gerald Hewitt. *The Political Theatre of David Edgar: Negotiation and Retrieval.* CUP, 2011.
Reinelt, Janelle. "Performing Race: Anna Deavere Smith's Fires in the Mirror." *Modern Drama*, vol. 39, 1996, pp. 609-617.
Reinelt, Janelle. "The Promise of Documentary." *Get Real: Documentary Theatre Past and Present*, edited by Alison Forsyth and Chris Megson. Palgrave Macmillan, 2009, pp. 6–23.
Reinelt, Janelle. "Toward a Poetics of Theatre and Public Events." *The Drama Review: TDR*, vol. 50, no. 3, 2006, pp. 69–87.
Reinelt, Janelle. "Stuff Happens (Review)." *Theater Journal*, vol. 57, no. 2, 2005, pp. 303-306.
Renov, Michael. "Re-Thinking Documentary: Towards a Taxonomy of Mediation." *Wide Angle*, vol. 8, no. ¾, 1986, pp. 71–77.
Robertson, Geoffrey. "2003-08-01: Morning." Automatically reformated transcripts of the Hutton Inquiry, http://www.hutton.softblade.com/transcripts.php?action=transcript&session=1&search=robertson.
Roesner, David. *Musicality in Theatre: Music as Model, Method and Metaphor in Theatre-making.* Routledge, 2016.
Rorty, Richard. "Realism and Reference." *The Monist*, vol. 59, no. 3, 1976, pp. 321–340.
Rosenthal, Daniel. *The National Theatre Story.* Oberon Books, 2013.
Rothberg, Michael. *Traumatic Realism: The Demands of Holocaust Representation.* University of Minnesota Press, 2000.
Rouse, John. "Brecht and the Contradictory Actor." *Critical Theory and Performance*, edited by Janelle G. Reinelt and Joseph R. Roach. The University of Michigan Press, [2007] 2010, pp. 295–310.

Rutherford, Malcom. "Arts – Scott Inquiry Takes Centre Stage." *Financial Times* (17 June 1994). Factiva Database.
Said, Edward W. "Introduction to the Fiftieth-anniversary Edition." *Mimesis: The Representation of Reality in Western Literature*, edited by Erich Auerbach, and translated by Williard R. Trask. PUP, 2003, pp. i–xxiv.
Saison, Maryvonne. *Les theatres du réel: pratiques de la représentation dans le théâtre contemporain.* L'Harmattan, 1998.
Sajnani, Nisha. "The Implicated Witness: Towards a Relational Aesthetic in Dramatherapy." *Dramatherapy*, vol. 34, no. 1, 2012, pp. 6–21.
Sakellaridou, Elizabeth. "Hyper-/Sur-/Realism and the Postmodern Stage." *Contemporary Drama in English*, vol. 3, 1996, pp. 47–58.
Saldaña, Johnny. *Ethnotheatre: Research from Page to Stage.* Left Coast Press, 2011.
Salverson, Julie. "Transgressive Storytelling or an Aesthetic of Injury: Performance, Pedagogy and Ethics." *Theatre Research in Canada*, vol. 20, no. 1, 1999, pp. 35–51.
Sanchez-Colberg, Ana. "Altered States and Subliminal Spaces: Charting the Road towards a Physical Theatre." *Performance Research*, vol. 1, no. 2, 1996, pp. 40–56.
Sauerberg, Lars Ole. *Fact into Fiction: Documentary Realism in the Contemporary Novel.* St. Martin's Press, 1991.
Saville, Alice. "London Road: On Ethics in Verbatim Theatre." *Exeunt*, 6 July 2015, http://exeuntmagazine.com/features/london-road- its-about-ethics- in-verbatim-theatre/.
Schechner, Richard. "6 Axioms for Environmental Theater." *The Drama Review : TDR*, 1968, pp. 41–64.
Schechner, Richard. "Anna Deavere Smith: Acting as Incorporation (1993)." *A Sourcebook on African-American Performance: Plays, People, Movements*, edited by Annemarie Bean. Routledge, 1999, pp. 265–266.
Schechner, Richard. "Believed-in Theatre: An Excerpt." *Seagull Theatre Quarterly*, vol. 18, 1998, pp. 6–12.
Schechner, Richard. "Believed- in Theatre." *Performance Research: A Journal of the Performative Arts*, vol. 2, no. 2, 1997, pp. 75–91.
Schechner, Richard. *Between Theater and Anthropology.* University of Pennsylvania Press, 1985.
Schechner, Richard. *Environmental Theater. Applause*, [1973] 1994.
Schechner, Richard. *Performance Studies: An Introduction (Third Edition).* Routledge, 2013.
Schechner, Richard. *Performance Theory.* Routledge, 1988.
Schechner, Richard. *Public Domain.* Discus/Avon, 1969.
Schlote, Christiane and Eckart Voigts-Virchow. "Introduction: 'The Creative Treatment of Actuality': New Documentarism." *Zeitschrift für Anglistik und Amerikanistik*, vol. 56, no. 2, 2008, pp. 107–116.
Schulze, Daniel. *Authenticity in Contemporary Theatre and Performance.* Bloomsbury Methuen Drama, 2017.
Schweitzer, Pam. *Reminiscence Theatre: Making Theatre from Memories.* Jessica Kingsley, 2007.
Scott, Robert Dawson. "Cherry Blossom Traverse in Edinburgh." *The Times*, 30 September 2008, p. 30.
Scott, William. *Documentary Expressions and Thirties America.* Oxford University Press, 1973.

Shand, Skip and Ric Knowles. 'The Process Issue.' *Canadian Theatre Review*, vol. 97, 1998, pp. 3–4.
Shaughnessy, Nicola. *Applying Performance: Live Art, Socially, Engaged Theatre and Affective Practice.* Palgrave Macmillan, 2012.
Shields, David. *Reality Hunger: A Manifesto.* Hamish Hamilton, 2010.
Shonkwiler, Alison, and Berge L. C. La. "Introduction: A Theory of Capitalist Realism." *Reading Capitalist Realism*, edited by Alison Shonkwiler and Leigh Claire La Berge. University of Iowa Press, 2014, pp. 1–25.
Shuster, Anthony. "The Girlfriend Experience – Lu Corfield (Interview, Summer 2012)." *Played in Britain: Modern Theatre in 100 Plays,* Kate Dorney and Frances Gray. Application for iPad.
Shuttleworth, Ian. "To Be Straight With You, Lyttleton Theatre, London." *Financial Times*, 30 October 2008, p. 2.
Sierz, Aleks. "A New National Service." *Financial Times*, 10 September 2004.
Sierz, Aleks. "BU21, Theatre 503." *Sierz.co.uk*, 20 March 2016, http://www.sierz.co.uk/reviews/bu21-theatre-503/.
Sierz, Aleks. *In-Yer-Face Theatre: British Drama Today.* Faber and Faber, 2001.
Sierz, Aleks. "Little Revolution, Almeida Theatre." Theartdesk.com, 4 September 2014, http://www.theartsdesk.com/theatre/little-revolution-almeida-theatre.
Sierz, Aleks. *Rewriting the Nation: British Theatre Today.* Methuen, 2011.
Sierz, Aleks. "Theatre in the 1990s." *Modern British Playwriting: The 1990s: Voices, Documents, New Interpretations*, edited by Aleks Sierz. Methuen Drama, 2012.
Sierz, Aleks. "Verbatim Drama: Jean Charles de Menezes (1/2)." 11 November 2009, http://www.theatrevoice.com/audio/verbatim-drama-jean-charles-de-menezes/.
Sierz, Aleks. "What Ever Happened to In-Yer-Face Theatre." *Coup de Théâtre*, vol. 29, 2015, pp. 17–36.
Slovo, Gillian. "Verbatim Theatre (Verbatim)." *The Tricycle: Collected Tribunal Plays 1994–2012.* Oberon Books, 2014, pp. 4–40.
Smith, Amanda. "Artworks Feature: David Hare and the Power of Language." *ABC,* 19 December 2010, https://www.abc.net.au/radionational/programs/archived/artworks/artworks-feature-david-hare-and-the-power-of/3105094.
Smith, Anna Deavere. "Introduction." *House Arrest and Piano*, Anna Deavere Smith. Anchor Books, 2004.
Smith, Anna Deavere. *Talk to Me: Listening between the Lines.* Random House, 2000.
Snow, Georgia. "National Theatre to Produce Virtual Reality Shows." *The Stage*, 9 June 2016, https://www.thestage.co.uk/news/2016/national-theatre-produce-virtual-reality-shows/.
Soncini, Sara. *Forms of Conflict: Contemporary Wars on the British Stage.* University of Exeter Press, 2015.
Soncini, Sara. "War in Words: The Tricycle Theatre's Re-voicing of the Bloody Sunday Inquiry." *Pólemos*, vol. 9, no. 2, 2015, pp. 393–409.
Sood, Sara Owsley and Athanasios V. Vasilakos. "Digital Theater: Dynamic Theater Spaces." *Handbook of Multimedia for Digital Entertainment and Arts,* edited by Borivoje Furht. Springer, 2009, pp. 423–446.
Soto-Morettini, Donna. "Trouble In the House: David Hare's Stuff Happens." *Contemporary Theatre Review*, vol. 15, no. 3, 2005, pp. 309–319.

Speer, Annika C. "The Feminist Potential of Docudrama: Destabilizing the Primacy of Primary Sources through Paula Kamen's Jane." *Frontiers: A Journal of Women Studies*, vol. 34, no. 3, 2013, pp. 1–26.

Spencer, Charles. "Fine Journalism and Wonderful Drama." *The* Telegraph, 13 September 2004, http://www.telegraph.co.uk/culture/theatre/drama/3623902/Fine-journalism-and-wonderful-drama.html.

Spencer, Charles. "London Road." *Theatre Record* (9 – 22 April 2011), p. 408.

Spencer, Charles. "Review: The Girlfriend Experience at the Royal Court." *The Telegraph*, 25 September 2008, p. 28.

Spencer, Charles. "Tragedy and Farce." *The Telegraph*, 6 November 2003, http://www.telegraph.co.uk/culture/theatre/drama/3605992/Tragedy-and-farce.html.

Spencer, Jenny. "Editor's Introduction." *Political and Protest Theatre after 9/11 : Patriotic Dissent*, edited by Jenny Spencer. Routledge, 2012, pp. 1–15.

Staff Reporter. "Former Norwich Pub Landlord's Prostitute Murders to Be Turned into a Musical." *Eastern Daily Express*, 27 January 2011, http://www.edp24.co.uk/news/former_norwich_pub_landlord_s_prostitute_murders_to_be_turned_into_a_musical_1_785879.

Stanislavski, Constantin. *An Actor Prepares*, translated by Elizabeth Reynolds Hapgood. Methuen, 1980.

Stephenson, Jenn. *Insecurity: Perils and Products of Theatres of the Real*. University of Toronto Press, 2019.

Stephenson, Jenn. "Verbatim without Words." Upsurges of the Real: A Performance Research Blog, 8 June 2016, https://realtheatre.wordpress.com/2016/06/08/verbatim-without-words/.

Stilz, Gerhard. "Naturalist Drama – 'Problem Plays' or 'Plays of Ideas'? Bernard Shaw's Mrs Warren's Profession and John Galsworthy's Justice," *A History of British Drama: Genres – Developments – Model Interpretations*, edited by Sibylle Baumbach, Birgit Neumann and Ansgar Nünning. Wissenschaftlicher Verlag Trier, 2011, pp. 237–250.

Stoller, Terry. "Crimes of the PM." *HotReviews.org*, May 2007, http://www.hotreview.org/articles/crimesofthepm.htm.

Stoller, Terry. *Tales of the Tricycle Theatre*. Methuen Drama, 2013.

Stourac, Richard and Kathleen McCreery. *Theatre as a Weapon: Workers' Theatre in the Soviet Union, Germany and Britain*, 1917–1934. Routledge and Kegan Paul, 1986.

Stowell, Sheila. "Rehabilitating Realism." *Journal of Dramatic Theory and Criticism*, vol. 6, no. 2, 1992, pp. 81–88.

Stucky, Nathan. "Toward an Aesthetic of Natural Performance." *Text and Performance Quarterly*, vol. 13, 1993, pp. 168–180.

Summerskill, Clare. "Hearing Voices: Creating Theatre from Stories Told by Mental Health Service Users." *Creative Positions in Adult Mental Health: Outside In-Inside Out*, edited by Sue McNab and Karen Partridge. Karnac Books, 2014, pp. 227–240.

Sutcliffe, Tom. "The Sheer Joy of Miró's 'First Room'." *Arts & Books Review* (22 April 2011), p. 2.

Szalwinska, Maxie. "Review: We Haven't Said a Porkie Pie Yet." *The Scotsman*, 14 August 2002, p. 15.

Taylor, Lib. "The Experience of Immediacy: Emotion and Enlistment in Fact-based Theatre." *Studies in Theatre & Performance*, vol. 31, no. 2, 2011, pp. 223–237.

Taylor, Lib. "Voice, Body and the Transmission of the Real in Documentary Theatre." *Contemporary Theatre Review*, vol. 23, no. 3, 2013, pp. 368–379.

Taylor, Millie. *Musical Theatre, Realism and Entertainment.* Ashgate, 2012.

Taylor, Paul. "Justifying War, Tricycle Theatre, London." *Independent,* 6 November 2003, p. 15.

Taylor, Paul. "Talking to Terrorists, Playhouse Oxford: When You're Forced to Listen." *Independent* (2 May 2005).

Thompson, James. *Applied Theatre: Bewilderment and Beyond.* Peter Lang, 2003.

Timmer, Nicoline. *Do You Feel It Too?: The Post-Postmodern Syndrome in American Fiction at the Turn of the Millennium.* New York: Rodopi, 2010.

Tinius, Jonas. "Lob des Realismus. By Bernd Stegemann." *Theatre Research International*, vol. 41, no. 1, 2016, pp. 86–87.

Tinius, Jonas. "Rehearsing Detachment: Refugee Theatre and Dialectical Fiction." *Cadernos de Arte e Antropologia*, vol. 5, no. 1, 2016.

Tomlin, Liz. *Acts and Apparitions: Discourses on the Real in Performance Practice and Theory,* 1990 – 2010. Manchester UP, 2013.

Tomlin, Liz. "A 'Political Suspension of the Ethical': *To Be Straight with You* (2007) and *An Evening with Psychosis* (2009)." *Performing Ethos: International Journal of Ethics in Theatre & Performance*, vol. 1, no. 2, 2010, pp. 167–180.

Tomlin, Liz. (ed.). *British Theatre Companies 1995–2014.* Methuen Drama, 2015.

Tomlin, Liz. "The Academy and the Marketplace: Avant-Garde Performance in Neoliberal Times." *Vanguard Performance Beyond Left and Right,* edited by Kimberly Jannarone. University of Michigan Press, 2015, pp. 264–282.

Tönnies, Merle. "Book Review. Jürs-Munsby, Karen, Jerome Carroll, and Steve Giles, eds. Postdramatic and the Political: International Perspectives on Contemporary Performance." *Journal of Contemporary Drama in English*, vol. 3, no. 2, 2015, pp. 355 –358.

Toth, Josh. *The Passing of Postmodernism: A Spectroanalysis of the Contemporary.* State University of New York Press, 2010.

Trezise, Bryoni and Caroline Wake. "On the Ethics of Non-Disclosure: A Roundtable with Urban Theatre Projects and Collaborators." *Performance Paradigm*, vol. 5, no. 2, 2009, pp. 1–16.

Troy-Donovan, Alice. "Interview Alecky Blythe: 'I'm Interested in Ordinary People in Extraordinary Situations'." *The Oxford Student,* 16 October 2014, http://oxfordstudent.com/2014/10/16/.

Tucker, David. "Introduction – 'an Anthropology of Ourselves' VS 'the Incomprehensibility of the Real': Making the Case for British Social Realism' *British Social Realism in the Arts since 1940*, edited by David Tucker. Palgrave Macmillan, 2011, pp. 1–16.

Turnbull, Olivia. "It's All about You: Immersive Theatre and Social Networking." *Journal of Contemporary Drama in English*, vol. 4, no. 1, 2016, pp. 150–163.

Turner, Cathy. "Framing the Site." *Studies in Theatre and Performance, Supplement*, vol. 5, 2000, pp. 23–40.

Tushingham, David. "Lloyd Newton: Dance about something." *Live 1. Food for the Soul: A New Generation of British Theatremakers*, edited by David Tushingham. Methuen Drama, 1994, pp. 43–53.

Twining, William. "The Hutton Inquiry: Some Wider Legal Aspects." *Hutton and Butler: Lifting the Lid on the Workings of Power,* edited by W. G. Runciman. OUP, 2004, pp. 29–50.

Ubersfeld, Anne. "The Pleasure of the Spectator, Pierre Bouillaguet and Charles Jose (trans.)." *Modern Drama*, vol. 25, no. 1, 1982, pp. 127–139.

Upton, Carole-Anne. "The Performance of Truth a nd Justice in Northern Ireland." *Get Real: Documentary Theatre Past and Present*, edited by Alison Forsyth and Chris Megson. Palgrave Macmillan, 2009, pp. 179–194.

Unwin, Stephen. *A Guide to the Plays of Bertolt Brecht.* Methuen, 2005.

Unwin, Stephen. *The Complete Brecht Toolkit.* Nick Hern Books, 2014.

Valentine, Alana. "Captivated by Reality: The Alex Buzo Lecture 2009.", http://performing.artshub.com.au/news-article/news/performing-arts/-captivated-by-reality-the-alex-buzo- lecture-2009-by-alana-valentine–179013.

Vaughan, Megan. "Periscope: A New Performance Platform?." *The Space* (2015), http://www.thespace.org/news/view/periscope-performance-art-megan-vaughan.

Voigts-Virchow, Eckart. "Shelley, Malley, Carey: A Performative Hermeneutics of Faking and Hoaxing." *Zeitschrift für Anglistik und Amerikanistik,* vol. 56, no. 2, 2008, pp. 175–191.

Wake, Caroline. "Book Review: Theatre of the Real, by Carol Martin (Basingstoke: Palgrave Macmillan, 2013)." *Performance Paradigm*, vol. 11, 2015, pp. 119–123.

Wake, Caroline. "Headphone Verbatim Theatre: Methods, Histories, Genres, Theories." *New Theatre Quarterly*, vol. 29, no. 4, 2013, pp. 321–335.

Wake, Caroline. "Review of Get Real: Documentary Past and Present and Dramaturgy of the Real on the World Stage." *Performance Paradigm*, vol. 7, 2011, pp. 1–11.

Wake, Caroline. "The Politics and Poetics of Listening: Attending Headphone Verbatim Theatre in Post-Cronulla Australia." *Theatre Research International*, vol. 39, no. 2, 2014, pp. 82–100.

Wake, Caroline. "Towards a Working Definition of Verbatim Theatre." *Verbatim: Staging Memory & Community,* edited by Paul Brown. Currency Press, 2010, pp. 2–5.

Wake, Caroline. "To Witness Mimesis: The Politics, Ethics, and Aesthetics of Testimonial Theatre in Through the Wire." *Modern Drama*, vol. 56, no.1, 2013, pp. 102–125.

Wake, Caroline. "Verbatim Theatre within a Spectrum of Practices." *Verbatim: Staging Memory & Community,* edited by Paul Brown. Currency Press, 2010, 6–8.

Wandor, Michelene. "Introduction." *Strike While the Iron is Hot: Three Plays on Sexual Politics*, edited by Michelene Wandor. The Journeyman Press, 1980, pp. 5–15.

Ward, Ian. "The Play of Terror." *Law and Art: Justice, Ethics and Aesthetics, Oren Ben-Dor.* Routledge, 2011, pp. 177–187.

Waters, Steve. "Political Playwriting: The Art of Thinking in Public." *Topoi,* vol. 30, 2011, pp. 137–144.

Waters, Steve. *The Secret Life of Plays.* Nick Hern Books, 2010.

Watson, Cynthia. "Transcript of Remarks by Secretary of Defense Donald Rumsfeld and Chairman of the Joint Chiefs of Staff General Richard B. Myers, USAF." *Stability, Security, Reconstuction and Transition Operations: A Guide to the Issues*, edited by Cynthia Watson. Praeger Security International, 2012, pp. 187–204.

Watt, David. "From Oral History to Performance." *Political Performances: Theory and Practice*, edited by Susan C. Haedicke, Deirdre Heddon, Avraham Oz and E.J. Westlake. Rodopi, 2009, pp. 189–214.

Weiss, Peter. "The Material and the Models: Notes Towards a Definition of Documentary Theatre." Translated by Heinz Bernard, *Theatre Quarterly*, vol. 1, no.1, 1971, pp. 41–43.
Wells, Liz (ed.). *Photography: A Critical Introduction*. Routledge, 2015.
Wessendorf, Markus. "Postmodern Drama Post-9/11: Adriano Shaplin's Pugilist Specialist and David Hare's Stuff Happens." *Contemporary Drama in English*, vol. 14, 2007, pp. 325–346.
Westlake, Martin. "Is That Really Me Up There?." *The Financial Times*, 17 February 2006, p. 7.
Wheatley, Thomas. "Thomas Wheatley interviewed by Derek Paget." University of Reading: Unpublished interview, 2008, https://www.reading.ac.uk/web/FILES/ftt/Thomas_Wheatley_2nd_April_2008.pdf.
White, Gareth. "On Immersive Theatre." *Theatre Research International*, vol. 37, no. 3, 2012, pp. 221–235.
White, Graham D. "Compelled to Appear: The Manifestation of Physical Space Before the Tribunal." *Contemporary Drama in English*, vol. 13, 2006, pp. 73–86.
Wilding, Jo. *Don't Shoot the Clowns: Taking a Circus to the Children of Iraq*. New Internationalist, 2006.
Wilkie, Fiona. "The Production of 'Site': Site-Specific Theatre." *A Concise Companion to Contemporary British and Irish Drama*, edited by Nadine Holdsworth and Mary Luckhurst. Blackwell Publishing, 2008, pp. 87–106.
Willett, John. *Brecht on Theatre: The Development of an Aesthetic*. Methuen, (1964) 1986.
Williams, David. "Putting Real People on Stage: Helgard Haug (Rimini Protokoll) in Conversation with Theatre Practitioners and Academics from Australia." *Performance Paradigm*, vol. 11, 2015, pp. 101–115.
Williams, Heathcote. *The Speakers*. Robin Clark, 1982.
Williams, Raymond. "A Lecture on Realism." Transcription of a lecture given at the London International Film School on 8–10 October 1976, http://www.afterall.org/journal/issue.5/lecture.realism.
Williams, Raymond. *Drama from Ibsen to Brecht: A Critical Account and Revaluation*. Penguin Books, (1952) 1976.
Williams, Raymond. *Keywords*. Fontana, 1976.
Williams, Raymond. *Marxism and Literature*. OUP, 1977.
Williams, Raymond. *Problems in Materialism and Culture: Selected Essays*. Verso Books, 1980.
Williams, Raymond. "Theatre as a Political Forum." *Visions and Blueprints: Avant-Garde Culture and Radical Politics in Early Twentieth-Century Europe*, edited by Edward Timms and Peter Collier. MUP, 1988, pp. 307–320.
Wilmer, S. E. "Renaming and Performative Reconstructions: The Uncanny Multiplication of Janez Janša." *Theatre Research International*, vol. 36, no. 1, 2011, pp. 47–63.
Wilson, Christina. "Challenging the 'Fetish of the Verbatim': New Aesthetics and Familiar Abuses in Christine Evans's Slow Falling Bird." *Imagining Human Rights in Twenty-First-Century Theater: Global Perspectives*, edited by Florian N. Becker, Paola S. Hernández and Brenda Werth. Palgrave Macmillan, 2013, pp. 121–136.
Winkler, Elizabeth Hale. *The Function of Song in Contemporary British Drama*. Associated University Presses, 1990.
Winston, Brian. *Claiming the Real: The Griersonian Documentary and its Legitimations*. British Film Institute, 1995.

Wood, Gaby. "Can David Hare Take Manhattan." *The Observer*, 12 November 2006, p. 4.
Woolf, Dennis. Beyond Belief: Scenes from the Shipman Inquiry. Official Programme. Manchester Library Theatre Company, 2004.
Worthen, William B. *Modern Drama and the Rhetoric of Theater.* University of California Press, 1992.
Worthen, William B. "'The Written Troubles of the Brain': Sleep No More and the Space of Character." *Theatre Journal*, vol. 64, no. 1, 2012, pp. 79–97.
Wright, Elizabeth. *Postmodern Brecht: A Re-Presentation.* Routledge, 1989.
Wroe, Nicholas. "Courtroom Dramas." *Guardian*, 24 July 2004, p. 16.
Wu, Duncan. *Making Plays: Interviews with Contemporary British Dramatists and Directors.* Macmillan Press, 2000.
Youker, Timothy. "Reading Collage Drama as Documentary Drama: Trash Archives and Dissensus in Charles L. Mee's Lives of the Artists." *Modern Drama*, vol. 58, no. 2, 2015, pp. 218–237.
Young, Stuart. "London Road (Review)." *Theater Journal*, vol. 64, no. 1, 2012, pp. 101–102.
Young, Stuart. "Playing with Documentary Theatre: Aalst and Taking Care of Baby." *New Theatre Quarterly*, vol. 25, no. 1, 2009, pp. 72–87.
Zeifman, Hersh. "An Interview with David Hare." *David Hare: A Casebook*, edited by Hersh Zeifman. Garland Publishing, 1994, pp. 3–22.
Žižek, Slavoj. "The Spectre of Ideology." *Mapping Ideology*, edited by Slavoj Žižek. Verso Books, 1994, pp. 1–33.
Žižek, Slavoj. *Welcome to the Desert of the Real! Five Essays on September 11 and Related Dates.* Verso Books, 2002.
Zola, Emile. *Naturalism in the Theatre*, translated by Albert Bermel. *The Theory of The Modern Stage*, edited by Eric Bentley. Penguin Books, 1968, 351–372.

Index

https://doi.org/10.1515/9783110715767-009

Printed and bound by CPI Group (UK) Ltd, Croydon, CR0 4YY
13/09/2026
14961092-0003